Springer Monographs in Mathematics

Kazuyuki Saitô • J.D. Maitland Wright

Monotone Complete C*-algebras and Generic Dynamics

Springer

Kazuyuki Saitô
Sendai, Japan

J.D. Maitland Wright
University of Aberdeen
Aberdeen, Scotland

Christ Church
University of Oxford
Oxford, UK

ISSN 1439-7382　　　　　　　ISSN 2196-9922　(electronic)
Springer Monographs in Mathematics
ISBN 978-1-4471-6773-0　　　ISBN 978-1-4471-6775-4　(eBook)
DOI 10.1007/978-1-4471-6775-4

Library of Congress Control Number: 2015958093

Mathematics Subject Classification: 46L05, 46L35, 37A20, 37A55, 46L36, 06A12, 54G05, 06F05

Springer London Heidelberg New York Dordrecht

Printed on acid-free paper

Springer-Verlag London Ltd. is part of Springer Science+Business Media (www.springer.com)

Abstract

This book is about monotone complete C^*-algebras, their properties, and the new classification theory (using spectroid invariants and a classification semigroup). A basic account of generic dynamics is included because of its important connections to these algebras. Each bounded, upward-directed net of real numbers has a limit. Monotone complete algebras of operators have a similar property. In particular, every von Neumann algebra is monotone complete, but the converse is false. The small von Neumann factors can be labelled by the set of real numbers. But there are many more ($2^{\mathbb{R}}$) small monotone complete C^*-algebras which are factors. The aim of this book is to give an account of monotone complete C^*-algebras which includes recent advances but also indicates the many mysteries and open problems which remain.

Preface

First, we wish to thank Professor Takesaki who encouraged us to write this book. Without him, we would never have started; without the unfailing support of Yuko and Harvinder, we would never have finished.

Monotone complete C^*-algebras are now entering a new and exciting stage of their development. The new classification semigroup and spectroid invariants are useful, but we are a very long way from a full understanding; many mysteries remain. The second author particularly wishes to thank Dennis Sullivan for stimulating discussions on generic dynamics and related topics.

We would both like to thank the many distinguished mathematicians who have shared their insights with us. There are too many to list them all, but among them are J.Bonet, J.K. Brooks, L.J. Bunce, Cho-Ho Chu, C.M. Edwards, P. de Lucia, Masanao Ozawa, A.Peralta, P. Ptak, B. Weiss, and K. Ylinen.

Sendai, Japan
Aberdeen, Scotland

Kazuyuki Saitô
J.D. Maitland Wright

Contents

Chapter 1
Introduction

This book is about monotone complete C^*-algebras, their properties and their classification. We also give a basic account of generic dynamics because of its useful connections to these algebras.

1.1 Monotone Complete Algebras of Operators

Fundamental to analysis is the completeness of the real numbers. Each bounded, monotone increasing sequence of real numbers has a least upper bound. Monotone complete algebras of operators have a similar property.

Let A be a C^*-algebra. Its self-adjoint part, A_{sa}, is a partially ordered, real Banach space whose positive cone is $\{zz^* : z \in A\}$. If each upward directed, norm-bounded subset of A_{sa}, has a least upper bound then A is said to be *monotone complete*. Every von Neumann algebra is monotone complete but the converse is false.

Recently there have been major advances in the theory of monotone complete C^*-algebras; for example the construction of classification semigroups [144]. This followed an important breakthrough in [66], which introduced huge numbers of new examples. But much remains to be discovered. The purpose of this book is to expound the new theory. We want to take readers from the basics to the frontiers of the subject. We hope they will be stimulated to work on the many fascinating open problems. Our intention is to strive for clarity rather than maximal generality. Our intended reader has a grounding in elementary functional analysis and point set topology and some exposure to the fundamentals of C^*-algebras, say, the first chapters of [161]. But prior knowledge of von Neumann algebras or operator systems is not essential. However, in this introduction, we may use terminology with which some readers are unfamiliar. If so, we apologise and reassure them that all necessary technicalities will be discussed later in the text.

© Springer-Verlag London 2015
K. Saitô, J.D.M. Wright, *Monotone Complete C*-algebras and Generic Dynamics*,
Springer Monographs in Mathematics, DOI 10.1007/978-1-4471-6775-4_1

Algebras of operators on Hilbert space, including C^*-algebras, von Neumann algebras and their generalisations, are the focus of intense research activity worldwide. They are fundamental to non-commutative geometry and intimately related to work on operator systems and operator spaces and have connections to many other fields of mathematics and quantum physics. But the first to be investigated (with a different name and a more "spatial" viewpoint) were the von Neumann algebras.

Monotone complete C^*-algebras arise in several different areas. There are close connections with operator systems, with operator spaces and with generic dynamics. In the category of operator systems, with completely positive maps as morphisms, each injective object can be given the structure of a monotone complete C^*-algebra in a canonical way. Injective operator spaces can be embedded as "corners" of monotone complete C^*-algebras, see Theorem 6.1.3 and Theorem 6.1.6 [38] and [25, 59, 60]. When a monotone complete C^*-algebra is commutative, its lattice of projections is a complete Boolean algebra. Up to isomorphism, every complete Boolean algebra arises in this way.

Let A be a monotone complete C^*-algebra then A is a von Neumann algebra precisely when it has a separating family of normal states. If a monotone complete C^*-algebra does not possess any normal states it is called *wild*.

The best known commutative example of a wild monotone complete C^*-algebra is straightforward to construct. Let $B(\mathbb{R})$ be the commutative C^*-algebra of all bounded, complex valued Borel measurable functions on $\mathbb{R}$. Let $M(\mathbb{R})$ be the ideal of all functions h in $B(\mathbb{R})$ such that $\{t \in \mathbb{R} : h(t) \neq 0\}$ is a meagre subset of $\mathbb{R}$. (Let us recall that a set is meagre if it is contained in the union of countably many nowhere dense sets; a set is nowhere dense if its closure has empty interior.) Then the quotient algebra $B(\mathbb{R})/M(\mathbb{R})$ is a commutative monotone complete C^*-algebra which has no normal states and so is not a von Neumann algebra. It turns out that if we replace $\mathbb{R}$ by any complete separable metric space, without isolated points, and perform the same construction then we end up with the same commutative monotone complete C^*-algebra.

A monotone complete C^*-algebra, like a von Neumann algebra, is said to be a *factor* if its centre is trivial. In other words, factors are as far as possible from being commutative. Just as for von Neumann algebras, monotone complete C^*-factors can be divided into Type I, Type II$_1$, Type II$_\infty$ and Type III. It turns out that all Type I factors are von Neumann algebras. So it is natural to ask: are *all* monotone complete C^*-factors, in fact, von Neumann algebras? The answer is "no" in general but to clarify the situation, we need some extra notions. Let H be a separable Hilbert space and $L(H)$ the bounded operators on H. A C^*-algebra A is said to be *separably representable* if there exists a $*$-isomorphism π from A into $L(H)$. It is known that if A is a monotone complete C^*-factor which is also a separably representable C^*-algebra then A must be a von Neumann algebra [179]. So where are the wild factors?

A (unital) C^*-algebra B is said to be *small* if there exists a unital complete isometry from B onto an operator system in $L(H)$, where H is separable. When an algebra is separably representable then it is small but the converse is false. In other words, there exist C^*-algebras which can be regarded as operator systems on a separable Hilbert space but which can only be represented as $*$-algebras of bounded operators on a larger Hilbert space. There do exist small Type III monotone complete C^*-factors which are not von Neumann algebras. In fact they exist in huge abundance. There are 2^c, where c is the cardinality of the real numbers. By contrast, there are only c small von Neumann algebras. (Each small von Neumann algebra is isomorphic to a $*$-subalgebra of $L(H)$ where the subalgebra is closed in the weak operator topology. In particular, each small von Neumann algebra is separably representable. This follows from [1].) Incidentally, if a small C^*-algebra is a wild factor then it is always of Type III.

One way to find a wild monotone complete C^*-factor is to start with a separable, simple, unital C^*-algebra and use a kind of "Dedekind cut" completion [173]. This approach will be discussed later. Another method is to associate a monotone complete C^*-algebra with a dynamical system. This "generic dynamics" approach is outlined below.

Monotone complete C^*-algebras are a generalisation of von Neumann algebras. The theory of the latter is now very well advanced. But it took many years before it was demonstrated that there were continuum many von Neumann factors of Type III [126], Type II$_1$ [100] and Type II$_\infty$ [148]. Then the pioneering work of Connes, Takesaki and other giants of the subject transformed our knowledge of von Neumann algebras, see [8, 30, 96, 162]. By comparison, the theory of monotone complete C^*-algebras is in its infancy with many fundamental questions unanswered. But great progress has been made in recent years. In the early study of monotone complete C^*-algebras the emphasis was on showing how similar they were to von Neumann algebras. Nowadays we realise how different they can be.

In 2001 Hamana [66] made a major breakthrough which implied that there are 2^c small monotone complete C^*-factors. In 2007 [144] we found a way to classify monotone complete C^*-algebras. This is set out in Chap. 3.

In [144] we introduced a quasi-ordering between monotone complete C^*-algebras. From this quasi-ordering we defined an equivalence relation and used this to construct, in particular, a classification semigroup $\mathcal{W}$ for small monotone complete C^*-algebras. This semigroup is abelian, partially ordered, and has the Riesz decomposition property. For each small monotone complete C^*-algebra A we assign a "normality weight", $w(A) \in \mathcal{W}$. If A and B are algebras then $w(A) = w(B)$, precisely when these algebras are equivalent. It turns out that algebras which are very different can be equivalent. In particular, the von Neumann algebras are equivalent to each other and correspond to the zero element of the semigroup. It might have turned out that $\mathcal{W}$ is very small and fails to distinguish between more than a few algebras. This is not so; the cardinality of $\mathcal{W}$ is 2^c, where $c = 2^{\aleph_0}$.

A natural reaction by anyone familiar with K-theory, is to construct the Grothendieck group of $\mathcal{W}$. But this group is trivial because each element of the semigroup is idempotent. However this implies that $\mathcal{W}$ has a natural structure as a semi-lattice. Furthermore, the Riesz Decomposition Property for $\mathcal{W}$ ensures that the semi- lattice is distributive.

As we shall see later, one of the useful properties of $\mathcal{W}$ is that it can sometimes be used to replace problems about factors by problems about commutative algebras [144].

To each monotone complete C^*-algebra we can associate a *spectroid* invariant, ∂A [144]. Just as a spectrum is a set which encodes information about an operator, a spectroid encodes information about a monotone complete C^*-algebra. It turns out that if $wA = wB$ then A and B have the same spectroid. So the spectroid may be used as a tool for classifying elements of $\mathcal{W}$.

Kaplansky wished to capture the algebraic essence of von Neumann algebras and to do it, introduced AW^*-algebras [90–92]. An AW^*-algebra may be defined as a unital C^*-algebra in which every maximal abelian $*$-subalgebra is monotone complete [146]. Every monotone complete C^*-algebra is easily seen to be an AW^*-algebra. Nobody has ever seen an AW^*-algebra which is *NOT* monotone complete. It is strongly suspected that *EVERY AW^*-algebra* is monotone complete. But in full generality this is a difficult open problem. But many positive results are known. In particular, all "small" AW^*-factors are known to be monotone complete. Since our interest is strongly focused on small C^*-algebras we shall postpone a discussion of AW^*-algebras until Chap. 8. (But this can be read now, without working through all the earlier chapters.) They will appear on our list of open problems, some of which have been unsolved for over 60 years. For a scholarly account of the classical theory of AW^*-algebras the reader may consult [13].

Generic dynamics is used in an essential way in this book but we shall not introduce this tool until Chap. 6. So some readers may prefer to turn immediately to Chap. 2 and postpone reading the introduction to generic dynamics.

1.2 Generic Dynamics

An elegant account of generic dynamics was given by Weiss [165]; the term occurred earlier in [157]. In these articles, the underlying framework is a countable group of homeomorphisms acting on a complete separable metric space with no isolated points (a perfect Polish space). The key result of [157] was a strong uniqueness theorem. As a consequence, the wild factor discovered by Dyer [36] and the factor found by Takenouchi [159] were shown to be isomorphic.

We devote a chapter to aspects of generic dynamics useful for monotone complete C^*-algebra theory, including some recent discoveries [145]. This is an elementary exposition. In this book, generic dynamics is only developed as far as we need it for applications to C^*-algebras. But this does require us to consider generic dynamics on compact non-metrisable separable spaces; a topic which has been little explored and gives rise to interesting open questions.

Let G be a countable group. Unless we specify otherwise, G will always be assumed to be infinite. Let X be a Hausdorff topological space with no isolated points. Further suppose that X is a Baire space i.e. such that the only meagre open set is the empty set. In other words, the Baire Category Theorem holds for X. We shall also suppose that X is completely regular. (These conditions are satisfied if X is compact or homeomorphic to a complete separable metric space or, more generally, a G_δ-subset of a compact Hausdorff space or is the extreme boundary of a compact convex set in a locally convex Hausdorff topological vector space.) A subset Y of X is said to be *generic* if $X \backslash Y$ is meagre.

Let ε be an action of G on X as homeomorphisms of X.

In classical dynamics we would require the existence of a Borel measure on X which was G-invariant or quasi-invariant, and discard null sets. In topological dynamics, no measure is required and no sets are discarded. In generic dynamics, we discard meagre Borel sets.

We shall concentrate on the situation where, for some $x_0 \in X$, the orbit $\{\varepsilon_g(x_0) : g \in G\}$ is dense in X. Of course this cannot happen unless X is separable. (A topological space is *separable* if it has a countable dense subset. This is a weaker property than having a countable base.) Let S be the Stone space of the (complete) Boolean algebra of regular open sets of X. Then, see below, the action ε of G on X induces an action $\hat{\varepsilon}$ of G as homeomorphisms of S; which will also have a dense orbit.

When, as in [165] and [157], X is a perfect Polish space, then S is unique; it can be identified with the Stone space of the regular open sets of $\mathbb{R}$. But if we let X range over all separable compact subspaces of the separable space, $2^{\mathbb{R}}$, then we obtain 2^c essentially different S; where S is compact, separable and extremally disconnected. For each such S, $C(S)$ is a subalgebra of ℓ^∞.

Let E be the relation of orbit equivalence on S. That is, sEt, if, for some group element g, $\hat{\varepsilon}_g(s) = t$. Then we can construct a monotone complete C^*-algebra M_E from the orbit equivalence relation. When there is a free dense orbit, the algebra will be a factor with a maximal abelian $*$-subalgebra, A, which is isomorphic to $C(S)$. There is always a faithful, normal, conditional expectation from M_E onto A. It can be shown that $wM_E = wA$. So some classification questions about factors can be replaced by questions about commutative algebras. *When E and F are orbit equivalence relations which coincide on a dense G_δ-subset of S then M_E is isomorphic to M_F.*

For $f \in C(S)$, let $\gamma^g(f) = f \circ \hat{\varepsilon}_{g^{-1}}$. Then $g \mapsto \gamma^g$ is an action of G as automorphisms of $C(S)$. Then we can associate a monotone complete C^*-algebra $M(C(S), G)$, the *monotone cross-product* with this action (see Chap. 7). When the action $\hat{\varepsilon}$ is free, then $M(C(S), G)$ is naturally isomorphic to M_E. In other words, the monotone cross-product does not depend on the group, only on the orbit equivalence relation.

In this book we shall consider 2^c algebras $C(S)$. Each is a subalgebra of ℓ^∞ and each takes different values in the weight semigroup $\mathcal{W}$. (Here $c = 2^{\aleph_0}$, the cardinality of $\mathbb{R}$.)

For general S there is no uniqueness theorem but we do show the following. Let G be a countably infinite group. Let α be an action of G as homeomorphisms of S and suppose this action has at least one orbit which is dense and free. Then, modulo meagre sets, the orbit equivalence relation obtained can also be obtained by an action of $\bigoplus \mathbb{Z}_2$ as homeomorphisms of S. This should be contrasted with the situation in classical dynamics. (e.g. It is shown in [31] that any action by an amenable group is orbit equivalent to an action of $\mathbb{Z}$. But, in general, non-amenable groups give rise to orbit equivalence relations which do not come from actions of $\mathbb{Z}$.)

On each of 2^c, essentially different, compact extremally disconnected spaces we construct a natural action of $\bigoplus \mathbb{Z}_2$ with a free, dense orbit. Let Λ be a set of cardinality 2^c, where $c = 2^{\aleph_0}$. Then by applying generic dynamics, as in [144], we can find a family of monotone complete C^*-algebras $\{B_\lambda : \lambda \in \Lambda\}$ with the following properties. Each B_λ is a monotone complete factor of Type III, and also a small C^*-algebra. For $\lambda \neq \mu$, B_λ and B_μ have different spectroids and so $wB_\lambda \neq wB_\mu$ and, in particular, B_λ is not isomorphic to B_μ. Furthermore each B_λ is generated by an increasing sequence of full matrix algebras.

Chapter 2
Order Fundamentals

This chapter presents basic material which will be needed later. Among the topics discussed are order limits, monotone σ-complete C^*-algebras and commutative algebras.

Our aim is to present an account of monotone complete C^*-algebras which can be followed without requiring the reader to constantly look up results elsewhere. So, particularly in this first chapter, we give proofs of some basic results which appear in standard texts. We shall also state a number of results without proof but with an indication of where proofs can be found.

The books [161] and [121] are our sources for much of the canonical theory of C^*-algebras; we have used their guidance for some basic results. There are many other excellent books in this area, an interesting recent example is [15] as well as the classic [34]. An elegant succinct introduction to operator algebras is to be found in [67].

In the 80 years of its development, operator algebra theory has woven together the thoughts of many brilliant contributors. In such a vast subject it is no longer practical to keep track of every individual contribution. If we fail to attribute results to their original discoverers this is not a slight, not an insult, but evidence of how fundamentally enmeshed in the general theory their work has become. In particular, none of the results are claimed as our own unless we specifically say so.

The first three sections of this chapter are "Order structures and order convergence", "Monotone σ-complete C^*-algebras" and "Commutative algebras". They are basic to all that follows. The final section, "Matrix algebras over a monotone complete C^*-algebra", is not needed until the later chapters.

© Springer-Verlag London 2015
K. Saitô, J.D.M. Wright, *Monotone Complete C*-algebras and Generic Dynamics*,
Springer Monographs in Mathematics, DOI 10.1007/978-1-4471-6775-4_2

2.1   Order Structures and Order Convergence

We are familiar with the fact that a bounded set of real numbers has a least upper
bound. Let P be a partially ordered set and S a subset of P. An *upper bound* for S
is an $x \in P$ such that $a \le x$ for each $a \in S$. We call y a *least upper bound* for S if
y is an upper bound for S and, whenever x is an upper bound for S, then $y \le x$. If
a least upper bound exists then it is unique. Lower bound and greatest lower bound
are defined analogously. We shall also use "supremum" and "least upper bound"
interchangeably; they mean the same. Similarly with "infimum" and "greatest lower
bound". If, for each x, y in P the set $\{x, y\}$ has a supremum and an infimum then P
is a *lattice*.

When $S \subset P$ and S has a supremum s, we write $s = supS$ or $s = \bigvee S$. Both these
notations are in common use and we shall make use of both of them.

We recall that S is *upward directed* if $a \in S$ and $b \in S$ implies there exists $c \in S$
such that $a \le c$ and $b \le c$. (Downward directed is defined similarly.)

Let A be any C^*-algebra (not necessarily with a unit element) and let A_{sa} be
the (real) Banach space of self-adjoint elements of A. We recall that the *positive
elements* of A are, by definition, those of the form zz^*. Let A^+ be the set of all
positive elements of A. Then A^+ is a cone, that is, if x and y are in A^+ and if λ and
μ are in $\mathbb{R}^+$ then $\lambda x + \mu y$ is in A^+. This cone A^+ is closed in the norm topology.
Furthermore, $A^+ \cap -A^+ = \{0\}$ and $A_{sa} = A^+ - A^+$. So we can define a partial
ordering on A_{sa} by $x \ge y$ precisely when $x - y \in A^+$. (We also use $y \le x$ to mean
$x \ge y$.) Then with this partial ordering, A_{sa} is a partially ordered Banach space with
the real numbers as scalars.

Let us put $A = C(T)$, the algebra of complex valued continuous functions on a
compact Hausdorff space T. Then A_{sa} can be identified with $C_\mathbb{R}(T)$, the real valued
continuous functions. Then $f \ge g$ if $f(t) \ge g(t)$ for each $t \in T$. It is easy to see that
this ordering makes A_{sa} into a lattice.

In fact this lattice property is equivalent to commutativity. Given a C^*-algebra A,
Sherman's Theorem [152] tells us that A_{sa} is a (vector) lattice if, and only if, A is
commutative. A striking theorem of Kadison [81] tells us that $L(H)$ is an anti-lattice.
That is, given $x, y \in L(H)_{sa}$, the pair $\{x, y\}$ does not have a supremum unless $x \le y$ or
$y \le x$. For these, and more general results, see [9, 29]. At first sight Kadison's result
seems puzzling, since we know the projections in $L(H)$ form a lattice, with respect
to the partial ordering induced by $\le$. But this apparent paradox is easily resolved.
Given projections p and q in $L(H)$ the set of projections above both of them has a
smallest element; but, in general, the set of all self-adjoint elements above both of
them does not have a smallest element.

Whenever J is a closed ideal of A then J^+ is a hereditary cone in A^+, in other
words, if $x \in A$, $0 \le x \le b$ and $b \in J^+$ then $x \in J^+$. (Our main reference for the
basic theory of C^*-algebras is Takesaki [161] but see, also, Pedersen [121] for his
elegant account of order properties in C^*-algebras.)

Let U be a subset of A_{sa}. Then U is a norm bounded, upward directed set if and only if $-U = \{-a : a \in U\}$ is a norm bounded, downward directed set. Furthermore U has a least upper bound x if, and only if, $x - U = \{x - y : y \in U\}$ has infimum 0.

If (a_n) is a monotone increasing sequence in A_{sa} then, clearly, $\{a_n : n = 1, 2 \ldots\}$ is an upward directed set; when this directed set has a supremum a we say that the sequence has supremum a. Similarly, a monotone decreasing sequence (b_n) has infimum b, when the downward directed set $\{b_n : n = 1, 2 \ldots\}$ has infimum b.

As well as sequences we shall also make use of nets, see below.

The set $\{t \in \mathbb{R} : t < 1\}$ is an upper bounded set of real numbers but it is not bounded. In general, given an upper bounded set U in A we can pick $y_0 \in U$ and define $U_0 = \{y \in U : y \geq y_0\}$. Then U_0 is norm bounded.

If U is upward directed then U_0 has the same set of upper bounds as U. But if U is *not* upward directed then this need not be true. (For a trivial example, recall that $L(H)$ is an anti-lattice. So we can find self-adjoint a and b for which $\{a, b\}$ has no supremum. Put $U = \{a, b\}$ and $U_0 = \{y \in U : y \geq b\} = \{b\}$.)

Definition 2.1.1 A C^*-algebra A is *monotone complete* if each norm bounded, upward directed subset of A_{sa} has a least upper bound.

By using the map $a \mapsto -a$ it is easy to see that A is monotone complete if each norm bounded, downward directed subset of A_{sa} has a greatest lower bound.

Definition 2.1.2 A C^*-algebra A is *monotone σ-complete* if each norm bounded, monotone increasing sequence has a least upper bound.

It is immediate that A is *monotone σ-complete* if each norm bounded, monotone decreasing sequence has a greatest lower bound.

We shall see, later, that all monotone complete C^*-algebras have a unit element. This is not true for monotone σ-complete C^*-algebras. However, suppose that A is monotone σ-complete and does not posses a unit. Then A^1 the algebra formed by adjoining a unit, will be shown to be monotone σ-complete.

When working with C^*-algebras things go much more smoothly when they possess a unit. But, particularly for dealing with ideals, we need to extend parts of the theory to the non-unital situation. Most of the (minor) contortions which this requires are dealt with in this chapter.

Let A be monotone (σ-)complete and let $\mathcal{T}$ be a Hausdorff topology for A. Let us call $\mathcal{T}$ *sequentially order compatible* if, whenever (a_n) is a norm bounded, monotone increasing sequence with least upper bound a, then $a_n \to a$ in the $\mathcal{T}$-topology. In $L(H)$ the strong operator topology is sequentially order compatible. When A is any von Neumann algebra, with predual A_*, then the $\sigma(A, A_*)$ topology is (sequentially) order compatible. Does every monotone complete C^*-algebra have a Hausdorff, locally convex vector space topology which is sequentially order compatible? If the answer were "yes" we could replace order considerations by topological arguments.

But the answer is "no". There are commutative counter examples:

Example 2.1.3

(a) Let $B[0, 1]$ be the space of all bounded Borel measurable, complex valued functions on the unit interval. Let M be the set of all f in $B[0, 1]$ for which $\{\lambda \in [0, 1] : f(\lambda) \neq 0\}$ is a meagre set. Then, when the algebraic operations are defined pointwise and $B[0, 1]$ is equipped with the supremum norm, it is a commutative C^*-algebra. Also M is a closed ideal and the quotient $B[0, 1]/M$ will be shown to be monotone complete in Chap. 4. We have already remarked, and will prove later, that this algebra has no normal states. It is known that if $\mathcal{T}$ is a Hausdorff, locally convex vector space topology for $B[0, 1]/M$ then $\mathcal{T}$ is not sequentially order compatible. This follows from [44] because this work shows that if $B[0, 1]/M$ is equipped with a Hausdorff topology $\mathcal{S}$ which is sequentially order compatible then the map $(x, y) \mapsto x - y$ is not jointly continuous. Hence such an $\mathcal{S}$ cannot be a locally convex vector topology. However, as we shall see later, the Wright Representation Theorem [171] does show that each monotone σ-complete C^*-algebra is the *quotient* of a monotone σ-complete C^*-algebra which *does* posses a sequentially order compatible topology. This is illustrated by the above example.

(b) Let $Bnd(\mathbb{R})$ be the commutative C^*-algebra of all bounded complex valued functions on $\mathbb{R}$. Let J be the ideal consisting of all $f \in Bnd(\mathbb{R})$ for which $\{r \in \mathbb{R} : f(r) \neq 0\}$ is a countable set. Then J is monotone σ-complete. It is not monotone complete and does not possess a unit element.

(c) Let $C[0, 1]$ be the C^*-algebra of continuous complex valued functions on the closed unit interval. Then this algebra is neither monotone complete nor monotone σ-complete. Now let U be the set of all $f \in C[0, 1]$ such that f is real valued, $f(0) = 0$ and $0 \leq f(\lambda) < 1$ for each $\lambda \in [0, 1]$. Then U is upward directed and the function with the constant value 1 is the least upper bound of U in $C[0, 1]$. We have $||f - 1|| = 1$ for each $f \in U$. So there does not exist a sequence in U which converges in norm to the least upper bound of U.

When A is not assumed to be monotone complete it may still have some norm bounded, downward directed subset which has a greatest lower bound e.g. in Example 2.1.3(c) put $D = \{1 - f : f \in U\}$. Then D has 0 as its greatest lower bound.

When A is a C^*-algebra without a unit, we define A^1 to be the algebra formed by adjoining a unit. Then A is a maximal ideal of A^1. When A does have a unit we put $A^1 = A$.

In the rest of this section A is a C^*-algebra which is not assumed to possess a unit and is not required to be monotone complete, unless this is stated explicitly.

Notation 2.1.4 *In an algebra A we use: "$(a_n) \uparrow a$" as an abbreviation for: (a_n) is a monotone increasing sequence in A_{sa} with least upper bound a in A_{sa}. We also use $(x_n) \downarrow x$ to indicate that (x_n) is a monotone decreasing sequence in A_{sa} with infimum x in A_{sa}. When we write "$(a_n) \uparrow$" that will mean that the sequence is* *monotone increasing. We shall also use a similar notation for sequences in more general partially ordered sets.*

We shall sometimes find it convenient to abuse our notation, mildly, by referring to a monotonic sequence or directed set as being "in A" when "in A_{sa}" would be more correct; similarly we sometimes refer to a supremum or infimum as being "in A".

Lemma 2.1.5 *Let A be any C^*-algebra. Let $(a_n) \uparrow a$ and $(b_n) \uparrow b$. Then $(a_n + b_n) \uparrow a + b$.*

Proof Let x be an upper bound for $(a_n + b_n)$. Then

$$a_r + b_m \leq a_{r+m} + b_{r+m} \leq x.$$

So $a_r \leq x - b_m$. Fix m. Then $x - b_m$ is an upper bound for (a_r). Thus $a \leq x - b_m$.

So, for all m, $b_m \leq x - a$. Hence $b \leq x - a$. It now follows that $a + b$ is the least upper bound of $(a_n + b_n)$.   □

Let Λ be a directed set. Let $(a_\lambda : \lambda \in \Lambda)$ be a net in A_{sa}. Then the net is increasing if $\lambda \leq \mu$ implies $a_\lambda \leq a_\mu$. We say the net has least upper bound a when its range $\{a_\lambda : \lambda \in \Lambda\}$ has a least upper bound a. It is clear that when $(a_\lambda : \lambda \in \Lambda)$ is increasing then $\{a_\lambda : \lambda \in \Lambda\}$ is an upward directed subset of A_{sa}. Similarly the net is decreasing if $\lambda \leq \mu$ implies $a_\lambda \geq a_\mu$ and it has infimum b if its range $\{a_\lambda : \lambda \in \Lambda\}$ has infimum b. When D is a downward directed set in A_{sa} then $(d : d \in D)$ is a decreasing net.

Lemma 2.1.6

(i) *Let $(a_\lambda : \lambda \in \Lambda)$ be an increasing net in A_{sa} with least upper bound a and $(b_\lambda : \lambda \in \Lambda)$ an increasing net in A_{sa} with least upper bound b. Then $(a_\lambda + b_\lambda : \lambda \in \Lambda)$ is an increasing net in A_{sa} with least upper bound $a + b$.*

(ii) *Let $(x_\lambda : \lambda \in \Lambda)$ be a decreasing net in A_{sa} with infimum x and $(y_\lambda : \lambda \in \Lambda)$ a decreasing net in A_{sa} with infimum y. Then $(x_\lambda + y_\lambda : \lambda \in \Lambda)$ is a decreasing net in A_{sa} with infimum $x + y$.*

Proof

(i) This is a straightforward modification of the proof of Lemma 2.1.5.

(ii) Put $x_\lambda = -a_\lambda$ and $y_\lambda = -b_\lambda$ and apply (i).   □

Lemma 2.1.7 *Let A be any C^*-algebra. Let (x_n) be a monotone increasing sequence in A_{sa}. Suppose this sequence converges in norm to x. Then the sequence has a supremum and this supremum is x.*

Proof For $m \geq n$, $x_m - x_n \geq 0$. Because the positive cone is closed in the norm topology,

$$x - x_n = lim_{m \to \infty}(x_m - x_n) \geq 0.$$

Let b be an upper bound for the sequence. Then $b - x = lim_{n \to \infty}(b - x_n) \geq 0$. So x is the least upper bound of the sequence.   □

The converse of this lemma is, of course, false. To see this, first take a separable, infinite dimensional, Hilbert space H. Then, in $L(H)$, take a monotone increasing sequence of finite rank projections converging to 1_H (the identity operator on H) in the strong operator topology. Then the sequence has 1_H as its supremum but the sequence is certainly not convergent in norm.

Corollary 2.1.8 *Let $\mathcal{T}$ be a Hausdorff locally convex vector topology for A_{sa} such that A^+ is closed in the $\mathcal{T}$ topology. Let $(x_\lambda : \lambda \in \Lambda)$ be an increasing net in A_{sa} such that $(x_\lambda : \lambda \in \Lambda)$ converges in the $\mathcal{T}$ topology to x. Then x is the least upper bound of $\{x_\lambda : \lambda \in \Lambda\}$.*

Proof Straightforward modification of the proof of the preceding lemma.   □

We shall see that if S is an upward directed set in A_{sa} with supremum s, then, for any z in A, zSz^* is upward directed with supremum zsz^*. We also obtain a weaker, but useful, result which can be used when S is not a directed set.

We call a subset $S \subset A_{sa}$ *order bounded in* A if there exist a and b in A_{sa} such that

$$a \leq x \leq b$$

for each x in S. Clearly an order bounded set is bounded in norm. When A is unital, norm bounded sets are order bounded. But when A is not unital, norm bounded sets need not be order bounded. For example, let A be the compact operators on a separable, infinite dimensional Hilbert space and take S to be the self-adjoint (compact) operators in the unit ball of A.

In the following lemma, there is no requirement that S be a directed set. Part of the proof is based on [61].

Lemma 2.1.9 *Let J be a (closed two sided) ideal in A. Let S be a norm bounded subset of J_{sa} with least upper bound s in J. If $z \in A^1$ is invertible then zSz^* is a norm bounded subset of J with least upper bound zsz^* in J. Furthermore, if S is order bounded in J_{sa}, and $c \in J^+$ then cSc has supremum csc in J^+.*

Proof We observe that J is a closed ideal of A^1.

Let $L = s - S = \{s - y : y \in S\}$. Then L is a subset of J^+ which is bounded in norm and has infimum 0. Since J is an ideal, zLz^* is a subset of J. We show that, in J, zLz^* has an infimum and that this infimum is 0.

Since $a \mapsto a^*$ is an anti-automorphism of A^1, z^* is invertible and $(z^{-1})^* = (z^*)^{-1}$. The map T_z defined by $T_z(x) = zxz^*$ for $x \in A$ is an order-isomorphism of A_{sa} onto itself with $T_z^{-1} = T_{z^{-1}}$ and $T_z(J) = J$. Hence zLz^* has infimum 0.

Now suppose S is order bounded in J_{sa}. Then there exists $a \in J_{sa}$ which is an upper bound for L. Let d be any element of L. So, there exists $a \in J_{sa}$ such that $d \leq a$ for all $d \in L$.

Then

$$d^2 \leq ||d||d \leq ||d||a \leq ||a||a.$$

Let $c \in J^+$. We shall show that cLc has an infimum in J and that this infimum is 0.
Let $x \in J$ be a lower bound for cLc and let d be any element of L. So $x \leq cdc$.
Let ε be a positive real number. Then

$$
\begin{aligned}
(c + \varepsilon 1)d(c + \varepsilon 1) &= cdc + \varepsilon(cd + dc) + \varepsilon^2 d \\
&= cdc + \varepsilon((c + d)(c + d)^* - c^2 - d^2) + \varepsilon^2 d \\
&\geq cdc - \varepsilon(c^2 + d^2) \\
&\geq x - \varepsilon(c^2 + ||a||a) \in J_{sa}.
\end{aligned}
$$

Hence it follows that

$$d \geq (c + \varepsilon 1)^{-1}(x - \varepsilon c^2 - \varepsilon ||a||a)(c + \varepsilon 1)^{-1} \text{ in } J_{sa}$$

for all $d \in L$. So, we have

$$0 \geq x - \varepsilon c^2 - \varepsilon ||a||a$$

for all ε and so $0 \geq x$ follows. So, $csc = \sup cSc$ in J_{sa}. $\qquad\square$

Proposition 2.1.10 *Let J be a (closed two sided) ideal in A. Let S be an upward directed subset of J_{sa} with least upper bound s in J. Then, for any $z \in A^1$, we have zSz^* is an upward directed subset of J with least upper bound zsz^* in J.*

Proof Let $a_0 \in S$ then $S_0 = \{a \in S : a \geq a_0\}$ is upward directed, order bounded and with supremum s. So, without loss of generality, we may assume that S is order bounded in J_{sa}. We use the same notation as in the preceding lemma.

Let $L = s - S = \{s - y : y \in S\}$. Then L is a subset of J^+ which is downward directed with infimum 0. It suffices to show that, in J, zLz^* has an infimum and that this infimum is 0.

We may assume that $z \neq 0$ for otherwise there is nothing to prove. Let $x = z + 2||z||$ and $y = z - 2||z||$. Then x and y are invertible elements of A^1. By the preceding lemma, xLx and yLy both have infimum 0. We shall show that $(x + y)L(x + y)^*$ also has infimum 0. To see this, we argue as follows.

For any f and g in A we find, by expanding $(f - g)(f - g)^*$, that $fg^* + gf^* \leq ff^* + gg^*$.

Let $d \in L$. Put $f = xd^{1/2}$ and $g = yd^{1/2}$. Then $xdy^* + ydx^* \leq xdx^* + ydy^*$. So $(x + y)d(x + y)^* \leq 2xdx^* + 2ydy^*$.

Let c be any lower bound for $(x + y)L(x + y)^*$. Then $c \leq 2xdx^* + 2ydy^*$ for every $d \in L$.

In Lemma 2.1.6 (ii) put $\Lambda = L$ and consider the nets $(2xdx^* : d \in L)$ and $(2ydy^* : d \in L)$. Then $(2xdx^* + 2ydy^* : d \in L)$ has infimum 0. So $c \leq 0$. Hence zLz^* has infimum 0 and so zSz^* has supremum zsz^*.   □

Suppose A does not possess a unit. Let S be an upward directed set with supremum s. When S is regarded as a subset of A^1 does it still have a supremum in A^1_{sa} and, if it does, is it s? Fortunately the answer is "yes". See the proposition below.

In contrast to this result, an upward directed set in A_{sa} may have a supremum in A^1_{sa} but fail to have a supremum in A_{sa}. To see this, let H be a separable, infinite-dimensional, Hilbert space. Let A be the algebra of compact operators on H. Then take an increasing sequence of finite rank projections converging (in the strong operator topology) to 1, the identity operator on H. This sequence has no supremum in A but, in A^1, it has 1 as its supremum.

Proposition 2.1.11 *Let A be a C^*-algebra without a unit element. Let S be an upward directed set in A_{sa} with supremum s in A_{sa}. Then S has supremum s in A^1.*

Proof Because A is a closed ideal in A^1 the quotient map $q : A^1 \mapsto A^1/A$ is a ∗-homomorphism and hence a positive map.

Let $a + \lambda 1$ be any upper bound for S in A^1, where $a \in A$. Since q maps each element of A to zero, it follows that $\lambda \geq 0$.

For any $z \in A$, since A is an ideal of A^1, $z(a + \lambda 1)z^*$ is an upper bound for zSz^* in A. So

$$zsz^* \leq z(a + \lambda 1)z^* \leq zaz^* + \lambda \|z\|^2 1.$$

Now let (z_α) be an approximate unit for A. Then $z_\alpha sz_\alpha^* \leq z_\alpha az_\alpha^* + \lambda 1$. Also $\|z_\alpha(a - s)z_\alpha^* - (a - s)\| \to 0$. Since the positive cone of A^1 is closed in the norm topology it follows that $0 \leq a - s + \lambda 1$. That is, $s \leq a + \lambda 1$.   □

Corollary 2.1.12 *Let A be monotone σ-complete. Let $(a_n) \uparrow a$. Let $\|a_n\| \leq 1$ and $a_n \geq 0$ for each n. Then $\|a\| \leq 1$.*

Proof In A^1, the unit 1 is an upper bound for the sequence. So, from the proposition, $a \leq 1$.   □

Lemma 2.1.13 *Let ϕ be a positive linear functional on a C^*-algebra A. Then ϕ is a bounded linear functional.*

Proof Let $A_1^+ = \{a \in A^+ : \|a\| \leq 1\}$. It suffices to show ϕ is bounded on A_1^+. Suppose this is false. Then for each n there is a_n in A_1^+ such that $n2^n < \phi(a_n)$. Using norm convergence, let $a = \sum_1^\infty \frac{1}{2^n} a_n$. Then, for each n,

$$n \leq \phi\left(\frac{1}{2^n} a_n\right) \leq \phi(a).$$

This is impossible.   □

Let ϕ be a positive linear functional on a C^*-algebra. Then ϕ is said to be *faithful* if $x \geq 0$ and $\phi(x) = 0$ implies $x = 0$.

Theorem 2.1.14 *Let A be a monotone σ-complete C^*-algebra. Let ϕ be a faithful, positive linear functional on A. Then A is monotone complete. In particular, let D be an upward directed, norm bounded set in A_{sa}. Then there is a monotone increasing sequence in D whose supremum is the supremum of D.*

Proof It suffices to prove this when D is a subset of A_1^+, the intersection of the closed unit ball with the cone of positive elements.

Let $\overline{D}$ be the set of all $x \in A_1^+$ for which there is a monotone increasing sequence in D with x as its supremum.

Since D is upward directed, it is easy to see that $\overline{D}$ is also upward directed. Let (d_n) be any sequence in $\overline{D}$. Then, for each n we have $(x_r^{(n)}) \uparrow d_n$ where each $x_r^{(n)}$ is in D. Since D is upward directed we can find a monotone increasing sequence in D, (y_n) such that $y_n \geq x_r^{(k)}$ for $n \geq k$ and $n \geq r$.

Let $(y_n) \uparrow d$. Then $d \in \overline{D}$ and $d \geq x_r^{(k)}$ for all r and all k. So $d \geq d_n$ for all n.

By the preceding lemma, ϕ is bounded. Let $\lambda = sup\{\phi(y) : y \in \overline{D}\}$. For each n let $d_n \in \overline{D}$ such that $\phi(d_n) \geq \lambda - 1/n$. Then there exists $d \in \overline{D}$ with $\phi(d) \geq \lambda$. So $\phi(d) = \lambda$.

Now let $b \in \overline{D}$. Then, because the set is upward directed, we can find $c \in \overline{D}$ such that $c \geq b$ and $c \geq d$. Thus $\lambda \geq \phi(c) \geq \phi(d) = \lambda$. So $\phi(c - d) = 0$. Since ϕ is faithful, $c = d$. So $d \geq b$. So d is an upper bound for D. Since d is the least upper bound of an increasing sequence from D, it is the least upper bound of D. □

Proposition 2.1.15 *Let A be monotone complete. Then A has a unit element.*

Proof Let $\Gamma = \{a \in A^+ : ||a|| < 1\}$. Then, see p.11 [121], Γ is upward directed and an approximate unit. Since it is norm bounded it has a supremum e in A. Then $0 \leq e$. By Proposition 2.1.11 $e \leq 1$ in A^1. So $||e|| \leq 1$. By spectral theory $e^2 \leq e$. So $z^* e^2 z \leq z^* e z \leq z^* z$ for all $z \in A$.

Let $x \in A_{sa}$. Since Γ is an approximate unit, and since

$$||x^2 - xax|| \leq ||x|| ||x - ax||$$

it follows that the net $(||x^2 - xax|| : a \in \Gamma)$ converges to 0. So the net $(xax : a \in \Gamma)$ converges in the norm topology to x^2

But, by Proposition 2.1.10, xex is the least upper bound of $x\Gamma x$. So, by Lemma 2.1.7, $xex = x^2$.

So we have

$$0 \leq (x - ex)^*(x - ex) = x^2 - xex - xex + xe^2x$$

$$= -x^2 + xe^2x \leq -x^2 + xex = 0$$

which implies that $||x - ex||^2 = 0$, that is, $x = ex$. Taking adjoints gives $x = xe$. So, e is a unit element of A. □

It is convenient to define an "order limit" for sequences which are not monotonic. In fact, there are at least two useful notions of order limit. For our current purposes we shall use the Kadison-Pedersen limit [86]. We shall use sequences but everything can easily be generalised to nets.

To show that our definition makes sense we need the following lemma.

Lemma 2.1.16 *Let A be any C^*-algebra. Let $(a_n) \uparrow a$, $(b_n) \uparrow b$, $(c_n) \uparrow c$ and $(d_n) \uparrow d$. Suppose that $a_n - b_n = c_n - d_n$. Then $a - b = c - d$.*

Proof We have $a_n + d_n = c_n + b_n$. By Lemma 2.1.5 we have $(a_n + d_n) \uparrow a + d$ and similarly $(c_n + b_n) \uparrow c + b$. Thus $a + d = c + b$. $\square$

Definition 2.1.17 Let A be any C^*-algebra. Let (x_n) be a sequence in A_{sa}. If $x_n = a_n - b_n$ where $(a_n) \uparrow a$ and $(b_n) \uparrow b$, then we define $LIMx_n = a - b$ and say that the order limit of (x_n) is $a - b$. When A is contained in a larger algebra we shall sometimes say (x_n) order converges to $a - b$ in A and write

$$LIM_A x_n = a - b.$$

It follows from Example 2.1.3 (a) that, in general, order limits do not correspond to convergence with respect to some locally convex vector topology for A. But part (iv) of the following lemma does say that the positive cone of A is "closed" with respect to order limits.

Lemma 2.1.18 *Let A be any C^*-algebra. Let (x_n) and (y_n) be sequences in A_{sa} such that $LIMx_n = x$ and $LIMy_n = y$. Let λ be any real number. Then*

(i) *$LIM\lambda x_n = \lambda LIMx_n$.*
(ii) *$LIM(x_n + y_n) = LIMx_n + LIMy_n$.*
(iii) *For any $z \in A^1$ we have $LIMzx_n z^* = z(LIMx_n)z^*$.*
(iv) *If $x_n \geq 0$ for each n then $LIMx_n \geq 0$.*

Proof Let $x_n = a_n - b_n$ where $(a_n) \uparrow a$ and $(b_n) \uparrow b$.

(i) It is immediate that $LIM(-x_n) = b - a = -LIMx_n$. For $\lambda \neq 0$ the map $x \to |\lambda|x$ is an order isomorphism of A_{sa}. So $(|\lambda|a_n) \uparrow |\lambda|a$ and $(|\lambda|b_n) \uparrow |\lambda|b$. Thus $LIM\lambda x_n = \lambda LIMx_n$.
(ii) Let $y_n = c_n - d_n$ where $(c_n) \uparrow c$ and $(d_n) \uparrow d$. By Lemma 2.1.5 $(a_n + c_n) \uparrow a + c$ and $(b_n + d_n) \uparrow b + d$. So $LIM(x_n + y_n) = a + c - b - d = LIMx_n + LIMy_n$.
(iii) We have $zx_n z^* = za_n z^* - zb_n z^*$. By Proposition 2.1.10 $za_n z^* \uparrow zaz^*$ and $zb_n z^* \uparrow zbz^*$. Since A is an ideal in A^1, zaz^* and zbz^* are in A. So $LIMzx_n z^* = zaz^* - zbz^* = z(LIMx_n)z^*$.
(iv) Finally, if $x_n \geq 0$ then $a_n \geq b_n$. So $a \geq b_n$ for all n. So $a \geq b$. Thus $LIMx_n = a - b \geq 0$. $\square$

Lemma 2.1.19 *Let A be any C^*-algebra. Let (x_n) be a sequence in A_{sa} which has an order limit x, that is $LIMx_n = x$. If (x_n) also converges in the norm topology then $\|x - x_n\| \to 0$.*

Proof Let c be the limit, in the norm topology, of (x_n). Let $y_n = x_n - c$. Then $||y_n|| \to 0$. It follows from Lemma 2.1.18 (ii) that $LIMy_n = x - c$.

Let $y_n = a_n - b_n$ where $a_n \uparrow a$ and $b_n \uparrow b$. Then, for any real number $\varepsilon > 0$, there is an m such that, when $n \geq m$,

$$||a_n - b_n|| < \varepsilon.$$

So, in A^1, we have $-\varepsilon 1 \leq a_n - b_n \leq \varepsilon 1$. Thus $a_n \leq b + \varepsilon 1$.

By applying Proposition 2.1.11, we find $a \leq b + \varepsilon 1$.

Since this holds for each positive ε it follows that $a \leq b$. Similarly we can show $b \leq a$. So $LIMy_n = a - b = 0$. So $x = c$. $\qquad\square$

In contrast to Lemma 2.1.7, we have NOT proved that $||x_n - c|| \to 0$ implies $LIMx_n = c$. But if we take a subsequence $(x_{n(j)})$ such that $||x_{n(j)} - c|| < \frac{1}{2^j}$ then this subsequence is order convergent and $LIM_{j\to\infty}x_{n(j)} = c$.

(Let $a_j = x_{n(j+1)} - x_{n(j)}$. Then $(\sum_1^m a_j^+)$ and $(\sum_1^m a_j^-)$ are monotone increasing and norm convergent.)

Let us recall that for any $z \in A$, $z = x + iy$ where x and y are self-adjoint. In particular, $x = 1/2(z + z^*)$ and $y = 1/2i(z - z^*)$. We call x the real part of z and y the imaginary part of z.

Definition 2.1.20 Let A be any C^*-algebra. Let (z_n) be a sequence in A where x_n is the real part of z_n and y_n is the imaginary part of z_n. We say that (z_n) is order convergent if $LIMx_n$ and $LIMy_n$ both exist. We define $LIMz_n$ to be $LIMx_n + iLIMy_n$.

Lemma 2.1.21 *Let A be any C^*-algebra. Let (z_n) and (w_n) be sequences in A such that $LIMz_n$ and $LIMw_n$ both exist. Then*

(i) $LIMz_n^* = (LIMz_n)^*$
(ii) $LIM(z_n + w_n) = LIMz_n + LIMw_n$.
(iii) *For any complex number ω, $LIM\omega z_n = \omega LIMz_n$.*
(iv) *When $LIMz_n = z$ and (z_n) converges in norm then $||z_n - z|| \to 0$.*

Proof (i) is an immediate consequence of the definitions. So, also, is $LIMiz_n = iLIMz_n$.

Let $\omega = \lambda + i\mu$. Then, by applying (i) and (ii) from Lemma 2.1.18 we find $LIM\omega z_n = \omega LIMz_n$.

By applying Lemma 2.1.18 again we get (ii).

For (iv) we observe that (z_n^*) is also convergent in norm. So we may apply Lemma 2.1.19 to the sequences $(\frac{1}{2}(z_n + z_n^*))$ and $(\frac{1}{2i}(z_n - z_n^*))$. $\qquad\square$

The following proposition is useful.

Proposition 2.1.22 *Let A be any C^*-algebra. Let (z_n) be a sequence in A which is order convergent to $z \in A$, that is $LIMz_n = z$. Let $w \in A^1$. Then $LIMwz_n = wLIMz_n$, where these order limits are in A.*

Proof By linearity and Lemma 2.1.21, it suffices to prove this when (z_n) is monotone increasing (with supremum $LIMz_n$ in A).

For any z we have the following polarisation identity:

$$\frac{1}{4} \sum_{k=0}^{3} i^k (w + i^k 1) z (w + i^k 1)^* = wz.$$

By Proposition 2.1.10

$$LIM(w + i^k 1) z_n (w + i^k 1)^* = (w + i^k 1)(LIMz_n)(w + i^k 1)^*.$$

By using the polarisation identity and the linearity results of Lemma 2.1.21 we find $LIMwz_n = wLIMz_n$. □

Corollary 2.1.23 *Let A and (z_n) be as in the preceding proposition. Let w and v be in A^1. Then $LIM(wz_n v^*) = w(LIMz_n)v^*$.*

Proof We have $vLIMz_n^* = LIMvz_n^*$. On taking adjoints $(LIMz_n)v^* = LIM(z_n v^*)$. Another application of the preceding proposition gives $w(LIMz_n)v^* = LIM(wz_n v^*)$. □

Let A be any C^*-algebra. Let Λ be a directed set. Let $(x_\lambda : \lambda \in \Lambda)$ be a net in A_{sa}. If $x_\lambda = a_\lambda - b_\lambda$ where $(a_\lambda : \lambda \in \Lambda)$ is an increasing net with supremum a and $(b_\lambda : \lambda \in \Lambda)$ is an increasing net with supremum b then $(x_\lambda : \lambda \in \Lambda)$ is said to order converge to $a-b$. We write $LIMx_\lambda = a-b$. Now consider a net $(z_\lambda : \lambda \in \Lambda)$ in A. Let $z_\lambda = x_\lambda + iy_\lambda$ where x_λ and y_λ are self-adjoint. We define the order limit $LIMz_\lambda$ to be $LIMx_\lambda + iLIMy_\lambda$ when $LIMx_\lambda$ and $LIMy_\lambda$ both exist. Then the preceding results on order convergence of sequences are easily extended to nets. The results from Lemma 2.1.16 to Corollary 2.1.23 were proved for sequences. From now onwards we will make use of them for nets as freely as we do for sequences.

Incidentally, we may always suppose that an order convergent net is bounded in norm. This follows from the observation that when $(a_\lambda : \lambda \in \Lambda)$ is an increasing net with supremum a we can fix λ_0 and work instead with $(a_\lambda : \lambda \geq \lambda_0)$, which is bounded in norm by $||a_{\lambda_0}|| + ||a||$.

The following technical lemma is sometimes needed.

Lemma 2.1.24 *Let A be any C^*-algebra. Let $(x_\lambda : \lambda \in \Lambda)$ be an increasing net with supremum x. Let $y \in A_{sa}$ be such that $x_\lambda y x_\lambda \geq 0$ for each λ. Then $xyx \geq 0$.*

Proof We have

$$-xyx \leq x_\lambda y x_\lambda - xyx = (x - x_\lambda)y(x - x_\lambda) - (x - x_\lambda)yx - xy(x - x_\lambda).$$

We shall suppose the net to be bounded in norm by K. We have

$$(x - x_\lambda)y(x - x_\lambda) \leq ||y||(x - x_\lambda)^2 \leq ||y|| \, ||x - x_\lambda||(x - x_\lambda).$$

So

$$-xyx \leq (||x|| + K)||y||(x - x_\lambda) - (x - x_\lambda)yx - xy(x - x_\lambda).$$

Since $LIM(x - x_\lambda) = 0$ it follows that $LIMw(x - x_\lambda)v^* = 0$ for any w and v in A^1. So, applying this to the above inequality, gives $xyx \geq 0$.   □

The following inequality will be used in a later section.

Lemma 2.1.25 *Let A be any C^*-algebra. Let b, z and d be in A with d self-adjoint and $z^*z \leq b^*b$. Let θ be a complex number of modulus 1. Then*

$$\|(b + \theta z)d^2(b + \theta z)^*\| \leq 4\|db^*bd\|.$$

Proof We have

$$\|(b+\theta z)d^2(b+\theta z)^*\| = \|d(b+\theta z)^*(b+\theta z)d\| \leq 2\|d(b^*b+z^*z)d\| \leq 4\|db^*bd\|.$$

                                                                                            □

2.1.1 Operator Monotone

Operator monotone functions are often useful. We recall some basic information but for more details see [121]. Let f be a continuous real valued function defined on an interval of $\mathbb{R}$. Then f is said to be *operator monotone (increasing)* if, for x and y in A_{sa}, $0 \leq x \leq y$ implies $f(x) \leq f(y)$, whenever the spectra of x and y belong to the interval of definition of f.

For $\varepsilon > 0$ let $f_\varepsilon(t) = (1 + \varepsilon t)^{-1}t = \frac{1}{\varepsilon}(1 - (1 + \varepsilon t)^{-1})$ for t in the open interval $(-1/\varepsilon, \infty)$. Then f_ε is operator monotone.

Let $f(t) = t^\beta$ for $t \geq 0$. Then f is operator monotone for $0 < \beta \leq 1$. But NOT operator monotone for $\beta > 1$. In particular, if A is a C^*-algebra such that whenever $0 \leq x \leq y$ then $x^2 \leq y^2$, it can be proved that the algebra must be commutative.

When A is unital, if x is positive and invertible and

$$0 \leq x \leq y$$

then $0 \leq y^{-1} \leq x^{-1}$. See [121].

2.1.2 Monotone Closed Subspaces

Let P be any partially ordered set and let C be a subset of P. Then C is a *monotone closed subset* of P if it satisfies the following two conditions. First, whenever L is an upward directed set in C which has a least upper bound b in P then b is in C. Secondly, whenever D is a downward directed set in C which has an infimum b in P then b is in C.

We have a sequential version of this. We call C a monotone σ-closed *subset* of P if it satisfies the following conditions. First, whenever $(b_n) \uparrow$ in C and $(b_n) \uparrow b$ in P, then $b \in C$. Secondly, whenever $(b_n) \downarrow$ in C and $(b_n) \downarrow b$ in P, then $b \in C$.

Now let S be any subset of P. The *monotone closure* of S in P is the intersection of all monotone closed subsets of P which contain S. Similarly for *monotone σ-closure*.

Let A be any C^*-algebra. Let us recall that a subset $S \subset A$ is *self-adjoint* if $z \in S$ implies $z^* \in S$. We call S a *$*$-subspace* of A (or equivalently, a *self-adjoint subspace*) if it is a (complex) vector subspace of A and also self-adjoint. When S is a $*$-subspace of A, let $S_{sa} = S \cap A_{sa}$ then $S = S_{sa} + iS_{sa}$.

Definition 2.1.26 Let A be any C^*-algebra and let V be a $*$-subspace of A. Then V is a monotone closed subspace of A if, V_{sa} is a monotone closed subspace of A_{sa}.

This definition makes sense for any A but if A is monotone complete then, clearly, a monotone closed subalgebra will also be monotone complete.

The sequential version of this is as follows.

Definition 2.1.27 Let A be any C^*-algebra and let V be a $*$-subspace of A. Then V is a *monotone σ-closed subspace* of A if, V_{sa} is a monotone σ-closed subspace of A_{sa}.

The following exercises help clarify some points concerning monotone (σ-) closures. They also introduce a technique which is frequently useful. When there is no risk of ambiguity, we abbreviate *monotone σ-closure* to σ-*closure*.

Exercise 2.1.28 Let B be a monotone σ-complete algebra and A a self-adjoint subspace of B. Show there exists a smallest σ-closed subset of B_{sa} which contains A_{sa}. Let V be this set. Show that V is a (real) vector subspace of B_{sa}.
Hint:

(i) Let V be the intersection of all *monotone* σ-closed subsets of B_{sa} which contain A_{sa}. Show V is σ-closed.
(ii) Fix $a \in A_{sa}$. Let $W_a = \{x \in V : a + x \in V\}$. Show W_a is σ-closed and contains A_{sa}.
(iii) Deduce from (ii) that $A_{sa} + V \subset V$.
(iv) Fix $b \in B_{sa}$. Let $W = \{y \in V : y + b \in V\}$. Deduce from (iii) that $A_{sa} \subset W$.
(v) From (i), deduce that $V = V \cap (-V)$.
(vi) Let r be a strictly positive real number. Show rV is σ-closed.

Exercise 2.1.29 Let B be a monotone complete algebra and A a self-adjoint subspace of B. Show there exists a smallest monotone closed subset of B_{sa} which contains A_{sa}. Let V be this set. Show that V is a (real) vector subspace of B_{sa}.

The notion of a commutant is familiar from von Neumann algebra theory. For our purposes we proceed as follows. Let A be a C^*-algebra. As usual, when A is not unital, A^1 is the algebra formed by adjoining a unit; when A does have a unit we put

$A^1 = A$. When S is a non-empty subset of A^1, the commutant of S relative to A is

$$S' \cap A = \{a \in A : as = sa \text{ for all } s \in S\}.$$

When there is no risk of ambiguity, we shall use S' instead of $S' \cap A$.

Proposition 2.1.30 *Let A be any C^*-algebra and S a self-adjoint subset of A. Then $S' \cap A$ is a monotone closed C^*-subalgebra of A.*

Proof Let $B = S' \cap A$. Then, because S is self-adjoint, it is straightforward to check that B is a C^*-subalgebra of A. Let $C^*(S, 1)$ be the C^*-subalgebra of A^1 generated by S and the unit element. Then

$$C^*(S, 1)' \cap A = S' \cap A.$$

Let L be an upward directed set of self-adjoint elements of $S' \cap A$ which has supremum b in A_{sa}. Let u be any unitary in $C^*(S, 1)$. Then every $x \in L$ commutes with u. So $uxu^* = x$.

By Proposition 2.1.10, ubu^* is the supremum, in A_{sa}, of uLu^*. But, $uLu^* = L$. So $ubu^* = b$. Thus $ub = bu$. Since each element of $C^*(S, 1)$ is the linear combination of at most four unitaries, it follows that $b \in C^*(S, 1)' \cap A = S' \cap A = B$. So B is a monotone closed subalgebra of A.   □

Corollary 2.1.31 *We have $S \subset S''$ and S'' is a monotone closed subalgebra of A.*

Corollary 2.1.32 *When C is any abelian $*$-subalgebra of A then C'' is abelian with $C \subset C''$. When C is a maximal abelian $*$-subalgebra of A then $C = C''$ and C is a monotone closed C^*-subalgebra of A.*

Proof Let y and z be in C''. Because C is abelian $C \subset C'$. So, for each x in C, $xy = yx$. Hence $y \in C'$. So y commutes with z. Thus C'' is abelian.

If C is maximal abelian it follows that $C = C''$. By applying Proposition 2.1.30 to C' we see that C is a monotone closed subalgebra of A.   □

2.1.3   Regular Subalgebras and Subspaces

Let B be a C^*-algebra and A a closed $*$-subalgebra. Also let D be a downward directed set of positive elements of A. Since B has more elements than A it is possible for D to have an infimum in B but fail to have an infimum in A. But even if D does have an infimum in A it may be different from the infimum in B.

Example 2.1.33 Let $A = C([0, 1])$ and let B be the algebra of all bounded (complex valued) functions on $[0, 1]$. Now let D be the set of continuous real functions f, where $0 \leq f \leq 1$ and $f(1) = 1$. Then D is downward directed and, in A_{sa}, it has a greatest lower bound; the function which takes the constant value 0. But in B, D has a different greatest lower bound. It is d where $d(1) = 1$ and $d(t) = 0$ for $0 \leq t < 1$.

We will be particularly interested in embeddings where this misbehaviour does not occur.

Definition 2.1.34 Let W be a partially ordered (real) vector space and V a subspace. Then V is a *regular subspace* of W if, for each $b \in W$, the set

$$L(b) = \{a \in V : a \leq b\}$$

has least upper bound b in W. When this condition is satisfied, by multiplying by -1 we find that for each $b \in W$, the set

$$U(b) = \{a \in V : b \leq a\}$$

has greatest lower bound b in W. When A and B are C^*-algebras, with A a $*$-subalgebra of B, and A_{sa} is a regular subspace of B_{sa} then we call A a *regular subalgebra* of B.

Proposition 2.1.35 *Let W be a partially ordered (real) vector space and V a regular subspace. Let D be a subset of V with infimum d in V. Then D has infimum d in W.*

Proof Let $b \in W$ be a lower bound of D. When $a \in L(b)$ then a is a lower bound, in V, for D. So $a \leq d$. Thus d is an upper bound of $L(b)$. So $b \leq d$ because V is a regular subspace of W. Thus d is the greatest lower bound of D in W. $\square$

The following technical lemma will be useful later. The proof is based on [61].

Lemma 2.1.36 *Let B be a unital C^*-algebra. Let A be a regular subalgebra of B with the unit of B in A. For each $y \in B_{sa}$ let $L(y) = \{a \in A_{sa} : a \leq y\}$. Then $L(y)$ is not bounded below and cannot be assumed to be upward directed. But $L(y)$ contains a bounded subset which also has supremum y.*

Proof Let us observe that, in any unital C^*-algebra, z is positive and invertible if, and only if, there is a positive real number ε such that $z \geq \varepsilon 1$. By spectral theory, when this occurs z^{-1} is positive.

First suppose that y is positive and invertible. Let

$$L_0 = \{a^{-1} \in A_{sa} : a \geq y^{-1}\}.$$

Since $a \geq y^{-1}$ implies $a^{-1} \leq y$, we see that $L_0 \subset L(y)$. Also $L_0 \subset A_{sa}^+$. So L_0 is bounded. Let $w \in B_{sa}$ be an upper bound of L_0. Then $w^{-1} \leq a$ for each a in $U(y^{-1})$. Since A is a regular subalgebra of B, we have $w^{-1} \leq y^{-1}$. Thus $y \leq w$. So L_0 has the same least upper bound, in B, as $L(y)$.

Now let $x \in B_{sa}$. Then, for some real number λ, $x + \lambda 1$ is positive and invertible. So there exists $L_0 \subset L(x + \lambda 1)$ where L_0 is bounded and has supremum $x + \lambda 1$. Then $L_0 - \lambda 1$ is a bounded subset of $L(x)$ whose supremum is x. $\square$

Definition 2.1.37 Let A be a unital C^*-algebra. A regular σ-completion of A is a pair (B, j) where B is a monotone σ-complete unital C^*-algebra and j is a unital $*$-isomorphism of A into B with the following properties. First, $j[A]$ is a regular subalgebra of B. Secondly, B is σ-generated by $j[A]$ that is, B is the smallest monotone σ-closed subset of B which contains $j[A]$.

In the above definition it would seem natural to put "σ-closed subspace of B" instead of "σ-closed subset of B". But, by Exercise 2.1.28, this would make no difference.

If, in Definition 2.1.37, B is monotone complete and B is monotone generated by $j[A]$ then (j, B) is said to be a regular completion of A.

For commutative algebras, the existence of regular (σ-)completions is discussed in Chap. 4. In fact, we can construct the regular σ-completion of every C^*-algebra, see Chap. 5. This implies, in particular, that each "small" C^*-algebra has a regular completion. ("Small" has a technical meaning. This class is large enough to contain most of the examples of primary interest to us. It contains some algebras which are too big to be represented as C^*-algebras on separable Hilbert spaces. See the discussion in Chap. 5.) Hamana [61] extended the results of [173] by showing that every C^*-algebra has a regular completion by first constructing its injective envelope. We shall discuss this in a later chapter. Incidentally, other σ-completions have been considered [172].

2.1.4   On ℓ^2-Summable Sequences

The following is straight forward, but will not be used until Sect. 5.8, so the reader can postpone reading it until later. See also [92] and [63].

A family $\{a_\lambda\}_{\lambda \in \Lambda}$ in A is ℓ^2-summable if $\left\{ \sum_{\lambda \in F(\Lambda)} a_\lambda^* a_\lambda : F \in F(\Lambda) \right\}$ is bounded in A_{sa}, where $F(\Lambda)$ is the class of all (non-empty) finite subsets of Λ.

Lemma 2.1.38 *Let A be a monotone σ-complete unital C^*-algebra and let (a_n) and (b_n) be ℓ^2-summable sequences in A. Then*

$$LIM \sum_{n \geq 1} b_n^* a_n \text{ does exist in } A.$$

Proof Let $\{x_n\}$ and $\{y_n\}$ be ℓ^2-summable sequences in A. Note that for any $x, y \in A$, $(x + y)^*(x + y) \leq 2x^*x + 2y^*y$ and $4y^*x = \sum_{j=0}^{3} i^j(x + i^j y)^*(x + i^j y)$. Hence the sequence $\{x_n + i^j y\}$ is also ℓ^2-summable for each j and, since

$$4 \sum_{k=1}^{n} y_k^* x_k = \sum_{j=0}^{3} i^j \left(\sum_{k=1}^{n} (x_k + i^j y_k)^* (x_k + i^j y_k) \right)$$

on putting $a_n(j) = \sum_{k=1}^{n} (x_k + i^j y_k)^* (x_k + i^j y_k)$ for each n and j, each $(a_n(j))(n = 1, 2, \cdots)$ is a norm bounded increasing sequence in A_{sa} and so the sequence $(\sum_{k=1}^{n} y_k^* x_k)(n = 1, 2, \cdots)$ converges to an order limit in A. The general case can be shown similarly.   □

We shall sometimes write this limit as $LIM_{n \to \infty} \sum_{k=1}^{n} y_k^* x_k = \sum_{n=1}^{\infty} y_n^* x_n$.

Corollary 2.1.39 *Let A be a monotone complete C^*-algebra and let $\{a_\lambda\}_{\lambda \in \Lambda}$ and $\{b_\lambda\}_{\lambda \in \Lambda}$ be ℓ^2-summable families in A. Then*

$$LIM_{F \in F(\Lambda)} \sum_{\lambda \in F} b_\lambda^* a_\lambda \ \text{does exist in } A.$$

Proof In the proof of Lemma 2.1.38, replace sequences by nets.   □

2.2   Monotone σ-Complete C^*-Algebras

In this section we discuss some basic properties of monotone σ-complete C^*-algebras. We shall begin by considering projections and use them to derive a polar decomposition theorem and also to show that when A is monotone σ-complete then so, also, is A^1. We shall consider homomorphisms and σ-ideals and formulate a fundamental representation theorem. Of course every monotone complete algebra is σ-complete. We saw in Sect. 2.1 that when a monotone σ-complete algebra has a faithful state then it is monotone complete.

We shall see later that the classical type theory of von Neumann algebras can readily be extended to monotone complete algebras. We shall also give a classification theory for monotone complete algebras which is totally different from the von Neumann theory.

Recall that we can embed a non-unital C^*-algebra A in the unification A^1 as a closed two-sided ideal. When A is unital, we put $A^1 = A$. In this section, we shall denote the unit of A^1 by 1. We also recall that for a, b in A_{sa}, $a \leq b$ if, and only if, $a \leq b$ in A_{sa}^1. Let us recall that, in any C^*-algebra A, a *projection* is a self-adjoint p such that $p = p^2$. Obviously 0 and, if the algebra is unital, 1 are projections but for some algebras these are the only ones. When there are non-trivial projections they inherit a partial ordering from the partial ordering of the self-adjoint elements. We shall show below that monotone σ-complete algebras have an abundance of projections and obtain a result which helps us to show that if A is monotone σ-complete then so, also, is A^1. We shall denote the set of projections in A by *ProjA*.

Lemma 2.2.1 *Let p and q be projections in A. Then $p \leq q$ if, and only if, $pq = p$.*

Proof Let $pq = p$. Then $qp = (pq)^* = p$. Also $(p-q)^2 = p - pq - qp + q = q - p$. So $q \geq p$.

Conversely, let $q \geq p$. Then,

$$0 = (1-q)q(1-q) \geq (1-q)p(1-q) \geq 0.$$

So $0 = ||(1-q)p(1-q)|| = ||p(1-q)||^2$. Thus $p = pq$.   □

Lemma 2.2.2 *Let p and q be projections in A. Let $p \leq Kq$ where K is a strictly positive real number. Then $p \leq q$.*

Proof We have $(1-q)p(1-q) \leq 0$. Then $||p(1-q)||^2 = ||(1-q)p(1-q)|| = 0$. So $p = pq$.   □

Let us recall that when A is a C^*-algebra without a unit and $a \in A$, then $\sigma(a)$ is defined to be the spectrum of a in A^1. The functional calculus for C^*-algebras, see [161], enables us to give a sensible meaning to $f(a)$ when a is a normal element of a C^*-algebra A and $f \in C(\sigma(a))$. Suppose $0 \leq a \leq 1$. So $\sigma(a) \subset [0,1]$. Then the Gelfand transform gives an isometric $*$-isomorphism of $C(\sigma(a))$ onto the norm closed $*$-subalgebra of A^1 generated by a and 1. The image of f is denoted by $f(a)$. In particular, a is $h(a)$ where $h(t) = t$ for all $t \in \sigma(a)$. In the future, when we wish to indicate that the functional calculus gives a required result we shall often write "by the functional calculus" or "by spectral theory".

Let $f \in C(\{0\} \cup \sigma(a))$ with $f(0) = 0$. Let f be real-valued. Then, by the Stone-Weierstrass Theorem, there is a sequence of polynomials (ϕ_n), with real coefficients, such that $||f - \phi_n|| \to 0$. Then $\phi_n(0) \to 0$. Let $\psi_n(t) = \phi_n(t) - \phi_n(0)$. Then each ψ_n is a polynomial with zero constant term and, as before, $||f - \psi_n|| \to 0$. It follows that $||f(a) - \psi_n(a)|| \to 0$.

Let r be a positive real number. Let (f_n) be a monotone increasing sequence in $C([0,r])$ with $f_n(0) = 0$ for each n and, for $0 < t \leq r, f_n(t) \to 1$. We shall normally use $f_n(t) = (\frac{1}{n}1 + t)^{-1}t$. An alternative which is sometimes useful is $f_n(t) = t^{1/n}$ but this is only useable if $r \leq 1$. Now let $g \in C([0,r])$ where $g(0) = 0$ and $g \geq 0$. Then (gf_n) is a monotone increasing sequence which converges pointwise to g. It now follows from Dini's Theorem that the sequence converges uniformly to g. Thus $||g(a)f_n(a) - g(a)|| \to 0$.

Proposition 2.2.3 *Let A be monotone σ-complete. Then, for each $z \in A$, there is a smallest projection p such that $pz = z$. Furthermore, p is the supremum of $((\frac{1}{n}1 + zz^*)^{-1}zz^*)$. Also, p is the smallest projection such that $pzz^* = zz^*$ and the smallest projection such that $p(zz^*)^{1/2} = (zz^*)^{1/2}$.*

Proof Let $a = zz^*$. Let $f_n(t) = (\frac{1}{n}1 + t)^{-1}t$. Then, by spectral theory, $(f_n(a))$ is a norm bounded, monotone increasing sequence in A_{sa}. So it has a least upper bound, p, in A. By Proposition 2.1.11 p is also the least upper bound in A^1. So $p \leq 1$.

Let $g \geq 0$ with $g(0) = 0$ and $g \in C([0, ||a||])$. Then, see the remarks above, $(g(a)f_n(a))$ converges in norm to $g(a)$. Also $(g(a)^{1/2}f_n(a)g(a)^{1/2})$ is monotone increasing with least upper bound $g(a)^{1/2}pg(a)^{1/2}$. But this sequence has norm limit $g(a)$.

So we have $g(a)^{1/2}(1-p)g(a)^{1/2} = 0$. So $||(1-p)^{1/2}g(a)^{1/2}|| = 0$. Thus $(1-p)^{1/2}g(a) = 0$.

For each n we may replace g by f_n. Then $((1-p)^{1/2}f_n(a)(1-p)^{1/2})(n = 1, 2 \ldots)$ is a sequence of zeroes with least upper bound $(1-p)^{1/2}p(1-p)^{1/2}$. So $(1-p)^{1/2}p(1-p)^{1/2} = 0$. That is, $p = p^2$.

We have

$$g(a) - pg(a) = (1-p)g(a) = (1-p)^{1/2}0 = 0.$$

On taking adjoints we also have $g(a) = g(a)p$.

In particular, by putting $g(t) = t$, we get $a = pa = ap$.

Now suppose that q is a projection such that $a = qa$. Then, whenever ψ is a polynomial with zero constant coefficient, $q\psi(a) = \psi(a)$.

Let $f \in C(\{0\} \cup \sigma(a))$ with $f(0) = 0$ and f real-valued. Then f is the norm limit of a sequence of real polynomials, each with zero constant coefficient.

So $qf(a) = f(a)$. (In particular, $qa^{1/2} = a^{1/2}$.) By taking adjoints, $f(a) = qf(a) = qf(a)q$.

By Proposition 2.1.10, qpq is the least upper bound of $(qf_n(a)q)$. So $qpq = p$. Multiplying by $(1 - q)$ gives $0 = p(1 - q) = p - pq$.

So, by the preceding lemma, $p \le q$.

Let e be a projection. Then $z = ez$ if and only if $(1 - e)zz^*(1 - e) = 0$. That is

$$(1 - e)a(1 - e) = 0.$$

Equivalently $(1 - e)a^{1/2} = 0$.

Now $(1 - e)a^{1/2} = 0$ implies $(1 - e)a = 0$. Conversely, if $a = ea$ then, as remarked above $ea^{1/2} = a^{1/2}$.

So $z = ez$ if and only if $ea = a$ and this holds if and only if $ea^{1/2} = a^{1/2}$.   □

Definition 2.2.4 Let A be a monotone σ-complete C^*-algebra and $z \in A$. Then the left projection of z is the smallest projection p such that $pz = z$. And we denote it by $LP(z)$.

Proposition 2.2.3 tells us that $LP(z) = LP(zz^*) = LP((zz^*)^{1/2})$. We define the right projection of z to be $LP(z^*)$ and denote it by $RP(z)$. So, whenever x is self-adjoint, $LP(x) = RP(x)$; we also call this the range projection of x. For a self-adjoint b the functional calculus gives meaning to $|b|$. But for a general z in A we define $|z|$ to be $(z^*z)^{1/2}$.

Corollary 2.2.5 *Let A be monotone σ-complete. Then, for each $z \in A$, $RP(z)$ exists. Also $RP(z) = RP(z^*z) = RP(|z|)$. Furthermore $RP(z)$ is the supremum of $((\frac{1}{n}1 + z^*z)^{-1}z^*z))$.*

Proof Apply Proposition 2.2.3 to z^*.   □

Corollary 2.2.6 *Let A be monotone σ-complete. Each countable set of projections has a least upper bound in ProjA; it also has a greatest lower bound in ProjA. In other words, ProjA is a σ-complete lattice.*

Proof Let (p_n) be an enumeration of a countable set of projections in *ProjA*. Let $a = \sum 1/2^n p_n$ where the sum converges in the norm topology.

Let p be a projection in *ProjA* such that $(1-p)a = 0$. Since $a \leq 1$ we have $a = pap \leq p$.

So $p_n \leq 2^n p$. So, by Lemma 2.2.2, p is an upper bound for $\{p_n : n = 1, 2 \ldots\}$. In particular, the range projection of a is an upper bound.

Now let q be any upper bound for $\{p_n : n = 1, 2 \ldots\}$ in *ProjA*. Then

$$(1-q)a = \sum 1/2^n (1-q)p_n = 0.$$

So the set of (projection) upper bounds of $\{p_n : n = 1, 2 \ldots\}$ is the same as the set of p such that $a = pa$.

It now follows from the preceding proposition that there is a smallest projection which is greater than each p_n. We denote this by $\bigvee\limits_{n=1}^{\infty} p_n$. It is straightforward to verify that

$$p - \bigvee_{n=1}^{\infty} (p - p_n) \text{ is the greatest lower bound of } (p_n).$$

In particular, each finite collection of projections in A has a supremum *in the set of projections*. So *ProjA* is a lattice. □

By Corollary 2.2.6, when A is monotone σ-complete then *ProjA* is a lattice. But, if A is not commutative, a pair of projections may fail to have a supremum in A_{sa}. See the discussion below.

Proposition 2.2.7 *Let (p_n) be a monotone increasing sequence of projections in a C^*-algebra A. If $LIMp_n$ exists in A it is a projection. If A is monotone σ-complete then*

$$LIMp_n = \bigvee_{n=1}^{\infty} p_n.$$

Proof Let $x = LIMp_n$. Then $xp_m = LIM(p_n p_m) = p_m$. So

$$x^2 = xLIMp_m = LIMxp_m = LIMp_m = x.$$

Then x is a projection and is the smallest projection which is greater than each p_n. □

By a slight modification of the above argument we have:

Proposition 2.2.8 *Let A be a monotone complete C^*-algebra. If $(p_\lambda : \lambda \in \Lambda)$ is an upward directed net of projections then $LIMp_\lambda$ is the smallest projection in A which is greater than each p_λ. Furthermore, ProjA is a complete lattice.*

Proof By adapting the above argument we can show that $LIMp_\lambda$ is the smallest projection in A which is greater than each p_λ.

Now let $Q = \{q_t : t \in T\}$ be a non-empty collection of projections in A. Let Λ be the collection of all finite, non-empty, subsets of T. For each $\lambda \in \Lambda$, let

$$p_\lambda = \bigvee \{q_t : t \in \lambda\}.$$

Then $LIMp_\lambda$ is the supremum of Q in $ProjA$. $\qquad\qquad\qquad\qquad\qquad$ □

Lemma 2.2.9 *Let A be a monotone σ-complete C^*-algebra which does not possess a unit. Let $\{a_r : r = 1, 2..\}$. be any countable subset of A_{sa}. Then there exists a projection p such that $pa_r = a_r = a_r p$ for each r.*

Proof Let p be the supremum in $ProjA$ of the range projections of each a_r. $\qquad$ □

Proposition 2.2.10 *Let A be a monotone σ-complete C^*-algebra. Then A^1 is monotone σ-complete.*

Proof We shall suppose that A does not possess a unit. For, otherwise $A^1 = A$ and there is nothing to prove. Let 1 be the unit of A^1.

Let $(\lambda_n 1 + x_n)$ be a norm-bounded monotone increasing sequence in $(A^1)_{sa}$, where each x_n is in A_{sa}.

By taking the quotient homomorphism onto A^1/A we find that (λ_n) is a bounded monotone increasing sequence in $\mathbb{R}$.

Hence it converges to λ_∞, say.

Let q be a projection in A which is bigger than the range projections of each x_r. Then

$$\lambda_n 1 + x_n = \lambda_n(1 - q) + \lambda_n q + x_n.$$

Then $\lambda_n q + x_n = q(\lambda_n 1 + x_n)q$. So $(\lambda_n q + x_n)$ is a norm-bounded monotone increasing sequence in A. So it has a supremum a in A_{sa} and hence, by Proposition 2.1.11, supremum a in $(A^1)_{sa}$.

We also have $(\lambda_n(1 - q))$ is a monotone increasing sequence in $(A^1)_{sa}$ with supremum $\lambda_\infty(1-q)$. (The sequence converges in the norm topology.) So $(\lambda_n 1 + x_n)$ has supremum $\lambda_\infty(1 - q) + a$. It follows that A^1 is monotone σ-complete. □

Let A be a C^*-algebra. Consider a monotone increasing sequence, $(x_n) \uparrow$ in A. Call the sequence *upper bounded* if there exists $c \in A_{sa}$ such that $x_n \leq c$ for all n. Then it is easy to check that every upper bounded monotone increasing sequence is bounded in norm. For $x_n - x_1 \leq c - x_1$ and so

$$||x_n|| \leq ||x_n - x_1|| + ||x_1|| \leq ||c - x_1|| + ||x_1||.$$

What about the converse? When A is *unital* it is easy to see that each norm bounded, monotone increasing sequence is upper bounded.

Call a C^*-algebra *pseudo monotone σ-complete* if each upper bounded, monotone increasing sequence has a least upper bound. Then, for unital algebras, it follows from the remarks above that each pseudo monotone σ-complete algebra is monotone σ-complete. But for non-unital algebras this breaks down.

Example 2.2.11 Let H be a separable, infinite dimensional Hilbert space and $K(H)$ the C^*-algebra of compact operators on H. Let (c_n) be a monotone increasing sequence in $K(H)_{sa}$ and suppose it has an upper bound c in $K(H)$. Then, in $L(H)$, (c_n) has a least upper bound x and $c_n \to x$ in the strong operator topology. Then $0 \leq x - c_1 \leq c - c_1$. Here $c - c_1$ is a positive compact operator. Since $K(H)$ is a closed ideal of $L(H)$, its self-adjoint part is an order ideal of $L(H)_{sa}$. So $x - c_1$ is compact. So x is a least upper bound of (c_n) in $K(H)$. Thus $K(H)$ is pseudo monotone σ-complete. But it is not monotone σ-complete. To see this take a projection $p \in L(H)$ such that the range of p is infinite dimensional. Let (p_n) be a monotone increasing sequence of finite rank projections such that $p_n \to p$ in the strong operator topology. Clearly this sequence is bounded in norm. If c were a compact operator which was an upper bound for this sequence then we would have $p \leq c$. Then, as before, this would imply that p is compact. But this is absurd.

Proposition 2.2.12 *Let A be a monotone σ-complete C^*-algebra. Let $b \in A_{sa}$ and $z \in A$ be such that $z^*z \leq b^2$. Then we can find $c \in A$ such that $z = cb$. Furthermore, $c = LIMz(\frac{1}{n}1 + b^2)^{-1}b$ and $||c|| \leq 1$.*

Proof By dividing by a suitable constant if necessary, we can suppose that $||b|| \leq 1$. Let $x_n = (\frac{1}{n}1 + b^2)^{-1}(\in A^1)$. Then, by Corollary 2.2.5 (x_nb^2) is a monotone increasing sequence in A with supremum $RP(b) = RP(b^2)$. Let $c_n = zx_nb$. We have the polarisation identity

$$c_n = \frac{1}{4}\sum_{k=0}^{3} i^k(b + i^kz^*)^*x_n(b + i^kz^*).$$

By applying Lemma 2.1.25 with b self-adjoint and $d = x_n^{1/2}$ we find that each of the monotone increasing sequences $((b + i^kz^*)^*x_n(b + i^kz^*))(n = 1, 2\ldots)$ is bounded in norm by $4||x_n^{1/2}b^2x_n^{1/2}|| = 4||b^2(\frac{1}{n}1 + b^2)^{-1}|| \leq 4$. Hence $LIMc_n$ exists. Let $c = LIMc_n$. Then

$$cb = LIMc_nb = LIMzx_nb^2 = zRP(b).$$

But

$$(1 - RP(b))z^*z(1 - RP(b)) \leq (1 - RP(b))bb(1 - RP(b)) = 0.$$

So $z(1 - RP(b)) = 0$. Thus $cb = z$.

It remains to show $||c|| \leq 1$. We observe that

$$cRP(b) = (LIMc_n)RP(b) = LIM(c_nRP(b)) = LIMc_n = c.$$

We have $bc^*cb = z^*z \le b^2$. So $b(1 - c^*c)b \ge 0$. This implies

$$(\frac{1}{n}1 + b)^{-1}b(1 - c^*c)b(\frac{1}{n}1 + b)^{-1} \ge 0.$$

We know that $((\frac{1}{n}1+b)^{-1}b) \uparrow RP(b)$. So, by Lemma 2.1.24, $RP(b)(1-c^*c)RP(b) \ge 0$. So $RP(b) - c^*c \ge 0$. Thus $||c|| \le 1$.   □

2.2.1 Open Problem

Let $a \in A_{sa}$. We showed that if A is monotone σ-complete then a has a range projection, that is a smallest projection p such that $pa = a = ap$. But an examination of the proof shows that we could prove the same result under a weaker hypothesis on A. It is sufficient to suppose that if (a_n) is a norm bounded, monotone increasing sequence in A_{sa} and the terms of the sequence are mutually commutative then (a_n) has a supremum in A_{sa}. Why did we not use this weaker hypothesis? Because no one has ever seen an algebra with this property which is not monotone σ-complete. Let A be a C^*-algebra where each maximal abelian $*$-subalgebra is monotone σ-complete. Is A, itself, monotone σ-complete? This is unknown.

We recently discovered [146] that if each maximal abelian $*$-subalgebra in A is monotone σ-complete and A is unital then A is a Rickart C^*-algebra. (Definition 3.3 in [13]). See also [3, 5].

So the question reduces to: are all Rickart C^*-algebras monotone σ-complete?

2.2.2 Polar Decomposition

Let us recall that, in any C^*-algebra, w is a *partial isometry* if ww^* is a projection.

Lemma 2.2.13 *Let w be a partial isometry. Then w^*w is also a projection.*

Proof We observe that $(w^*w)^3 = w^*(ww^*)^2w = w^*(ww^*)w = (w^*w)^2$. So, by spectral theory, w^*w is a projection.   □

In any C^*-algebra, a projection p is (Murray-von Neumann) equivalent to q if there exists a w in the algebra such that $p = w^*w$ and $q = ww^*$. We shall write $p \sim q$ to indicate that two projections are equivalent. We shall show that when A is monotone σ-complete then, for any z, we can find a partial isometry u such that $z = u|z|$ and $u^*u = LP(z)$ and $uu^* = RP(z)$.

Let us recall [161] that in any C^*-algebra A, for each x and y we have

$$\sigma(xy) \cup \{0\} = \sigma(yx) \cup \{0\}.$$

Lemma 2.2.14 *Let A be any C^*-algebra and $z \in A$. Then for any $f \in C(\sigma(zz^*) \cup \{0\})$*

$$zf(z^*z) = f(zz^*)z.$$

Proof By induction on n, this holds when $f(t) = t^n$. Hence, by linearity, it holds for all polynomials. So, by Stone-Weierstrass, it holds when f is real valued. For the general situation, split f into real and imaginary parts.   □

Theorem 2.2.15 *Let A be monotone σ-complete. For any $z \in A$ there is a partial isometry u such that $z = u|z|$. Also $u^*u = LP(z)$ and $uu^* = RP(z)$.*

Proof Let $u_n = z(\frac{1}{n}1 + |z|^2)^{-1}|z|$.

In Proposition 2.2.12 put $b = |z|$. Then $LIMu_n = u$ where $z = u|z|$.

So $u_m^*u_n = |z|(\frac{1}{m}1 + |z|^2)^{-1}z^*z(\frac{1}{n}1 + |z|^2)^{-1}|z|$.

We recall that $z^*z = |z|^2$ and, by Corollary 2.2.5, $LIM(\frac{1}{n}1 + |z|^2)^{-1}|z|^2 = RP(z)$.

We have

$$u^*u = LIM_{m \to \infty}(LIM_{n \to \infty}u_m^*u_n) = RP(z)^2 = RP(z).$$

Now consider

$$u_nu_m^* = z(\frac{1}{n}1 + |z|^2)^{-1}|z||z|(\frac{1}{m}1 + |z|^2)^{-1}z^*$$

$$= z(\frac{1}{n}1 + z^*z)^{-1}z^*z(\frac{1}{m}1 + z^*z)^{-1}z^*.$$

We now apply Lemma 2.2.14

$$u_nu_m^* = (\frac{1}{n}1 + zz^*)^{-1}zz^*(\frac{1}{m}1 + zz^*)^{-1}zz^*.$$

First keep m fixed and let $n \to \infty$. Then let $m \to \infty$. We find that

$$uu^* = LP(z).$$

□

2.2.3 Sequentially Closed Subspaces

Let L be a vector subspace of a C^*-algebra A. We recall that L is said to be a self-adjoint subspace, or $*$-subspace if $z \in L$ implies $z^* \in L$. We defined (monotone) σ-subspaces in 2.1.27. When we refer to an ideal of a C^*-algebra we always mean a norm-closed, self-adjoint, two sided ideal, unless we specifically state otherwise.

Definition 2.2.16 Let A be a monotone σ-complete C^*-algebra. Then a σ-ideal is an ideal of A which is a monotone σ-subspace of A. Similarly, a σ-subalgebra is a $*$-subalgebra which is a monotone σ-subspace of A.

When A is monotone σ-complete and L is a closed $*$-subalgebra which is a σ-subspace then it is easy to check that for any $a \in L_{sa}$ the range projection of a, obtained in A, is the same as its range projection constructed in L.

For the rest of this section, J is a closed ideal of a C^*-algebra A and ρ is the quotient homomorphism onto A/J. It is easy to see that each self-adjoint element of the quotient is $\rho(a)$ for some $a \in A_{sa}$. That is $(A/J)_{sa} = \rho[A_{sa}]$. Let d be a positive element of A/J then, since ρ is surjective, $d^{1/2} = \rho(w)$ for some w. Hence $ww^* \in A^+$ and $\rho(ww^*) = d$. So $\rho[A^+] = (A/J)^+$.

The following "intermediate value theorem" is a key technical result.

Lemma 2.2.17 Let $a \in A^+$. Let $0 \le y \le \rho(a)$. Then there exist z in A^+ such that $\rho(z) = y$ and $0 \le z \le a$.

Proof See Proposition I.5.10 in [121]. For our purposes, we only need this result for the special case where A is monotone σ-complete. Making this assumption we can argue as follows.

Since $\rho(a) - y$ is positive, it is $\rho(w)$ for some positive w. Let $x = a - w$. Then $x \le a$ and $\rho(x) = y$.

By C^*-algebra spectral theory, there exist unique positive x^+ and x^- such that $x = x^+ - x^-$ and $x^+ x^- = 0$. Then $\rho(x^+)$ and $\rho(x^-)$ provide the unique decomposition for $\rho(x) = y$. So $\rho(x^+) = y$ and $\rho(x^-) = 0$.

We have $x^+ \le x^- + a$. So, by Proposition 2.2.12, there exists s, with $||s|| \le 1$, such that $(x^+)^{1/2} = s(x^- + a)^{1/2}$.

Put $u = sa^{1/2}$. Then $u^* u = a^{1/2} s^* s a^{1/2} \le a$.

Since $\rho(a) = \rho(a + x^-)$ we have $\rho(a^{1/2}) = \rho((a + x^-)^{1/2})$. From this it follows that

$$\rho(u^* u) = \rho((a + x^-)^{1/2}) \rho(s^* s) \rho((a + x^-)^{1/2})$$
$$= \rho(x^+)^{1/2} \rho(x^+)^{1/2} = \rho(x^+) = y.$$

Put $z = u^* u$.   □

Corollary 2.2.18 Let (b_n) be a monotone increasing sequence of positive elements of A/J which is bounded above by $\rho(c)$, with $c \ge 0$. Then there exists a monotone increasing sequence, (x_n) in A^+ such that c is an upper bound for this sequence and $\rho(x_n) = b_n$ for each n.

Proof By the lemma there exists x_1 in A^+ such that $\rho(x_1) = b_1$ and $x_1 \le c$.

Then $0 \le b_2 - b_1 \le \rho(c - x_1)$ with $c - x_1 \ge 0$.

So we may apply the lemma again to find $x \in A^+$ such that $\rho(x) = b_2 - b_1$ and $x \le c - x_1$.

Now let $x_2 = x + x_1$. Then $0 \leq x_1 \leq x_2 \leq c$ and $\rho(x_2) = b_2$. Proceeding inductively, we obtain (x_n).   □

Lemma 2.2.19 *Let A be monotone σ-complete. Let J be a σ-ideal and ρ the quotient homomorphism from A onto A/J. Let $(x_n) \uparrow x$ and $(a_n) \uparrow a$. Let $\rho(x_n) = \rho(a_n)$ for each n. Then $\rho(x) = \rho(a)$.*

Proof Since $x_n - a_n \in J_{sa}$ we have $RP(x_n - a_n) \in J$. Let q be the supremum of these range projections in $Proj J$.

By Corollary 2.2.6 we can find a projection e in A such that it is greater than q, $RP(x)$, $RP(a)$ and, also, greater than $RP(x_n)$ and $RP(a_m)$ for each n and m.

Then $(e - q)x_n = (e - q)a_n$.

But $((e - q)x_n(e - q)) \uparrow (e - q)x(e - q)$ and $(e - q)a_n(e - q) \uparrow (e - q)a(e - q)$.

So $\rho((e - q)x(e - q)) = \rho((e - q)a(e - q))$.

Since ρ is a homomorphism and $\rho(q) = 0$ it follows that $\rho(exe) = \rho(eae)$. So $\rho(x) = \rho(a)$. See [27].   □

Definition 2.2.20 (i) Let A and B be monotone σ-complete C^*-algebras and ϕ : $A \mapsto B$ a positive linear map. Then ϕ is said to be σ-*normal* if, whenever $(a_n) \uparrow a$ in A_{sa} then $\phi(a_n) \uparrow \phi(a)$ in B_{sa}. If a σ-normal map is a homomorphism we call it a σ-*homomorphism*. (ii) Let A and B be monotone complete. The map ϕ is said to be *normal* if, whenever $D \subset A_{sa}$ is a upward directed set with supremum a in A, then $\phi[D]$ has supremum b in B.

Proposition 2.2.21 *Let A be a monotone σ-complete C^*-algebra with a closed ideal J. Let $\rho : A \mapsto A/J$ be the quotient homomorphism. Then A/J is monotone σ-complete and ρ is a σ-homomorphism if, and only if, J is a σ-ideal.*

Proof Let J be a σ-ideal.

(i) Let $(a_n) \uparrow a$. Clearly $\rho(a)$ is an upper bound of $(\rho(a_n))$. We shall show that this is the least upper bound. It suffices to prove this when (a_n) is a sequence in A^+; for, otherwise, we could work with $(a_n - a_1)$.

Let d be any upper bound of $(\rho(a_n))$. Then, see the remarks preceding Lemma 2.2.17, $d = \rho(c)$ for some $c \geq 0$.

By Corollary 2.2.18 we can find $(x_n) \uparrow$ in A^+ such that $\rho(x_n) = \rho(a_n)$ and $x_n \leq c$.

Let $(x_n) \uparrow x$. Then by Lemma 2.2.19 we have $\rho(a) = \rho(x) \leq \rho(c) = d$.

This tells us that ρ is a σ-homomorphism provided we can show that A/J is monotone σ-complete.

(ii) First suppose that A is unital.

Let (c_n) be a monotone increasing sequence in A/J which is bounded in norm. Since A/J has a unit, the sequence is bounded above by a multiple of the unit element. Let $b_n = c_n - c_1$. Then it follows from Corollary 2.2.18 that there is $(z_n) \uparrow z$ in A^+ such that $\rho(z_n) = b_n$. It now follows from the argument above, that $(b_n) \uparrow \rho(z)$. So $(c_n) \uparrow \rho(z) + c_1$. Thus A/J is monotone σ-complete.

(iii) Now suppose that A is not unital. By Proposition 2.2.10, A^1 is monotone σ-complete. It is easy to see that J is a closed ideal of A^1. By applying

Proposition 2.1.11, it is straightforward to verify that J is a σ-ideal of A^1. It now follows from the preceding paragraph that A^1/J is monotone σ-complete. We abuse our notation by using ρ for the quotient homomorphism of A^1 onto A^1/J. Then $\rho[A] = A/J$.

Let (b_n) be a norm bounded, monotone increasing sequence in A/J. We wish to show it has a supremum in A/J. Without loss of generality, we assume $b_1 \geq 0$. Let d be the sequence's supremum in A^1/J. Then $d = \rho(x)$ for some positive x in A^1. By Corollary 2.2.18 we can find $(y_n) \uparrow$ in A^1 such that $\rho(y_n) = b_n$ and $y_n \leq x$. For each n, $b_n = \rho(a_n)$ for some $a_n \in A$. Then $y_n - a_n \in J$. Since $J \subset A$ it follows that $y_n \in A$.

The sequence (y_n) is norm bounded and A is monotone σ-complete. So $(y_n) \uparrow y$ for some y in A. By Proposition 2.1.11, y is the supremum of the sequence in A^1. By applying part (i), we have $\rho(y) = d$. So d is in $\rho[A] = A/J$. Thus A/J is monotone σ-complete.

(iv) Conversely, let ρ be a σ-homomorphism onto a monotone σ-complete algebra. Then it is clear that the kernel of ρ is a σ-ideal.   □

Let H be a Hilbert space and $L(H)$ the C^*-algebra of all bounded operators on H. Then $L(H)$ is monotone complete. Let $(a_\lambda : \lambda \in \Lambda)$ be a norm bounded, increasing net in $L(H)_{sa}$. Then this net converges in the strong operator topology to $a \in L(H)$, that is, $a_\lambda \xi \to a\xi$ (in the norm topology of H) for each $\xi \in H$. Equivalently, $a_\lambda \to a$ in the weak operator topology. By Corollary 2.1.8 it follows that a is the least upper bound of $(a_\lambda : \lambda \in \Lambda)$.

Definition 2.2.22 Let B be a C^*-subalgebra of $L(H)$. The *Pedersen envelope* of B in $L(H)$ is the smallest σ-subspace of $L(H)$ which contains B. We shall denote it by B^σ.

By a theorem of Pedersen [118] (see also Corollary 5.3.8), B^σ is a C^*-algebra. It is now clear that B^σ is a σ-subalgebra of $L(H)$. Then we can easily find a sequentially order compatible, Hausdorff, locally convex topology for B^σ; either the weak operator topology or the strong operator topology will do.

Let A be an "abstract" C^*-algebra. For simplicity, let A have a unit element. Let us recall that by taking the direct sum of all the Gelfand-Naimark-Segal representations of A, we obtain the universal representation (π_u, H_u). Then π_u is an (isometric) $*$-isomorphism of A onto $\pi_u[A]$. Let A'' be the closure of $\pi_u[A]$ in the strong operator topology of $L(H_u)$; then A'' is the double commutant of $\pi_u[A]$ in $L(H_u)$. Then we can identify A with $\pi_u[A]$. The second dual of A, as a Banach space, can be identified with A''.

Definition 2.2.23 We define the *Pedersen-Baire envelope of* A to be the Pedersen envelope of $\pi_u[A]$ in $L(H_u)$. We shall denote it by A^∞. So $A^\infty = \pi_u[A]^\sigma$.

Then A^∞ is a σ-subalgebra of $L(H_u)$. We shall return to this topic in Chap. 5, where we shall also consider other envelopes.

The Wright Representation Theorem [171] below tells us that *each* monotone σ-complete C^*-algebra is the quotient of a σ-subalgebra of some $L(H)$.

Theorem 2.2.24 *Let A be a unital monotone σ-complete C^*-algebra. There exists a σ-homomorphism $\rho : A^\infty \mapsto A$ such that $\rho(a) = a$ for each $a \in A$. Let J be the kernel of ρ. Then J is a σ-ideal of A^∞ and $A^\infty/J = A$.*

It can be extended to non-unital algebras. More importantly, by taking a larger envelope than A^∞ we get a representation which works well for general monotone complete algebras. We shall postpone proofs until later, when we have more tools at our disposal.

We have seen above that the quotient of a monotone σ-complete algebra by a σ-ideal is always monotone σ-complete. But sometimes the quotient may turn out to be monotone complete. For example, in a later section, we shall show the following. Let X be a compact Hausdorff space and $B(X)$ the algebra of bounded Borel measurable functions on X. Let J be the set of $f \in B(X)$ for which $\{x \in X : f(x) \neq 0\}$ is a meagre set. Then $B(X)$ is a commutative monotone σ-complete algebra and J is a σ-ideal. So $B(X)/J$ is, necessarily, monotone σ-complete. But more is true, this quotient algebra is monotone complete.

Another, very familiar, example occurs if we take $X = [0, 1]$ and J to be those f in $B([0, 1])$ such that $\int |f| d\mu = 0$. For then $B([0, 1])/J$ is a (commutative) von Neumann algebra and so monotone complete.

The following propositions will be used later.

Proposition 2.2.25 *Let A be monotone σ-complete and let e be a projection in A. Then the map*

$$x \mapsto exe$$

is σ-normal from A onto eAe. Also eAe is monotone σ-closed in A (and so the natural inclusion map is σ-normal from eAe into A).

Proof Suppose (a_n) is a norm-bounded, monotone increasing sequence in $eA_{sa}e$ with supremum a in A_{sa}. By Proposition 2.1.22,

$$(1 - e)a = (1 - e)LIMa_n = LIM(1 - e)a_n.$$

But $(1 - e)a_n = 0$ for all n. So $(1 - e)a = 0$. So $a = ea$. Taking adjoints gives $a = ae$. Hence $a \in eAe$. It follows that eAe is monotone σ-complete and is a σ-subspace of A.

Now let (a_n) be any norm-bounded, monotone increasing sequence in A_{sa}. Let its supremum in A be a. Then (ea_ne) is monotone increasing and its supremum in A_{sa} is

$$LIMea_ne = e(LIMa_n)e = eae.$$

Since eAe is a σ-subspace of A, eae is also the supremum of (ea_ne) in eAe. □

Proposition 2.2.26 *When A is monotone complete the map $x \mapsto exe$ is normal. Also eAe is monotone closed in A (and so the natural inclusion map is normal from eAe into A).*

The proof is a straightforward generalisation of Proposition 2.2.25.

2.3 Basics for Commutative Algebras

Each maximal abelian $*$-subalgebra of a monotone complete C^*-algebra is, itself, a (commutative) monotone complete C^*-algebra.

There is a duality between compact Hausdorff spaces and commutative (unital) C^*-algebras. i.e. for each compact Hausdorff space X, $C(X)$ is a commutative C^*-algebra; conversely, given a commutative unital C^*-algebra A it is isomorphic to $C(\Omega)$ where Ω is its spectrum.

What property of Ω corresponds to $C(\Omega)$ being monotone complete? The answer is that Ω must be *extremally disconnected* i.e. the closure of each open subset is open.

Throughout this section, indeed throughout this book, *we shall always suppose that* the topologies we use are Hausdorff unless we explicitly state the contrary. In this section X is a compact (Hausdorff) space. Let Y be a topological space where Y is not assumed to be compact. We use $C_b(Y)$ to denote the algebra of *bounded* complex valued continuous functions on Y and $C_b(Y)_{sa}$ the (real) algebra of bounded real valued continuous functions on Y.

We recall that in any topological space, a subset is *clopen* if it is both closed and open. It is easy to see that $f \in C(X)$ is a projection if, and only if, it is the characteristic function of a clopen set. Let us recall that a topological space is *zero-dimensional* if its topology has a base of clopen sets.

Lemma 2.3.1 *Let X be a compact space. For each pair of distinct points t_1 and t_2, let there exist a clopen K such that $t_1 \in K$ and $t_2 \notin K$. Then X is zero-dimensional. Furthermore given disjoint closed sets E and F, we can find a clopen K such that $E \subset K$ and $F \subset X \backslash K$.*

Proof

(i) Let $t_0 \in X$ and F a closed subset of X with $t_0 \notin F$. For each $s \in F$ there is a clopen Q_s with $s \in Q_s$ and $t_0 \notin Q_s$. By compactness, there are finitely many points $\{s_1, \ldots, s_n\}$ such that the corresponding clopen sets cover F. We now put $K = \cap \{X \backslash Q_k : k = s_1, \ldots, s_n\}$. Then K is a clopen neighbourhood of t_0 and K is disjoint from F. Hence the clopen subsets form a base for the topology.

(ii) For each $t \in E$ there is a clopen K_t with $t \in K_t$ and F disjoint from K_t. Since E is compact there are finitely many points $\{t_1, \ldots, t_n\}$ such that the corresponding clopen sets cover E. Let $K = K_{t_1} \cup K_{t_2} \cup \ldots \cup K_{t_n}$. $\qquad\square$

Proposition 2.3.2 *Let X be a compact (Hausdorff) space such that C(X) is σ-complete. Then X is zero-dimensional.*

Proof Suppose s and t are distinct points. Then there exist disjoint open sets U and V where $s \in U$ and $t \in V$. By Urysohn's lemma, there exist continuous functions $a : X \mapsto [0, 1]$ and $b : X \mapsto [0, 1]$ such that $a(s) = 1$ and $a[X \backslash U] = \{0\}$ and also $b(t) = 1$ and $b[X \backslash V] = \{0\}$.

Since U and V are disjoint $ab = 0$. Let p be the range projection of a. Then the sequence $((\frac{1}{n}1 + a)^{-1}a)$ is monotone increasing to p. So $pb = 0$.

Since $0 \le a \le 1$ it follows that $a = pa \le p$. Now let K be the clopen set such that $p = \chi_K$. Since $a(s) = 1$ it follows that $\chi_K(s) = 1$. That is, $s \in K$.

But $pb = 0$. So $p(t)b(t) = 0$. Since $b(t) = 1$ we have $p(t) = 0$. That is $t \notin K$.

We can now apply the preceding lemma to deduce that X is zero-dimensional. □

Let X be a compact zero-dimensional space. We call $s \in C(X)$ *simple* if it takes only finitely many values. This is equivalent to

$$s = \sum_{i=1}^{n} \lambda_i \chi_{K_i}$$

where $\{K_1, K_2, \ldots K_n\}$ is a finite collection of pairwise disjoint clopen sets; equivalently s is a finite linear combination of orthogonal projections. A straightforward application of the Stone-Weierstrass Theorem shows that the simple continuous functions are norm-dense in $C(X)$. In other words, each element of $C(X)$ is the limit in norm of a sequence of finite linear combinations of projections.

Let Y be any non-empty set and $Bnd(Y)$ the set of all bounded complex valued functions. Then with the natural pointwise operations of addition, multiplication etc., it becomes a *-algebra. We define a norm by

$$\|f\| = \sup\{|f(y)| : y \in Y\}.$$

Then $Bnd(Y)$ becomes a commutative C^*-algebra. It is monotone complete, in fact a von Neumann algebra. The self-adjoint part of $Bnd(Y)$ consists of the bounded real valued functions on Y. Given f, g in $Bnd(Y)_{sa}$ we define $f \vee g$ by

$$(f \vee g)(y) = \max(f(y), g(y)) = 1/2(f(y) + g(y) + |f(y) - g(y)|).$$

We define $f \wedge g$ in a similar way. Then $Bnd(Y)_{sa}$ is a vector lattice.

When Y is a topological space and f, g are continuous real valued functions then $f \vee g$ and $f \wedge g$ are continuous functions.

Let X be compact and $C(X)$ monotone σ-complete. Let (f_n) be a sequence in $C(X)_{sa}$ which is bounded above. Let

$$g_n = f_1 \vee f_2 \vee \ldots \vee f_n.$$

Then (g_n) is a monotone increasing sequence in $C(X)_{sa}$ with the same set of upper bounds as (f_n). So every upper bounded sequence has a least upper bound in $C(X)_{sa}$.

Clearly the supremum of (g_n) in $Bnd(X)_{sa}$ is the function which is the pointwise limit of the sequence i.e. $\underline{g}$ where $\underline{g}(x) = \sup(g_n(x))$ for each x in X. But, in general, $\underline{g}$ is not continuous; although it is always lower semicontinuous. So the supremum of (g_n) in $C(X)_{sa}$ is g, where g is the smallest continuous function in $C(X)_{sa}$ for which $g \geq \underline{g}$.

For any topological space Y, a real valued function $f : Y \mapsto \mathbb{R}$ is *lower semicontinuous* if $\{y \in Y : f(y) > t\}$ is open for each real number t. So each (real valued) continuous function is also lower semicontinuous. The characteristic function of $U \subset Y$ is lower semicontinuous precisely when U is open. Clearly rf is lower semicontinuous, when f is, whenever r is a positive real number.

We define $u : Y \mapsto \mathbb{R}$ to be upper semicontinuous if $-u$ is lower semicontinuous. This is equivalent to $\{y \in Y : u(y) < t\}$ is open for each real number t.

Lemma 2.3.3 *Let $\{f_\lambda\}$ be an upper bounded family of continuous, real-valued functions on a topological space Y. Let $f(y) = \sup f_\lambda(y)$ for each $y \in Y$. Then f is lower semicontinuous.*

Proof We observe that $\{y \in Y : f(y) > t\} = \bigcup \{y \in Y : f_\lambda(y) > t\}$. $\square$

The same argument shows that if we weaken the hypotheses to each f_λ being lower semicontinuous we still obtain the same conclusion.

When Y is completely regular then there is a converse to this lemma. Let us recall that Y is *completely regular* if, whenever F is a closed subset of Y and $y_0 \in Y \backslash F$ then there is a continuous function $g : Y \mapsto [0, 1]$ with $g(y_0) = 1$ and $g[F] = 0$. We are only considering Hausdorff spaces; sometimes Hausdorff completely regular spaces are referred to as Tychonov spaces. Any subset of a compact (Hausdorff) space is completely regular in the relative topology. Conversely, it can be shown that every (Hausdorff) completely regular space arises in this way.

Lemma 2.3.4 *Let Y be a (Hausdorff) completely regular space. Let $f : Y \mapsto \mathbb{R}$ be a positive, upper bounded, lower semicontinuous function. Let*

$$V(f) = \{a \in C_b(Y)_{sa} : 0 \leq a \leq f\}.$$

Then, for each y in Y, $f(y) = sup\{a(y) : a \in V(f)\}$.

Proof Let $v(y) = sup\{a(y) : a \in V(f)\}$. Then, by the preceding lemma, v is lower semicontinuous. Also $v \leq f$. We wish to show that $v = f$.

Suppose this is false. Then for some y_0 we can find a real number t such that $v(y_0) < t < f(y_0)$.

Let $F = \{y \in Y : f(y) \leq t\}$. Then F is closed, because f is lower semicontinuous. Also, $y_0 \notin F$.

By complete regularity, there is a continuous $g : Y \mapsto [0, 1]$ with $g(y_0) = 1$ and g vanishing on F.

Let $h = tg$. Then $h \in V(f)$ and so $v(y_0) \geq h(y_0) = t$. But this is a contradiction. So $v = f$.   □

Corollary 2.3.5 *Let Y be a completely regular space. Let f, g be bounded, lower semicontinuous functions. Then so, also, are $f + g$, $f \vee g$ and $f \wedge g$. If f and g are positive and lower semicontinuous, so, also, is fg.*

Proof Let $\lambda(f) = inf\{f(y) : y \in Y\}$. Let

$$V(f) = \{a \in C_b(Y) : \lambda(f) \leq a(y) \leq f(y) \text{ for each } y \in Y\}.$$

Then, by applying Lemma 2.3.4 to $f - \lambda(f)1$,

$$f(y) = sup\{a(y) : a \in V(f)\}.$$

We define $\lambda(g)$ and $V(g)$ similarly.

Consider $\{a \wedge b : a \in V(f) \text{ and } b \in V(g)\}$. It is straightforward to check that this set is upward directed. Let

$$w(y) = sup\{a(y) \wedge b(y) : a \in V(f) \text{ and } b \in V(g)\}.$$

Clearly $w \leq f \wedge g$. By Lemma 2.3.3, w is lower semicontinuous.

Fix $y \in Y$ and fix $b \in V(g)$. Then $w(y) \geq a(y) \wedge b(y)$ for all $a \in V(f)$. So $w(y) \geq f(y) \wedge b(y)$.

So $w(y) \geq sup\{f(y) \wedge b(y) : b \in V(g)\} = f(y) \wedge g(y)$. Then $f \wedge g = w$ and so $f \wedge g$ is lower semicontinuous.

Similar arguments work for $f + g$ and $f \vee g$. For fg we use positivity of f and g to ensure that $\{ab : a \in V(f), b \in V(g)\}$ is upward directed.

Let $g \geq 0$. Then $\lambda(g) \geq 0$. So, for each $y \in Y$, and each $b \in V(g)$ we have $b(y) \geq 0$. Similarly, $f \geq 0$ implies $a(y) \geq 0$ for each $a \in V(f)$.

Consider a_1, a_2 in $V(f)$ with $a_1 \geq a_2$ and b_1, b_2 in $V(g)$ with $b_1 \geq b_2$. Then

$$a_1(y)b_1(y) \geq a_2(y)b_1(y) \geq a_2(y)b_2(y).$$

So $\{ab : a \in V(f), b \in V(g)\}$ is upward directed. The pointwise supremum is fg. So fg is lower semicontinuous.   □

Proposition 2.3.6 *Let $C(X)$ be monotone complete. Then X is extremally disconnected.*

Proof Let U be an open subset of X. We wish to show that clU is open. By Proposition 2.3.2, X is zero-dimensional i.e. has a base of clopen sets. So U is the union of its clopen subsets.

Let $V = \{\chi_L : L \text{ is a clopen subset of } U\}$. Then V is an upward directed set of projections. Since $C(X)$ is monotone complete, V has a least upper bound in $C(X)_{sa}$ which is a projection. So this least upper bound must be of the form χ_K where K is clopen. Clearly $U \subset K$. Hence $clU \subset K$.

Then $K \backslash clU$ is an open set. Suppose it is not empty. Since X is zero-dimensional, this implies there exists a non-empty clopen $Q \subset K \backslash clU$. Then $K \backslash Q$ is a clopen set containing clU. Then $\chi_{K \backslash Q}$ is an upper bound for V. This implies $\chi_{K \backslash Q} \geq \chi_K$. Hence $K \backslash Q \supset K$. But this implies Q is empty, which is a contradiction. So clU is the clopen set K. $\qquad\qquad\qquad\qquad\qquad\qquad\qquad\qquad\qquad\qquad\qquad\qquad\qquad\qquad\qquad\qquad$ $\square$

To show that $C(X)$ is monotone complete if, and only if, X is extremally disconnected we need the converse of the above proposition. See [33, 156]. We give a proof here but later on, when we have some more machinery, we will give another proof which may be more transparent.

Theorem 2.3.7 *Let X be a compact Hausdorff space. Then $C(X)$ is monotone complete if, and only if, X is extremally disconnected.*

Proof In one direction this is already proved. So we assume that X is extremally disconnected and wish to show that $C(X)$ is monotone complete.

Let V be a norm bounded, upward directed set in $C(X)_{sa}$ we shall show that V has a least upper bound in $C(X)_{sa}$. It suffices to prove this when V is bounded below by 0.

Let $f(x) = sup\{a(x) : a \in V\}$. Then f is a non-negative, bounded lower semicontinuous function.

For each $t \in \mathbb{R}^+$ let $F_t = \{x : f(x) \leq t\}$. Because f is lower semicontinuous, each F_t is closed.

Let C_t be the interior of F_t. Then $F_t \backslash C_t$ is a closed nowhere dense set.

Since X is extremally disconnected, clC_t is clopen. But $clC_t \subset clF_t = F_t$. So C_t is clopen.

Let M be the union of $F_t \backslash C_t$ where t ranges over the (countable) set of all non-negative rational numbers. So M is a meagre set.

Since f is bounded above, $F_t = X$ for large enough t. Also, $s < t$ implies $F_s \subset F_t$ and hence $C_s \subset C_t$.

Let $g(x) = inf\{t \in \mathbb{Q}^+ : x \in C_t\}$ where $\mathbb{Q}^+$ is the set of non-negative rationals. Then

$$\{x \in X : g(x) < s\} = \bigcup\{C_t : t \in \mathbb{Q}^+ \,\&\, t < s\}, \text{ which is open;}$$

$$\{x \in X : g(x) \leq r\} = \bigcap\{C_t : t \in \mathbb{Q}^+ \,\&\, t > r\}, \text{ which is closed.}$$

So, whenever $r < s$, $\{x \in X : r < g(x) < s\}$ is open. Hence g is a continuous function from X into $\mathbb{R}^+$.

Now fix x_0 in X and let $g(x_0) = r$. Then $x_0 \in C_t$ for $t > r$ and $t \in \mathbb{Q}^+$. So $f(x_0) \leq t$ for $t > r$. Hence $f(x_0) \leq g(x_0)$. So $f \leq g$.

Now let $x \in X \backslash M$. Let $f(x) = r$. Then $x \in F_t$ for all rational t with $t \geq r$. Since $x \notin M$, in fact, $x \in C_t$ for all rational t with $t \geq r$. Hence $g(x) \leq r = f(x)$.

In summary, $g \geq f$ and $g(x) = f(x)$ except on a meagre set.

Let $h \in C(X)$ with $h \geq f$. Then $g \geq g \wedge h \geq f$. So $\{x \in X : (g - g \wedge h)(x) > 0\}$ is a meagre open set. It now follows from the Baire Category Theorem that this set is empty. Thus $g \leq h$.

So g is the supremum of V.   □

Exercise 2.3.8 Show that for any compact Hausdorff X, the construction of the function g in the above proof gives an upper semicontinuous function with $g \geq f$ and $g = f$ except on a meagre set.

2.3.1 Extension Theorems

Let A be a commutative monotone complete C^*-algebra. Then it is sometimes helpful to think of A_{sa} as an analogue of the real numbers. Both have a Dedekind complete order structure, that is, bounded sets have a supremum. So we may regard A as an infinite dimensional generalisation of the complex numbers. This viewpoint is fruitful for a number of extension theorems. See below.

Proposition 2.3.9 *Let A be a commutative monotone complete C^*-algebra. Let B be a unital C^*-algebra and C a self-adjoint subspace of B with $1 \in C$. Let $\phi_0 : C \mapsto A$ be a positive linear map. Then there exists a positive linear map $\phi : B \mapsto A$ which is an extension of ϕ_0.*

Proof The restriction of ϕ_0 to C_{sa} is a real linear map from a real vector space into A_{sa}. It suffices to find a positive, real-linear extension $\mu : B_{sa} \mapsto A_{sa}$. (Because we can then define ϕ on the whole of B by $\phi(a + ib) = \mu(a) + i\mu(b)$ whenever a and b are self-adjoint. Then ϕ is complex linear.)

Let $\mathcal{E}$ be the collection of all pairs (V, ψ) where (i) V is a (real) vector subspace of B_{sa}, (ii) $C_{sa} \subset V$, (iii) $\psi : V \mapsto A_{sa}$ is a positive (real) linear map which coincides with ϕ_0 on C_{sa}.

Partially order $\mathcal{E}$ by $(V_1, \psi_1) \preceq (V_2, \psi_2)$ when $V_1 \subset V_2$ and ψ_2 extends ψ_1. Just as in the classical Hahn-Banach Theorem, a straightforward Zorn's Lemma argument shows that $\mathcal{E}$ has a maximal element (M, μ).

Suppose there exists a positive $b \in B_{sa} \backslash M$. If $x \leq b$ then $x \leq \|b\| 1$.

Then $U = \{\mu(x) \in A_{sa} : x \in M \text{ and } x \leq b\}$ is an upper bounded set in the boundedly complete lattice A_{sa}. So U has a supremum u. Since $0 \in U$ it follows that $u \geq 0$.

Let L be the (real) linear span of M and b. That is $L = \{m + \lambda b : m \in M \text{ and } \lambda \in \mathbb{R}\}$.

Let $\eta(m + \lambda b) = \mu(m) + \lambda u$. Then η is an extension of μ to a real-linear map. It remains to show that η is positive.

Let $m + \lambda b \geq 0$.

Case (i) $\lambda > 0$. Then $-\frac{1}{\lambda} m \leq b$. So, from the definition of U, $\mu(-\frac{1}{\lambda} m) \leq u$. Hence $0 \leq \mu(m) + \lambda u = \eta(m + \lambda u)$.

Case (ii) $\lambda < 0$. So $-\frac{1}{\lambda}m \geq b$. So $\mu(-\frac{1}{\lambda}m)$ is an upper bound for U. Thus $\mu(-\frac{1}{\lambda}m) \geq u$. Thus $\mu(m) + \lambda u \geq 0$.

So (L, η) is in $\mathcal{E}$, which is a contradiction. It follows that every positive element of B must be in M. So, by linearity, $B_{sa} = M$.   □

Corollary 2.3.10 *Let H be a Hilbert space of arbitrary dimension and let A be a commutative, unital closed $*$-subalgebra of $L(H)$. Suppose that A is monotone complete. Then there exists a conditional expectation from $L(H)$ onto A.*

Proof In the proposition, let $C = A$ and let ϕ_0 be the identity map from C to A. Then the extension ϕ is a positive linear idempotent map from $L(H)$ onto A; hence ϕ is a conditional expectation from $L(H)$ onto A.   □

Proposition 2.3.11 *Let A and B be commutative, unital C^*-algebras with A monotone complete. Let C be a (unital) $*$-subalgebra of B and let $\gamma : C \mapsto A$ be a $*$-homomorphism mapping the unit of C to the unit of A. Then there exists a $*$-homomorphism $\beta : B \mapsto A$ whose restriction to C coincides with γ.*

Proof (Sketch) A proof of this can be found in [168]. When D is a self-adjoint, unital subspace of B, let $P(D_{sa}, A_{sa})$ be the (convex) set of those positive, real-linear maps from D_{sa} to A_{sa}, which map unit to unit. The key observation is that when D_{sa} is a closed subalgebra of B_{sa}, then the extreme points of $P(D_{sa}, A_{sa})$ coincide with the homomorphisms of D_{sa} to A_{sa} (see the exercise below). See [124]. We adapt the method of proof of Proposition 2.3.9 but this time requiring our extensions to be extremal.

The homomorphism γ is an extreme point of $P(C_{sa}, A_{sa})$. Let $E^{\#}$ be the collection of all pairs (V, ψ) in E such that ψ is an extreme point of $P(V, A_{sa})$ and ψ is an extension of γ. Again, by a Zorn's Lemma argument, $E^{\#}$ has a maximal element (M, μ). So here μ is an extreme point of $P(M, A_{sa})$. If $M = B_{sa}$ then μ is an extreme point of $P(B_{sa}, A_{sa})$. So μ is an homomorphism and its "complexification" is a $*$-homomorphism of B into A, which extends γ.

Otherwise, there is a positive $b \in B_{sa} \backslash M$. We now define (L, η) as in Proposition 2.3.9. Suppose $\eta = \frac{1}{2}(\alpha + \beta)$ where α and β are in $P(L, A_{sa})$. Then $\mu = \frac{1}{2}(\alpha|_M + \beta|_M)$. Since μ is extreme, $\mu(m) = \alpha(m) = \beta(m)$ for all $m \in M$.

If $m \leq b$ then, by positivity, $\alpha(b) \geq \alpha(m) = \mu(m)$. So $\alpha(b) \geq u = \eta(b)$. Similarly $\beta(b) \geq \eta(b)$. So $\alpha = \beta$. Thus (L, η) is in $E^{\#}$. But this contradicts maximality.   □

Exercise 2.3.12 Let B and A be commutative, unital C^*-algebras. Let $P(B_{sa}, A_{sa})$ be the convex set of those positive linear maps from B_{sa} into A_{sa} which map the unit element of A to the unit element of B. (i) Let h be an extreme point of $P(B_{sa}, A_{sa})$. Show that h is an homomorphism. (ii) Let h be an homomorphism of B_{sa} to A_{sa} where $h(1) = 1$. Show that h is an extreme point of $P(B_{sa}, A_{sa})$.

Hint: (i) Fix $b \in B$ with $0 \leq b \leq 1$. Let $g(x) = h(xb) - h(x)h(b)$. Show $h \pm g$ are in $P(B_{sa}, A_{sa})$. Then use the extremality of h to show that $g = 0$. (ii) Reduce to the situation where $A = \mathbb{C}$. Suppose $h = \frac{1}{2}(\alpha + \beta)$ where α, β are states. Show that α, β both vanish on the maximal ideal which forms the kernel of h.

These two propositions are "best possible" in the sense that the hypothesis that A is monotone complete is essential. This follows from the following result.

Proposition 2.3.13 *Let A be a unital C^*-subalgebra of a monotone complete C^*- algebra B. Let $\Gamma : B \mapsto A$ be a positive linear map such that $\Gamma a = a$ for each a in A. Then A is monotone complete.*

Proof Let $(a_\lambda : \lambda \in \Lambda)$ be a norm bounded increasing net in A. Let $x \in A$ be any upper bound of this net. Let b be the supremum of this net in B. Then, for each λ,

$$x \geq b \geq a_\lambda.$$

Hence, by applying Γ to the above inequality,

$$x \geq \Gamma b \geq a_\lambda.$$

So Γb is the least upper bound of $(a_\lambda : \lambda \in \Lambda)$ in A. $\square$

We note that there is no assumption here that A is commutative. We can weaken the hypotheses by not requiring A to be a sub-algebra; merely a self-adjoint subspace with $1 \in A$. Then the same argument shows that A_{sa} is monotone complete in an obvious sense. This argument can easily be extended further to categories of partially ordered sets (see Lemma 1 [168]).

When A is a commutative monotone complete C^*-algebra, Vincent-Smith [164] obtained far reaching generalisations of the Hahn-Banach Theorem, where vector spaces are replaced by modules over A. See also [48, 109, 160].

2.4   Matrix Algebras over a Monotone Complete C^*-Algebra

When considering completely positive maps, we shall need to know that matrix algebras over monotone complete C^*-algebras are, themselves, monotone complete. The reader can accept this fact and move on, or else, read this section. For the convenience of the reader, we begin by recalling some basic C^*-algebra facts.

Definition 2.4.1 Let A be a unital C^*-algebra. For any $n \in \mathbb{N}$, let $M_n(A)$ be the set of all $n \times n$ matrices $\mathbf{x} = (x_{ij})_{1 \leq i,j \leq n}$ (we shall sometimes write (x_{ij}) if no confusion is likely to occur) with elements from A in each position. Let $M_n(A)$ be given the natural $*$-algebra operations: $\mathbf{x} + \mathbf{y} = (x_{ij} + y_{ij})$, $\lambda \mathbf{x} = (\lambda x_{ij})$, $\mathbf{xy} = (w_{ij})$ with $w_{ij} = \sum_{k=1}^{n} x_{ik} y_{kj} (1 \leq i,j \leq n)$ and $\mathbf{x}^* = (z_{ij})$ with $z_{ij} = x_{ji}^*(1 \leq i,j \leq n)$ for any $\mathbf{x} = (x_{ij})$, $\mathbf{y} = (y_{ij})$ in $M_n(A)$ and $\lambda \in \mathbb{C}$, the field of complex numbers. When equipped with these operations $M_n(A)$ becomes a $*$-algebra over $\mathbb{C}$.

Let A be a unital C^*-algebra. As was pointed out in [161], Chapter 4, Section 3, we can suppose that A acts on a Hilbert space H; that is, we may identify A with a unital subalgebra of $L(H)$. Then $M_n(A)$ acts on $\tilde{H} = H \oplus \cdots \oplus H$ (n-times) as a $*$-subalgebra of $L(\tilde{H})$ defined by

$$\mathbf{a}\underline{\xi} = \left(\sum_{k=1}^n a_{1k}\xi_k, \cdots, \sum_{k=1}^n a_{nk}\xi_k \right) \text{ for } \underline{\xi} = (\xi_1, \cdots, \xi_n) \in \tilde{H}$$

for each $\mathbf{a} = (a_{ij}) \in M_n(A)$.

Then clearly, $\mathbf{a}$ is a bounded linear operator on $\tilde{H}$ such that $sup_{1 \leq i,j \leq n} \|a_{ij}\| \leq \|\mathbf{a}\| \leq \sum_{1 \leq i,j \leq n} \|a_{ij}\|$. So $M_n(A)$ acts on $\tilde{H}$ as a unital C^*-algebra on $\tilde{H}$. Since any $*$-isomorphism between C^*-algebras is an isometry, $M_n(A)$ has a unique C^*-norm, $\| \cdot \|$, defined by this operator norm of $\mathbf{a}$ for each $\mathbf{a} \in M_n(A)$. It follows that $\|\cdot\|$ does not depend on the choice of H. We call this norm, the matrix norm of $\mathbf{a} = (a_{ij})_{1 \leq i,j \leq n}$ in $M_n(A)$. It is also clear that with this norm $\| \cdot \|$, $M_n(A)$ is a unital C^*-algebra.

Definition 2.4.2 Let B be a unital C^*-algebra or more generally a unital $*$-algebra. A system $\{u_{ij} : 1 \leq i,j \leq n\}$ of elements in B is said to be an $n \times n$ system of matrix units for B if $u_{ij}{}^* = u_{ji}$, $u_{ij}u_{k\ell} = \delta_{j,k}u_{i\ell}$ for all i, j, k and ℓ, where $\delta_{j,k}$ is the Kronecker's delta and $\sum_{i=1}^n u_{ii} = 1$.

We note that $\{u_{ii} : 1 \leq i \leq n\}$ is an orthogonal family of projections in B that is also a homogeneous partition of 1, that is, $u_{11} \sim u_{ii}$ via the partial isometry u_{i1} for each i.

Take any $n \in \mathbb{N}$ and let $\mathbf{e}_{ij}^{(n)}$ be the matrix with 1 (the unit of A) in the (i,j) position and zero otherwise. Then the system $\{\mathbf{e}_{ij}^{(n)} : 1 \leq i,j \leq n\}$ is an $n \times n$ system of matrix units for $M_n(A)$. We call this system the *standard* $(n \times n)$ system of matrix units in $M_n(A)$. If no confusion is likely, we shall sometimes write $\{\mathbf{e}_{ij} : 1 \leq i,j \leq n\}$.

Let $\{e_{ij}^{(n)} : 1 \leq i,j \leq n\}$ be the standard $n \times n$ system of matrix units in $M_n(\mathbb{C})$, which we shall sometimes write $\{e_{ij}\}$ if no confusion is likely to occur. Then clearly $M_n(\mathbb{C}) = \sum_{1 \leq i,j \leq n} \mathbb{C}e_{ij}$ which is an n^2 dimensional vector space over $\mathbb{C}$.

Let us define the map

$$\Pi : M_n(A) \ni \mathbf{t} = (t_{ij})_{1 \leq i,j \leq n} \longmapsto \sum_{i,j=1}^n t_{ij} \otimes e_{ij} \in A \otimes M_n(\mathbb{C}).$$

Then clearly Π is a $*$-isomorphism from $M_n(A)$ onto $A \otimes M_n(\mathbb{C})$ such that

$$\Pi(\mathbf{e}_{ij}) = 1 \otimes e_{ij} \text{ for every pair } i,j \text{ with } 1 \leq i,j \leq n.$$

We also have that for $\mathbf{x} \in M_n(A)$, $\mathbf{x}\mathbf{e}_{ij} = \mathbf{e}_{ij}\mathbf{x}$ for all pairs i and j with $1 \leq i,j \leq n$ if, and only if, $\mathbf{x} = (\delta_{i,j}a)_{1 \leq i,j \leq n}$ for some $a \in A$. This is easy to check by the map

Π defined above. So,

$$M_n(A) \cap \{e_{ij} : i,j = 1, \cdots, n\}' = \{(\delta_{i,j}a) : a \in A\} \cong A.$$

Furthermore, for every $\mathbf{b} = (b_{ij}) \in M_n(A)$ and for each pair i and j with $1 \leq i, j \leq n$,

we have $\mathbf{e}_{1i}\mathbf{b}\mathbf{e}_{j1} = \begin{pmatrix} b_{ij} & 0 & \cdots & 0 \\ 0 & 0 & \cdots & 0 \\ \vdots & \vdots & \ddots & \vdots \\ 0 & 0 & \cdots & 0 \end{pmatrix}$ and we have $\Pi(\mathbf{e}_{11}M_n(A)\mathbf{e}_{11}) = A \otimes e_{11}$. Hence it

follows that

$$\sum_{k=1}^{n} \mathbf{e}_{ki}\mathbf{b}\mathbf{e}_{jk} = \begin{pmatrix} b_{ij} & 0 & \cdots & 0 \\ 0 & b_{ij} & \cdots & 0 \\ \vdots & \vdots & \ddots & \vdots \\ 0 & 0 & \cdots & b_{ij} \end{pmatrix} \in M_n(A) \cap \{e_{rs} : 1 \leq r, s \leq n\}'.$$

In this section, instead of using "i" for the square root of -1 we shall write "$\sqrt{-1}$"; to avoid any possible confusion with the use of "i" as an integer subscript.

Lemma 2.4.3 *Let B be a unital C^*-algebra or more generally, a $*$-algebra and let $\{f_{ij}\}_{1 \leq i,j \leq n}$ be any $n \times n$ system of matrix units for B.*

(1) *Then, for any pair i and j with $1 \leq i, j \leq n$, we have for every $t \in B$, by polarization*

$$4f_{1i}tf_{j1} = \sum_{k=0}^{3} \sqrt{-1}^k (f_{j1} + \sqrt{-1}^k f_{i1})^* t(f_{j1} + \sqrt{-1}^k f_{i1})$$

and each term of the above equality belongs to $f_{11}Bf_{11}$. For each pair i and j with $1 \leq i, j \leq n$ and k, put

$$t(i,j;k) = (f_{j1} + \sqrt{-1}^k f_{i1})^* t(f_{j1} + \sqrt{-1}^k f_{i1}).$$

Then $4f_{1i}tf_{j1} = \sum_{k=0}^{3} \sqrt{-1}^k t(i,j;k)$ and the map $B \ni t \longmapsto t(i,j;k) \in f_{11}Bf_{11}$ is a positive normal linear map when B is a monotone complete C^*-algebra.

(2) *Let $C = \{f_{ij} : 1 \leq i,j \leq n\}' \cap B$. Take any $x \in B$ and put, for each k, l with $1 \leq k, l \leq n$, $x_{kl} = \sum_{r=1}^{n} f_{rk}xf_{lr}$. Then $x_{kl} \in C$ for all k and l and C is a unital $*$-subalgebra of B. If B is a C^*-algebra, then so is C and when B is monotone $(\sigma\text{-})complete$, so is C. The map $\Phi : B \ni x \longmapsto (x_{ij})_{1 \leq i,j \leq n} \in M_n(C)$ is a unital $*$-isomorphism.*

(3) *The map $\psi : C \ni x \longmapsto f_{11}x \in f_{11}Bf_{11}$ is a $*$-isomorphism.*

(4) *Let $\phi : B \ni x \longmapsto (f_{1i}xf_{j1})_{1\le i,j\le n} \in M_n(f_{11}Bf_{11})$. Then ϕ is also a $*$-isomorphism from B onto $M_n(f_{11}Bf_{11})$ and we have the following commutative diagram:*

$$
\begin{array}{ccccc}
B & = & B & \xrightarrow{\;\Phi\;} & M_n(C) \\
\phi \downarrow & & & & \downarrow \Pi_C \\
M_n(f_{11}Bf_{11}) & \xleftarrow[\Pi^{-1}]{} & f_{11}Bf_{11} \otimes M_n(\mathbb{C}) & \xleftarrow[\psi \otimes \iota_n]{} & C \otimes M_n(\mathbb{C}),
\end{array}
$$

where Π is the canonical map from $M_n(f_{11}Bf_{11})$ onto $f_{11}Bf_{11} \otimes M_n(\mathbb{C})$ and Π_C is the canonical map from $M_n(C)$ onto $C \otimes M_n(\mathbb{C})$.

Proof By Propositions 2.1.10 and 2.2.26, (1) is clear.

Since the system $\{f_{ij} : 1 \le i,j \le n\}$ is self-adjoint, C is a unital C^*-subalgebra of B. Now take any $x \in B$. Since $x_{kl} = \sum_{r=1}^{n} f_{rk}xf_{lr}$ for each pair k and l with $1 \le k$, $l \le n$, calculation shows

$$
f_{ij}x_{kl} = f_{ij}\left(\sum_{r=1}^{n} f_{rk}xf_{lr}\right) = f_{ik}xf_{lj} \text{ and } x_{kl}f_{ij} = \left(\sum_{r=1}^{n} f_{rk}xf_{lr}\right)f_{ij} = f_{ik}xf_{lj}
$$

and so $x_{kl} \in C$ follows.

Consider the map $\Phi : B \ni x \longmapsto (x_{ij})_{1\le i,j\le n} \in M_n(C)$. It is clear that Φ is an injective $*$-linear map from B to $M_n(C)$. Moreover, for each $x \in B$,

$$
x = \sum_{i,j=1}^{n} f_{ii}xf_{jj} = \sum_{i,j=1}^{n}\left(\sum_{k=1}^{n} f_{ki}xf_{jk}\right)f_{ij} = \sum_{i,j=1}^{n} x_{ij}f_{ij}
$$

that is, $B = \sum_{i,j=1}^{n} Cf_{ij}$ and the map is surjective.

We also have that for every pair x and y in B and i and j with $1 \le i,j \le n$,

$$
\sum_{r=1}^{n} x_{ir}y_{rj} = \sum_{r=1}^{n}\left(\sum_{k=1}^{n} f_{ki}xf_{rk}\right)\left(\sum_{l=1}^{n} f_{lr}yf_{jl}\right) = \sum_{r=1}^{n}\sum_{k=1}^{n} f_{ki}xf_{rr}yf_{jk} = (xy)_{ij}
$$

which implies that the map Φ is a homomorphism. Hence it is a $*$-isomorphism. This is (2).

Furthermore, the map $\psi : C \ni x \longmapsto f_{11}x \in f_{11}Bf_{11}$ is a $*$-isomorphism. In fact, for any $x \in C$, $xf_{11} = 0$ implies $f_{1j}x = f_{11}f_{1j}x = f_{11}xf_{1j} = 0$ for all j. Hence $f_{jj}x = 0$ for all j and so $x = \sum_{j=1}^{n} f_{jj}x = 0$ follows, which implies that the map is injective.

Take any $y \in f_{11}Bf_{11}$ and put $z = \sum_{r=1}^{n} f_{r1}yf_{1r}$. Then as was shown above, $z \in C$ and $zf_{11} = f_{11}yf_{11} = y$. So the map is surjective. Since it is clear that this map is a $*$-homomorphism, it is a $*$-isomorphism. This is (3).

Now take any $x \in B$ and as before put, for each pair i and j with $1 \leq i, j \leq n$, $x_{ij} = \sum_{k=1}^{n} f_{ki} x f_{jk}$, we have $x_{ij} f_{11} = f_{1i} x f_{j1} \in f_{11} B f_{11}$. Let us define the map $\phi : B \longmapsto M_n(f_{11} B f_{11})$ by $\phi(x) = (f_{1i} x f_{j1})_{1 \leq i,j \leq n}$. Then, as before, ϕ is a $*$-isomorphism from B onto $M_n(f_{11} B f_{11})$. This comes from the fact that $\psi \otimes \iota_n$ is a $*$-isomorphism from $C \otimes M_n(\mathbb{C})$ onto $f_{11} B f_{11} \otimes M_n(\mathbb{C})$. Since the map $\phi = \Pi^{-1} \circ (\psi \otimes \iota_n) \circ \Pi_C \circ \Phi$, where Π is the canonical map from $M_n(f_{11} B f_{11})$ onto $f_{11} B f_{11} \otimes M_n(\mathbb{C})$ and Π_C is the canonical map from $M_n(C)$ onto $C \otimes M_n(\mathbb{C})$, this induces a commutative diagram in the statement of (4).   □

Corollary 2.4.4

(1) *For any $b \in A$, $b \geq 0$ if, and only if,* $\begin{pmatrix} b & 0 & \cdots & 0 \\ 0 & 0 & \cdots & 0 \\ \vdots & \vdots & \ddots & \vdots \\ 0 & 0 & \cdots & 0 \end{pmatrix} \geq 0$ *in $M_n(A)$.*

(2) *For $\mathbf{a} = (a_{ij}) \in M_n(A)$, if $\mathbf{a} \geq 0$, then $a_{kk} \geq 0$ for each k with $1 \leq k \leq n$.*

Proof Since the map $A \ni a \longmapsto \begin{pmatrix} a & 0 & \cdots & 0 \\ 0 & 0 & \cdots & 0 \\ \vdots & \vdots & \ddots & \vdots \\ 0 & 0 & \cdots & 0 \end{pmatrix} \in e_{11} M_n(A) e_{11}$ is a $*$-isomorphism,

(1) follows.

If $\mathbf{a} \geq 0$, we have, as before, $e_{1k} \mathbf{a} e_{k1} = \begin{pmatrix} a_{kk} & 0 & \cdots & 0 \\ 0 & 0 & \cdots & 0 \\ \vdots & \vdots & \ddots & \vdots \\ 0 & 0 & \cdots & 0 \end{pmatrix} \geq 0$, which means by (1)

that $a_{kk} \geq 0$ for every k with $1 \leq k \leq n$.   □

More generally, we have the following:

Lemma 2.4.5 *Let A be a unital C^*-algebra and let $n \in \mathbb{N}$. Take any $\mathbf{s} = (s_{ij})_{1 \leq i,j \leq n} \in M_n(A)$. Then $\mathbf{s}$ is non-negative if, and only if,*

$$\sum_{i,j=1}^{n} x_i^* s_{ij} x_j \geq 0 \text{ for every } \{x_1, \cdots, x_n\} \subset A. \tag{*}$$

Furthermore, for any n-family $\{a_1, \cdots, a_n\} \subset A$, the matrix $\mathbf{b} = (a_i^* a_j)_{1 \leq i,j \leq n}$ is non-negative.

See Lemma IV.3.1 and Lemma IV.3.2 in [161].

Now we come to the main result of this section, which was originally proved in [61] by making use of the regular completion (see later arguments) of $M_n(A)$. Here we shall show this in an elementary way by applying Lemma 2.4.5. In [63], it was treated in a more general setting where $n = \infty$. We shall discus this more general situation later.

Theorem 2.4.6 *Let the C^*-algebra A be monotone complete (monotone σ-complete). Then for each $n \in \mathbb{N}$, the matrix algebra $M_n(A)$ is also monotone complete (respectively, monotone σ-complete).*

Proof Suppose that A is monotone σ-complete. Let $(\mathbf{a}_m)$ be any norm bounded increasing sequence of elements in $M_n(A)_{sa}$ with $\mathbf{a}_m = (a_{ij}(m))_{1 \leq i,j \leq n}$ for some $a_{ij}(m) \in A$ for each such m and i,j. Take any n-family $\{x_1, \cdots, x_n\} \subset A$. By Lemma 2.4.5, each sequence $\left(\sum_{i,j=1}^n x_i^* a_{ij}(m) x_j \right)$ in A_{sa} is norm bounded and increasing. Since A is monotone σ-complete, there exists an $a(x_1, \cdots, x_n) \in A_{sa}$ such that $\sum_{i,j=1}^n x_i^* a_{ij}(m) x_j \uparrow a(x_1, \cdots, x_n)$ in A_{sa}.

As in Lemma 2.4.3 (1), by polarization, for any $\mathbf{b}$ we have:

$$4\mathbf{e}_{1i}\mathbf{b}\mathbf{e}_{j1} = \sum_{k=0}^3 \sqrt{-1}^k (\mathbf{e}_{j1} + \sqrt{-1}^k \mathbf{e}_{i1})^* \mathbf{b}(\mathbf{e}_{j1} + \sqrt{-1}^k \mathbf{e}_{i1})$$

where each term lies in $\mathbf{e}_{11} M_n(A) \mathbf{e}_{11}$. This tells us that for each i,j and each k with $1 \leq i,j \leq n$ and $k = 0, 1, 2, 3$, there exist four norm bounded increasing sequences $(a_{ij}(m; k) : m = 1, 2, \cdots)$ of elements in A_{sa} such that

$$a_{ij}(m) = \sum_{k=0}^3 \sqrt{-1}^k a_{ij}(m; k) \text{ for each } i,j \text{ with } 1 \leq i,j \leq n \text{ and } m \in \mathbb{N}.$$

Since A is monotone σ-complete, we have $a_{ij}(k)$ $(k = 0, 1, 2, 3; 1 \leq i,j \leq n)$ such that $a_{ij}(m; k) \uparrow a_{ij}(k)$ in A_{sa} as $m \to \infty$. Let $a_{ij} = \sum_{k=0}^3 \sqrt{-1}^k a_{ij}(k)$ for each pair i and j with $1 \leq i,j \leq n$.
We shall show $\sum_{i,j=1}^n x_i^* a_{ij} x_j = a(x_1, \cdots, x_n)$.

By our definition of order limit, $LIM_{m \to \infty} a_{ij}(m) = \sum_{k=0}^3 \sqrt{-1}^k a_{ij}(k)$ in A. By Proposition 2.1.22, we also have $LIM_{m \to \infty} \sum_{i,j=1}^n x_i^* a_{ij}(m) x_j = \sum_{i,j=1}^n x_i^* a_{ij} x_j$ and it follows that $a(x_1, \cdots, x_n) = \sum_{i,j=1}^n x_i^* a_{ij} x_j$.

Hence we have $\sum_{i,j=1}^n x_i^* a_{ij}(m) x_j \leq \sum_{i,j=1}^n x_i^* a_{ij} x_j$ for all $\{x_1, \cdots, x_n\} \subset A$ and m. As was noted before, $\mathbf{a} = (a_{ij})_{1 \leq i,j \leq n} \in M_n(A)$ is self-adjoint. Since $\sum_{i,j=1}^n x_i^* (a_{ij} - a_{ij}(m)) x_j \geq 0$ for all $\{x_1, \cdots, x_n\} \subset A$, we have $\mathbf{a}_m \leq \mathbf{a}$ for all m. Now take any $\mathbf{b} \in M_n(A)_{sa}$ and suppose that $\mathbf{a}_m \leq \mathbf{b}$ for all m. Then for every n-family $\{x_1, \cdots, x_n\} \subset A$, we have $\sum_{i,j=1}^n x_i^* a_{ij}(m) x_j \leq \sum_{i,j=1}^n x_i^* b_{ij} x_j$ for all m, which implies that $\sum_{i,j=1}^n x_i^* a_{ij} x_j \leq \sum_{i,j=1}^n x_i^* b_{ij} x_j$, because $\sum_{i,j=1}^n x_i^* a_{ij}(m) x_j \uparrow \sum_{i,j=1}^n x_i^* a_{ij} x_j$ in A_{sa}. Hence it follows that $\mathbf{a} \leq \mathbf{b}$ and so $\mathbf{a}_m \uparrow \mathbf{a}$ in $M_n(A)_{sa}$. By a similar argument, we can show $M_n(A)$ is monotone complete when A is monotone complete. $\square$

Chapter 3
Classification and Invariants

In this chapter we show how monotone complete C^*-algebras, of bounded cardinality, can be classified by a semigroup which we construct. We introduce the spectroid invariant for monotone σ-complete C^*-algebras and show how it can also be defined for elements of the classification semigroup. In later chapters this theory will be applied to exhibit huge numbers of examples. We also indicate how aspects of this theory can be extended to more general classes of partially ordered set. We begin with a brief discussion of C^*-algebras which are not "too large" and cardinalities.

3.1 C^*-Algebras of Small Size

Each monotone complete C^*-algebra is unital. Also, adjoining a unit to a monotone σ-complete C^*-algebras gives a unital monotone σ-complete C^*-algebra. In this chapter we shall assume all C^*-algebras considered are unital, unless we specifically state otherwise. Let H be a Hilbert space and $L(H)$ the (von Neumann) algebra of all bounded operators on H.

We recall from Chap. 1 that if M is a von Neumann algebra with a separable state space then, see [1], M is isomorphic to a von Neumann subalgebra of $L(H)$, where H is separable. But there do exist monotone complete C^*-algebras which have separable state spaces but which are *NOT* isomorphic to C^*-subalgebras of $L(H)$ [178].

We need a few basic notions on complete positivity and complete isometry.

Let A and B be unital C^*-algebras and let $\Phi : A \mapsto B$ be a linear map. Let $\Phi_n : M_n(A) \mapsto M_n(B)$ be defined by $\Phi_n((a_{ij})) = [\Phi(a_{ij})]$.

Then Φ is said to be *completely positive* if Φ_n is positive for every n. Similarly, Φ is said to be a *complete isometry* if Φ_n is an isometry for each n. If Φ maps the unit of A to the unit of B then Φ is said to be *unital*.

© Springer-Verlag London 2015

K. Saitô, J.D.M. Wright, *Monotone Complete C*-algebras and Generic Dynamics*,
Springer Monographs in Mathematics, DOI 10.1007/978-1-4471-6775-4_3

Our standard reference for operator spaces and operator systems is [38]. There are interesting and important connections between this theory and monotone complete C^*-algebras. But for the beginner, most of these considerations can be postponed until later.

We recall that a C^*-algebra A is said to be *small* if there exists a unital complete isometry Φ from A into $L(H)$ where H is separable. It follows from Corollary 5.1.2 [38] that when such a Φ exists, then Φ is a completely positive isometry onto an operator system in $L(H)$, and its inverse is completely positive. Saitô's Theorem [137] tells us that A is a small C^*-algebra if, and only if, the state space of $M_n(A)$ is separable for each n.

A C^*-algebra B is *almost separably representable* if the state space of B is separable. This turns out to be equivalent to: there exists a separable H and a completely positive unital

$$\Gamma : B \mapsto L(H)$$

such that Γ is an isometric order isomorphism of B_{sa} into $L(H)_{sa}$ [178].

Clearly each small C^*-algebra is almost separably representable. We conjecture that the converse is true; it is certainly true in most cases of interest to us. But in full generality this seems a delicate question. We shall discuss this later.

Lemma 3.1.1 *Let A be a monotone σ-complete C^*-algebra with a separable state space. Then A is monotone complete.*

Proof Let (ϕ_n) be a dense sequence of states. Then $\sum_1^\infty 1/2^n \phi_n$ is a faithful state. So, by Theorem 2.1.14, A is monotone complete. $\square$

We use $\#S$ or $card(S)$ to denote the cardinality of a set S. We frequently use "c" to denote the cardinal of the continuum of real numbers, that is, $c = 2^{\aleph_0} = \#\mathbb{R}$.

Proposition 3.1.2 *Whenever A is a C^*-algebra with a separable state space then $\#A = \#\mathbb{R} = c$.*

Proof Let H be a separable Hilbert space. Because H has a countable orthonormal basis, each operator on H can be represented by an infinite $\mathbb{N} \times \mathbb{N}$ matrix over the complex numbers. So $L(H)$ injects into $\mathbb{C}^{\mathbb{N} \times \mathbb{N}}$ which has cardinality c. Since there is a completely positive injection of A into $L(H)$, A also has cardinality c. $\square$

Proposition 3.1.3 *Let A be any C^*-algebra of cardinality c. Then it has a faithful representation on a Hilbert space of dimension c.*

Proof We may assume that the algebra is unital, if not we can adjoin a unit without increasing the cardinality.

For each $a \in A\backslash\{0\}$, there is a pure state ϕ_a such that $\phi_a(aa^*) \neq 0$. By the Gelfand-Naimark-Segal process and the fact that the state is pure, there is a surjection from A onto the corresponding Gelfand-Naimark-Segal Hilbert space $H(\phi_a)$. So $\#H(\phi_a) \leq \#A = c$. Let H be the Hilbert space direct sum of $\{H(\phi_a) : a \in A\backslash\{0\}\}$. So H has an orthonormal basis of cardinality not exceeding

$c \times c = c$. Since each element of the Hilbert space is orthogonal to all but countably many basis elements, $\#H \leq c \times c^{\aleph_0} = c$. The natural representation of A on H is faithful. $\qquad\qquad\qquad\qquad\qquad\qquad\qquad\qquad\qquad\qquad\qquad\qquad\qquad\qquad$ $\square$

3.2 Classification Semigroup

In this section we construct a classification semigroup, $\mathcal{W}$, for monotone complete C^*-algebras. We shall see that $\mathcal{W}$ is abelian and equipped with a zero element. It is partially ordered with 0 as its smallest element.

Each element of $\mathcal{W}$ is idempotent, that is $x + x = x$. We show that the semigroup has the Riesz Decomposition Property.

From these facts it follows that, in its natural partial ordering, $\mathcal{W}$ is a distributive semi-lattice.

To avoid some set theoretic difficulties we fix a large Hilbert space $H^\#$ and, for the rest of this section, only consider algebras which are isomorphic to subalgebras of $L(H^\#)$. For most of our applications it suffices to take an $H^\#$ which has an orthonormal basis of cardinality $c = 2^{\aleph_0} = \#\mathbb{R}$. This ensures that each C^*-algebra of cardinality $\#\mathbb{R}$ has a faithful representation in $L(H^\#)$. In particular, all small C^*-algebras are included.

Many of these constructions can be generalised to other classes of partially ordered sets. But for this section, we shall only discuss monotone complete C^*-algebras.

Let A and B be monotone complete C^*-algebras and let $\phi : A \mapsto B$ be a positive linear map. We recall that ϕ is *faithful* if $x \geq 0$ and $\phi(x) = 0$ implies $x = 0$.

Let A and B be monotone complete C^*-algebras and let $\phi : A \mapsto B$ be a positive linear map. We also recall that the map ϕ is said to be *normal* if, whenever D is a downward directed set of positive elements of A_{sa}, ϕ maps the infimum of D to the infimum of $\{\phi(d) : d \in D\}$ in B_{sa}. That is,

$$\bigwedge\{\phi(d) : d \in D\} = \phi\left(\bigwedge\{d : d \in D\}\right).$$

When defining the classification semigroup we shall use positive linear maps which are faithful and normal. By varying the conditions on ϕ we would get a slightly different theory. (For example we could, alternatively, require ϕ to be completely positive or replace normality by a sequential condition.)

Let Ω be the class of all monotone complete C^*-algebras which are isomorphic to norm closed *-subalgebras of $L(H^\#)$; and let $\Omega^\#$ be the set of all C^*-subalgebras of $L(H^\#)$ which are monotone complete (in themselves) (They cannot be monotone closed subalgebras of $L(H^\#)$ unless they are von Neumann algebras.). So every $A \in \Omega$ is isomorphic to an algebra in $\Omega^\#$.

Definition 3.2.1 We define a relation between algebras in Ω by $A \precsim B$ if there exists a positive linear map $\phi : A \mapsto B$ which is faithful and normal.

Let π be an isomorphism of A onto B. Then π and π^{-1} are both normal so $A \precsim B$ and $B \precsim A$. Now suppose π is an isomorphism of A onto a subalgebra of B. Then π need not be normal. It will only be normal if its range is a monotone closed subalgebra of B. In particular, if A is a monotone closed subalgebra of B, then by taking the natural injection as ϕ, we see that $A \precsim B$.

Lemma 3.2.2 *Let A, B, C be in Ω. If $A \precsim B$ and $B \precsim C$ then $A \precsim C$. Also $A \precsim A$.*

Proof There exists a normal, faithful positive linear map $\phi : A \mapsto B$ and there exists a normal, faithful positive linear map $\psi : B \mapsto C$.

Then $\psi \circ \phi : A \mapsto C$ is a normal, faithful positive linear map.

The identity map from A to A is a surjective isomorphism and so, by the remarks above, $A \precsim A$.   □

Lemma 3.2.3 *Let A and B be in Ω. Then $A \oplus B$ is also in Ω.*

Proof Let $(a_n \oplus b_n : n = 1, 2, \cdots)$ be any norm bounded increasing sequence in $(A \oplus B)_{sa}$. Since for $x \oplus y \in A \oplus B$, $x \oplus y \geq 0$ if, and only if, $x \geq 0$ and $y \geq 0$, (b_n) is norm bounded and increasing. Hence there exists $b \in B_{sa}$ such that $b_n \uparrow b$ in B_{sa}. Similarly, we have $a \in A_{sa}$ such that $a_n \uparrow a$ in A_{sa}. By applying Lemma 2.1.5, it is easy to check that $a_n \oplus b_n \uparrow a \oplus b$ and so $A \oplus B$ is monotone σ-complete.

By replacing sequences by nets, the same argument will show $A \oplus B$ is monotone complete.

Since $H^{\#}$ is an infinite dimensional Hilbert space it is isomorphic to the direct sum of two isomorphic copies of itself, $H_1 \oplus H_2$. Then A is isomorphic to $A_1 \subset L(H_1)$ and B is isomorphic to $B_2 \subset L(H_2)$. Then $A \oplus B$ is isomorphic to $A_1 \oplus B_1$ and clearly $A_1 \oplus B_1$ can be identified with a subalgebra of $L(H_1 \oplus H_2) = L(H^{\#})$. So $A \oplus B$ is in Ω.   □

It is clear that Lemma 3.2.3 implies that Ω is closed under the taking of finite direct sums. In fact more is true. We can generalise to infinite sums. But, before doing so, we shall clarify what "infinite sums" we have in mind. The infinite product of a family of C^*-algebras is defined in ([121], page 6, 1.2.4).

In particular, let $(A_n)_{n=1}^{\infty}$ be a sequence of C^*-algebras. Then the direct product is

$$\prod_{n=1}^{\infty} A_n = \{(a_n) : a_n \in A_n \ \ ||(a_n)|| = \sup_n ||a_n|| < \infty\}.$$

We recall that when the algebraic operations are defined in the obvious way, $\prod_{n=1}^{\infty} A_n$ is a C^*-algebra. Now consider $\mathbf{x} = (x_n)$ and $\mathbf{y} = (y_n)$ in $\prod_{n=1}^{\infty} A_n$. Then $\mathbf{x} \geq \mathbf{y}$ if, and only if, $x_n \geq y_n$ for every n. Using this, a straightforward argument shows that when each A_n is monotone complete then so, also, is their direct product.

Lemma 3.2.4 *Let (A_n) be a sequence of algebras in Ω. Then the (infinite) direct product $\prod A_n$ is in Ω.*

Proof The proof is essentially the same as in the preceding lemma, except we split $H^{\#}$ into a countable direct sum of isomorphic copies of itself. □

Provided $H^{\#}$ has an orthonormal basis of cardinality c or more, we can generalise Lemma 3.2.4 and show that the infinite direct product of continuum many algebras from Ω, is in Ω.

Lemma 3.2.5 *Let $A_1 \precsim B_1$ and $A_2 \precsim B_2$ where A_1, B_1, A_2 and B_2 are in Ω. Then $A_1 \oplus A_2 \precsim B_1 \oplus B_2$.*

Proof By hypothesis, there exist positive, faithful, normal linear maps $\phi_j : A_j \mapsto B_j$ for $j = 1, 2$.

We define $\psi : A_1 \oplus A_2 \mapsto B_1 \oplus B_2$ by $\psi(a_1 \oplus a_2) = \phi_1(a_1) \oplus \phi_2(a_2)$. Then it is straightforward to verify that ψ is a positive, faithful, normal linear map. □

Definition 3.2.6 We now define a relation between algebras in Ω by $A \sim B$ if $A \precsim B$ and $B \precsim A$.

Lemma 3.2.7 *The relation $\sim$ is an equivalence relation.*

Proof By Lemma 3.2.2, $A \sim A$. Again by Lemma 3.2.2, if $A \sim B$ and $B \sim C$ then $A \sim C$. □

Lemma 3.2.8 *If $A_1 \sim B_1$ and $A_2 \sim B_2$ then $A_1 \oplus A_2 \sim B_1 \oplus B_2$.*

Proof This follows from Lemma 3.2.5. □

We adopt the following temporary notation. For each $A \in \Omega^{\#}$ we define $[A]$ to be the set of all B in $\Omega^{\#}$ such that $B \sim A$.

Let $\mathcal{W} = \{[A] : A$ is a norm closed $*$-subalgebra of $L(H^{\#})$, and A is a monotone complete C^*-algebra$\}$.

For each $B \in \Omega$, there is an isomorphism π from B onto $A \in \Omega^{\#}$; we define $w(B)$ to be $[A]$. It is clear that w is well defined. In particular $w(A) = w(B)$ if, and only if, $A \sim B$. So, by Lemma 3.2.8, $w(A_1) = w(B_1)$ and $w(A_2) = w(B_2)$ implies $w(A_1 \oplus A_2) = w(B_1 \oplus B_2)$. It follows that we can define an operation $+$ on $\mathcal{W}$ by setting $w(A) + w(B) = w(A \oplus B)$. The associativity of taking direct sums immediately implies that $+$ is associative on $\mathcal{W}$. So $(\mathcal{W}, +)$ is a semigroup; we shall abuse our notation and use $\mathcal{W}$ for both the semigroup and the underlying set.

Proposition 3.2.9 *The semigroup $\mathcal{W}$ is abelian and has a zero element. The zero element is $w(\mathbb{C})$, where $\mathbb{C}$ is the one dimensional algebra, the complex numbers.*

Proof Consider $w(A)$ and $w(B)$. Define $\phi : A \oplus B \mapsto B \oplus A$ by $\phi(a \oplus b) = b \oplus a$. Then ϕ is a surjective $*$-isomorphism. Hence $w(A \oplus B) = w(B \oplus A)$. So

$$w(A) + w(B) = w(B) + w(A).$$

Fix A. Let $\phi : A \mapsto A \oplus \mathbb{C}$ be defined by $\phi(a) = a \oplus 0$. Then ϕ is positive, normal and faithful. So $A \precsim A \oplus \mathbb{C}$.

Now consider $\psi : A \oplus \mathbb{C} \mapsto A$ defined by $\psi(a \oplus \lambda) = a + \lambda 1$, where 1 is the unit element of the algebra A. Then ψ is positive and normal. Suppose that $\psi(aa^* \oplus \lambda\bar{\lambda}) = 0$. Then $aa^* + |\lambda|^2 1 = 0$. So $a = 0$ and $\lambda = 0$. i.e. ψ is faithful. So $A \oplus \mathbb{C} \precsim A$. Hence

$$w(A) = w(A \oplus \mathbb{C}) = w(A) + w(\mathbb{C}).$$

□

We shall denote the zero element of $\mathcal{W}$ by 0.

Proposition 3.2.10 *Each element of $\mathcal{W}$ is idempotent, that is, $w(A) + w(A) = w(A)$.*

Proof Let $\phi : A \oplus A \mapsto A$ be defined by $\phi(a \oplus b) = a + b$. Then ϕ is a faithful, normal positive linear map. So $A \oplus A \precsim A$.

Now consider $\psi : A \mapsto A \oplus A$ defined by $\psi(a) = a \oplus a$. Then ψ is a faithful, normal positive linear map. So $A \precsim A \oplus A$.   □

From Lemma 3.2.2, we see that if $A_1 \sim A_2 \precsim B_1 \sim B_2$ then $A_1 \precsim B_2$. So, without ambiguity, we may define $w(A) \leq w(B)$ if, and only if $A \precsim B$.

Lemma 3.2.11 *The relation $\leq$ is a partial ordering of the semigroup $\mathcal{W}$. Then $0 \leq w(A)$ for all elements of $\mathcal{W}$. Also, $w(A_1) \leq w(B_1)$ and $w(A_2) \leq w(B_2)$ implies*

$$w(A_1) + w(A_2) \leq w(B_1) + w(B_2).$$

Proof For any A consider the positive linear map $\phi : \mathbb{C} \mapsto A$ defined by $\phi(\lambda) = \lambda 1$. Then ϕ is faithful and normal. So $\mathbb{C} \precsim A$ hence $0 \leq w(A)$. The second part of the lemma follows from Lemma 3.2.5.   □

Corollary 3.2.12 *In the partially ordered semigroup $\mathcal{W}$, $w(A) + w(B)$ is the least upper bound of $w(A)$ and $w(B)$.*

Proof Since $w(A) \leq w(A)$ and $0 \leq w(B)$, we have $w(A) + 0 \leq w(A) + w(B)$. Similarly $w(B) \leq w(A) + w(B)$.

Now suppose that $w(X)$ is an upper bound for $w(A)$ and $w(B)$. Then

$$w(A) + w(B) \leq w(X) + w(X) = w(X).$$

□

Corollary 3.2.13 *We have $w(A) \leq w(B)$ if and only if $w(A) + w(B) = w(B)$.*

Proof This is because $w(A) \leq w(B)$ if and only if $w(B)$ equals the least upper bound of $w(A)$ and $w(B)$.   □

Proposition 3.2.14 *Let A be an algebra in Ω. Then $w(A) = 0$ if, and only if, A is a von Neumann algebra with a faithful normal state.*

Proof Let 1 be the unit element of A. Then $\lambda \to \lambda 1$ is a positive, faithful normal map from $\mathbb{C}$ to A. Hence $\mathbb{C} \precsim A$. So $A \sim \mathbb{C}$ if, and only if, $A \precsim \mathbb{C}$. But this is equivalent to the existence of a faithful normal state on A. By a well known theorem of Kadison ([84] or Exercise 7.6.3 in [88]) the existence of a faithful normal state on a monotone complete C^*-algebra implies that A is a von Neumann algebra. Also, if A is a von Neumann algebra with a faithful normal state, then $A \precsim \mathbb{C}$. □

Corollary 3.2.15 *Let A be a von Neumann algebra which has a faithful state. Then* $w(A) = 0$.

Proof When a von Neumann algebra A possesses a faithful state, say φ, then, by a well known theorem of Takesaki, the state φ can be split into the sum of a positive normal functional η and a completely singular positive functional σ (see Theorem III.3.8 in [161]).

Suppose there exists a non-zero projection p such that $\eta(p) = 0$. Then, because σ is completely singular, there is a non-zero projection q such that $q \leq p$ and $\sigma(q) = 0$. But then $\varphi(q) = 0$. Since φ is faithful this is a contradiction. So η is a faithful normal positive functional on A. □

From the above we get:

Theorem 3.2.16 *The semigroup $\mathcal{W}$ is a partially ordered, abelian semigroup with zero. Each element of $\mathcal{W}$ is idempotent. The natural partial ordering induced on $\mathcal{W}$ by $\precsim$ coincides with the partial ordering defined by:* $x \leq y$ if, and only if $x + y = y$.

A join semi-lattice is a partially ordered set in which each pair of elements has a least upper bound. See [55].

Corollary 3.2.17 *The semigroup $\mathcal{W}$ is a join semi-lattice with a least element 0.*

Proposition 3.2.18 *Let $(A_n)(n = 1, 2, \ldots)$ be a sequence of algebras in Ω. Let their (infinite) direct product be $\prod A_n$. Then $w(\prod A_n)$ is the least upper bound of the countable set $\{w(A_n) : n = 1, 2, \ldots\}$. In other words, in $\mathcal{W}$ each countable set has a supremum.*

Proof Let $w(X)$ be an upper bound for $\{w(A_n) : n = 1, 2 \ldots\}$. Then for each n there exists a faithful positive normal linear map ϕ_n from A_n into X. By multiplying by a suitable constant, if necessary, we can suppose $\|\phi_n\| \leq 1$.

For each $\mathbf{x} \in \prod A_n$ with $\mathbf{x} = (x_n)(n = 1, 2, \cdots)$, let $\Phi(\mathbf{x}) = \sum_{n=1}^{\infty} 2^{-n} \phi_n(x_n)$. Then $\Phi(\mathbf{x}^*\mathbf{x}) = 0$ implies that $\phi_n(x_n^*x_n) = 0$ which implies $x_n = 0$, for all n. So Φ is faithful.

Let $(\mathbf{x}^\alpha)$ be a downward directed net with infimum 0 in $\prod A_n$ where $\mathbf{x}^\alpha \leq 1$ for all α. Choose $\epsilon > 0$. Then for large enough N, and all α,

$$0 \leq \Phi(\mathbf{x}^\alpha) \leq \sum_{n=1}^{N} 2^{-n} \phi_n(x_n^\alpha) + \sum_{n=N+1}^{\infty} 2^{-n} 1 < \sum_{n=1}^{N} 2^{-n} \phi_n(x_n^\alpha) + \epsilon 1.$$

Hence $\bigwedge_\alpha \Phi(\mathbf{x}^\alpha) \le \sum_{n=1}^N \phi_n(x_n^{\alpha_n}) + \epsilon$. Where $\alpha_1, \alpha_2, \dots, \alpha_N$ are arbitrary elements of the index set for the net. Hence $\bigwedge_\alpha \Phi(\mathbf{x}^\alpha) \le \epsilon 1$. Since ϵ was arbitrary it follows that Φ is normal. So $w(\prod A_n) \le w(X)$. It is easy to see that $w(A_q) \le w(\prod A_n)$ for each q. The result follows.   □

Lemma 3.2.19 *Let $\phi : A \mapsto B$ be a positive linear map which is normal. Then there exists a projection $e \in A$ such that ϕ vanishes on $(1 - e)A(1 - e)$ and its restriction to eAe is faithful.*

Proof Let K be the hereditary cone $\{x \in A^+ : \phi(x) = 0\}$. Let $K_0 = \{x \in K : ||x|| < 1\}$. Then, see ([121] page 11) or ([161] pp25–27), K_0 is an upward directed set. Let p be the supremum of K_0 in A. In fact, ([121], Theorem 1.4.2) or ([161], Theorem 1.7.4) K_0 is an increasing approximate unit for the hereditary subalgebra B generated by K so its supremum, p, is a unit for this algebra. So p is a projection. (See the proof of Proposition 2.1.15.)

Let $z \in (1 - p)A(1 - p)$ and $\phi(zz^*) = 0$. Then zz^* is in K. So $zz^* = pzz^* = p(1 - p)zz^* = 0$. So ϕ is faithful when restricted to $(1 - p)A(1 - p)$. On putting $e = 1 - p$ the lemma follows.   □

We call e the support projection of ϕ.

Lemma 3.2.20 *Let e be any projection in A. Then*

$$eAe \oplus (1 - e)A(1 - e) \sim A.$$

Proof If $e = 0$ or $e = 1$ then the statement is trivially true. So we suppose that neither e nor $1 - e$ is zero. Let $\phi : A \mapsto eAe \oplus (1-e)A(1-e)$ be defined by $\phi(x) = exe \oplus (1-e)x(1-e)$. Then ϕ is positive, linear and normal. (Use Lemmas 2.1.9, 2.1.5 and Proposition 2.2.26.)

Suppose $\phi(zz^*) = 0$. Then $ezz^*e = 0$ and $(1 - e)zz^*(1 - e) = 0$. So $ez = 0$ and $(1 - e)z = 0$. Thus $z = 0$. So ϕ is faithful.

Hence $w(A) \le w(eAe \oplus (1 - e)A(1 - e))$.

By Proposition 2.2.26, the natural embedding of eAe in A is an isomorphism onto a monotone closed subalgebra. Hence $w(eAe) \le w(A)$. Similarly $w((1 - e)A(1 - e)) \le w(A)$. Since $w(A) + w(A) = w(A)$, we have $w(eAe \oplus (1-e)A(1-e)) \le w(A)$. The result is proved.   □

The next result shows that the ordered semigroup $\mathcal{W}$, has the Riesz Decomposition Property. This will then imply that, regarded as a join semi-lattice, it is distributive. This is useful because there is a well developed structure theory for distributive join semi-lattices which can then be applied to $\mathcal{W}$.

Theorem 3.2.21 *Let a, b, c be elements of $\mathcal{W}$ such that $a \le b + c$. Then $a = a_1 + a_2$ such that $0 \le a_1 \le b$ and $0 \le a_2 \le c$.*

Proof Let $a = w(A)$, $b = w(B)$, and $c = w(C)$. Then there exists a faithful normal positive linear map $\phi : A \mapsto B \oplus C$.

For each y in the algebra B and for each z in the algebra C, let $\pi_1(y \oplus z) = y$. Then π_1 is the canonical projection from $B \oplus C$ onto the first component B. Similarly, define $\pi_2 : B \oplus C \mapsto C$.

Let $\phi_j = \pi_j \circ \phi$ for $j = 1, 2$. Then each ϕ_j is positive, linear and normal. Also $\phi(x) = \phi_1(x) \oplus \phi_2(x)$ for each $x \in A$. First let us suppose that $\phi_1 = 0$. Then $a \leq c$. So, on putting $a = a_2$ and $a_1 = 0$, we are done.

Hence we shall now suppose that $\phi_1 \neq 0$. Then, by Lemma 3.2.19, it has a non-zero support projection e. Then $\phi_1|eAe$ is a faithful normal map into B. Also $\phi(1 - e) = 0 \oplus \phi_2(1 - e)$. If this vanishes then, since ϕ is faithful, $e = 1$. So ϕ_1 would be faithful on A, which implies $a \leq b$. On putting $a = a_1$ and $a_2 = 0$ we would be finished. We now suppose that $\phi(1 - e) = 0 \oplus \phi_2(1 - e) \neq 0$. Since ϕ_1 vanishes on $(1-e)A(1-e)$, we have that $0 \oplus \phi_2|(1-e)A(1-e) = \phi|(1-e)A(1-e)$ which is faithful.

Hence $eAe \precsim B$ and $(1 - e)A(1 - e) \precsim C$. On putting $a_1 = w(eAe)$ and $a_2 = w((1 - e)A(1 - e))$, we find that $0 \leq a_1 \leq b$ and $0 \leq a_2 \leq c$.

By appealing to Lemma 3.2.20 we have $a_1 + a_2 = a$. $\qquad\square$

Corollary 3.2.22 *Regarded as a join semi-lattice, $\mathcal{W}$ is distributive.*

Proof When we interpret '+' as the lattice operation '$\vee$' this is just a straightforward translation of the statement of the theorem. $\qquad\square$

The well established theory of distributive join semi-lattices can now be applied to $\mathcal{W}$. See [55]. Since we wish to keep this chapter of reasonable length we shall not pursue this here. But distributivity is a key property which, in particular, leads to an elegant representation theory akin to the Stone representation for Boolean algebras.

The classification given here maps each small von Neumann algebra, equivalently, each von Neumann algebra acting on a separable Hilbert space, to the zero of the semigroup. It could turn out that $\mathcal{W}$ is very small and fails to distinguish between many algebras. We shall see later that this is far from true. Even when w is restricted to special subclasses of algebras (e.g. subalgebras of ℓ^∞), we can show that its range in $\mathcal{W}$ is huge, 2^c, where $c = 2^{\aleph_0}$. In the next section we shall introduce the spectroid of an algebra and show that it is, in fact, an invariant for elements of $\mathcal{W}$.

The following technical result will be useful to us later.

Lemma 3.2.23 *Let A be a monotone complete C^*-algebra. Let M be a maximal abelian $*$-subalgebra of A. Let $\Gamma : A \mapsto M$ be a faithful, normal idempotent map from A onto M. Then $wA = wM$.*

Proof By Corollary 2.1.32, M is a monotone closed subalgebra of A. So the natural embedding of M in A implies $M \precsim A$. On the other hand, the map Γ gives $A \precsim M$. So $A \sim M$. Thus $w(A) = w(M)$. $\qquad\square$

3.3   Spectroid and Representing Functions

Although our main interest is focused on monotone complete C^*-algebras, in this section we shall also use the larger class of monotone σ-complete C^*-algebras.

Let A be a monotone σ-complete C^*-algebra. When $\{a_k : k \in K\}$ is a countable, lower bounded, downward directed subset of A_{sa}, we denote the infimum of $\{a_k : k \in K\}$ by $\bigwedge_{k \in K} a_k$.

For any non-empty set J we let $F(J)$ be the collection of all finite subsets of J, including the empty set. In particular we note that $F(\mathbb{N})$, where $\mathbb{N}$ is the set of natural numbers, is countable.

Definition 3.3.1 A representing function for a monotone σ-complete C^*-algebra, A, is a function $f : F(\mathbb{N}) \mapsto A^+$ such that

(i) $f(k) \geq 0$ and $f(k) \neq 0$ for all k.
(ii) f is downward directed, that is, when k, l are finite subsets of N, then $f(k \cup l) \leq f(k)$ and $f(k \cup l) \leq f(l)$.
(iii) $\bigwedge_{k \in F(\mathbb{N})} f(k) = 0$ in A_{sa}.

Let T be a set of cardinality $2^{\aleph_0}$. Let $\mathbf{N} : T \mapsto \mathcal{P}(\mathbb{N})$ be an injection and let $\mathbf{N}(t)$ be infinite for each t. We could, for example, take T to be the collection of all infinite subsets of $\mathbb{N}$ and let $\mathbf{N}(t) = t$ for each t. But we do not demand that $\{\mathbf{N}(t) : t \in T\}$ contains every infinite subset of $\mathbb{N}$. We shall regard T and the function $\mathbf{N}$ as fixed until further notice.

Definition 3.3.2 Let A be a monotone σ-complete C^*-algebra and let $f : F(\mathbb{N}) \longmapsto A$ be a representing function. Then let $R_{(T,\mathbf{N})}(f)$ be the subset of T defined by

$$\{t \in T : \bigwedge_{k \in F(\mathbf{N}(t))} f(k) = 0\}.$$

The set $R_{(T,\mathbf{N})}(f)$ is said to be represented by f in A, modulo $(T, \mathbf{N})$.

Any subset of T which can be represented in A is said to be a representing set of A (modulo $(T, \mathbf{N})$).

Definition 3.3.3 Let A be a monotone σ-complete C^*-algebra. Then the spectroid of A (modulo $(T, \mathbf{N})$), written $\partial_{(T,\mathbf{N})} A$, is the collection of all sets which can be represented in A, modulo $(T, \mathbf{N})$, by some representing function $f : F(\mathbb{N}) \mapsto A^+$, that is,

$$\partial_{(T,\mathbf{N})} A = \{R_{(T,\mathbf{N})}(f) : f \text{ is a representing function for } A\}.$$

When it is clear from the context which $(T, \mathbf{N})$ is being used, we shall sometimes write ∂A.

We use Card(S) or $\#S$ to denote the cardinality of a set S.

Proposition 3.3.4 *Let* $(T, \mathbf{N})$ *be fixed and let A be any monotone σ-complete C^*-algebra of cardinality c. Then $\partial_{(T,\mathbf{N})}A$ is of cardinality not exceeding c.*

Proof Each element of $\partial_{(T,\mathbf{N})}A$ arises from a representing function for A. But the cardinality of all functions from $F(\mathbf{N})$ into A is $\text{Card}(A^{F(\mathbf{N})}) = c^{\aleph_0} = c$. So $\text{Card}(\partial_{(T,\mathbf{N})}A) \leq c$. □

Corollary 3.3.5 *Let* $(T, \mathbf{N})$ *be fixed and let A be a small monotone complete C^*-algebra of cardinality c. Then $\partial_{(T,\mathbf{N})}A$ is of cardinality not exceeding c.*

Lemma 3.3.6 *Let* $(T, \mathbf{N})$ *be fixed and let S be the set of all spectroids, modulo* $(T, \mathbf{N})$, *of monotone σ-complete C^*-algebras of cardinality c. Then $\text{Card}(S) \leq 2^c$.*

Proof Each subset of cardinality $\leq c$ is the range of a function from $\mathbb{R}$ into $\mathcal{P}(T)$. So $\text{Card}(S) \leq \text{Card}(\mathcal{P}(\mathbb{R})^{\mathbb{R}}) \leq \text{Card}(\mathcal{P}(\mathbb{R} \times \mathbb{R})) = 2^c$. □

Suppose that A and B are monotone σ-complete C^*-algebras and $\phi : A \mapsto B$ is a faithful positive linear map. Let us recall that ϕ is σ-normal if, whenever $(a_n)(n = 1, 2, \ldots)$ is a monotone decreasing sequence in A with $\bigwedge_{n=1}^{\infty} a_n = 0$ then $\bigwedge_{n=1}^{\infty} \phi(a_n) = 0$.

Lemma 3.3.7 *Let A be monotone σ-complete and D a downward directed subset of A_{sa}, which is countable and bounded from below. Then there exists a monotone decreasing sequence $(a_n)(n = 1, 2, \ldots)$ in D such that $d \in D$ implies $d \geq a_n$ for some n. Furthermore $\bigwedge D = \bigwedge_1^{\infty} a_n$.*

Proof Since D is countable, we can write $D = \{b_n : n = 1, 2, \cdots\}$. Let $a_1 = b_1$. Since D is downward directed, we can find a_2 in D such that $a_2 \leq a_1$ and $a_2 \leq b_2$. We can now find a_3 in D such that $a_3 \leq a_2$ and $a_3 \leq b_3$. Continuing in this way we find a sequence $(a_n)(n = 1, 2, \ldots)$ with the required properties. □

Corollary 3.3.8 *Let A and B be monotone σ-complete C^*-algebras. Let $\phi : A \mapsto B$ be a positive, faithful σ-normal linear map. Let D be a downward directed subset of non-negative elements in A which is countable. Then $\bigwedge\{d : d \in D\} = 0$ if, and only if $\bigwedge\{\phi(d) : d \in D\} = 0$.*

Definition 3.3.9 Let A and B be monotone σ-complete C^*-algebras. If there exists a positive, faithful σ-normal linear map $\phi : A \mapsto B$ we write $A \precsim_{\sigma} B$. Then the relation $\precsim_{\sigma}$ is a quasi-ordering of the class of monotone σ-complete C^*-algebras.

When $A \precsim_{\sigma} B$ and $B \precsim_{\sigma} A$ we say that A and B are σ-normal equivalent and write $A \sim_{\sigma} B$. This is an equivalence relation on the class of monotone σ-complete C^*-algebras. Clearly, if A and B are monotone complete C^*-algebras and $A \precsim B$ then $A \precsim_{\sigma} B$. So if $A \sim B$ it follows that $A \sim_{\sigma} B$.

Proposition 3.3.10 *Let* $(T, \mathbf{N})$ *be fixed and let A and B be monotone σ-complete C^*-algebras. Let $A \precsim_{\sigma} B$. Then $\partial_{(T,\mathbf{N})}(A) \subset \partial_{(T,\mathbf{N})}(B)$.*

Proof Each element of ∂A is of the form $R_{(T,\mathbf{N})}(f)$ where f is a representing function for A. It is straightforward to verify that ϕf is a representing function for B. Since

ϕ is faithful it follows from Corollary 3.3.8 that $R_{(T,N)}(f) = R_{(T,N)}(\phi f)$. Thus $\partial A \subset \partial B$. $\square$

It is clear that the spectroid is an isomorphism invariant but from Proposition 3.3.10 it is also invariant under σ-normal equivalence.

Corollary 3.3.11 *Let* (T, N) *be fixed and let A and B be monotone σ-complete C^*-algebras. Let* $A \sim_\sigma B$. *Then* $\partial_{(T,N)}(A) = \partial_{(T,N)}(B)$.

Corollary 3.3.12 *Let A and B be monotone complete C^*-algebras with* $w(A) = w(B)$. *Then* $\partial_{(T,N)}(A) = \partial_{(T,N)}(B)$ *for any given* (T, N).

So the spectroid is an invariant for the semigroup W and we may talk about the spectroid of an element of the semigroup.

For the rest of this section we shall consider only monotone complete C^*-algebras, although some of the results (and proofs) are still valid for monotone σ-complete C^*-algebras. Let $\mathcal{C}$ be the class of all monotone complete C^*-algebras of cardinality c. Let W be the semigroup constructed in Sect. 3.2. We shall now assume that w has been restricted to the class of all monotone complete C^*-algebras of cardinality not exceeding c and, from now on, use W to denote the semigroup $\{w(A) : A \in \mathcal{C}\}$. (So, in effect we are taking a sub-semigroup of the one constructed in Sect. 3.2, and abusing our notation by giving it the same name.)

Theorem 3.3.13 *Let* (T, N) *be fixed and consider only spectroids modulo* (T, N). *Let* $\{A_\lambda : \lambda \in \Lambda\}$ *be a collection of small monotone complete C^*-algebras such that the union of their spectroids has cardinality* 2^c. *Then there is a subcollection* $\{A_\lambda : \lambda \in \Lambda_0\}$ *where* Λ_0 *has cardinality* 2^c *and* $\partial(A_\lambda) \neq \partial(A_\mu)$ *whenever* λ *and* μ *are distinct elements of* Λ_0.

Proof Let us define an equivalence relation on Λ by $\lambda \approx \mu$ if, and only if, $\partial(A_\lambda) = \partial(A_\mu)$. By using the Axiom of Choice we can pick one element from each equivalence class to form Λ_0. Clearly $\partial(A_\lambda) \neq \partial(A_\mu)$ whenever λ and μ are distinct elements of Λ_0. Also $\bigcup\{\partial(A_\lambda) : \lambda \in \Lambda_0\}$ is equal to the union of all the spectroids of the original collection. So

$$2^c = \#\left(\bigcup\{\partial(A_\lambda) : \lambda \in \Lambda_0\}\right).$$

By Corollary 3.3.5, $\#\partial(A_\lambda) \leq c$ for each $\lambda \in \Lambda_0$. Hence $2^c \leq c \times \#(\Lambda_0)$. It follows that we cannot have $\#(\Lambda_0) \leq c$. So $c \times \#(\Lambda_0) = \#(\Lambda_0)$. So $2^c \leq \#(\Lambda_0)$. From Lemma 3.3.6, we get $\#\Lambda_0 \leq 2^c$. So $\#\Lambda_0 = 2^c$. $\square$

Corollary 3.3.14 *Given the hypotheses of the theorem, whenever* λ *and* μ *are distinct elements of* Λ_0 *then* $wA_\lambda \neq wA_\mu$. *So* A_λ *is not equivalent to* A_μ. *In particular, they cannot be isomorphic.*

Proof Apply Corollary 3.3.12. $\square$

We have seen that the small monotone complete C^*-algebras can be classified by elements of W and also by their spectroids. Since w maps every small von Neumann algebra with a faithful state to the zero of the semigroup, this classification might be very coarse, possibly $\mathcal{W}$ might be too small to distinguish between more than a few classes of algebras. But we shall see later that this is far from the truth. By applying Theorem 3.3.13 for appropriate $(T, \mathbf{N})$ we shall see that $\#\mathcal{W} = 2^c$.

We can define the spectroid in much greater generality. In particular we can do this for order feasible sets.

Definition 3.3.15 A partially ordered set M is said to be *order feasible* if it has a smallest element 0 and each monotone decreasing sequence in M has a greatest lower bound.

For example, the positive cone A^+ of monotone (σ-)complete C^*-algebra and more generally, the positive cone V^+ of a monotone (σ-)complete operator system V are order feasible.

Quasi-orderings of classes of order feasible sets can be defined and associated weight semigroups can be constructed. However, the Riesz Decomposition property may fail for these generalized weight semigroups unless extra conditions are imposed.

We constructed the classification semigroup $\mathcal{W}$ in [144]. Representing functions for Boolean algebras were used by Monk and Solovay [104]; their definition was extended to monotone complete C*-algebras by Hamana in [66]. See also [103]. The spectroid invariant was introduced by us in [144]. Its extension to order feasible sets is outlined in [182].

Chapter 4
Commutative Algebras: Constructions and Classifications

The second chapter gave the foundations of the theory of monotone complete C^*-algebras. We then introduced a classification semigroup and spectroid invariants. But, up to now, we have seen few concrete examples of wild monotone complete C^*-algebras. A good place to start is by finding commutative examples. This is what we shall do in this chapter. In Sect. 4.2 we give general constructions for commutative algebras. In Sect. 4.3 we show that ℓ^∞ has $2^{\mathbb{R}}$ subalgebras $\{A_t : t \in 2^{\mathbb{R}}\}$, where each A_t is a (small) wild, commutative monotone complete C^*-algebra. Furthermore, when $r \neq s$, then A_r and A_s take different values in the semigroup $\mathcal{W}$ and have different spectroids. In a later chapter we shall use group actions on commutative algebras to construct huge numbers of small wild factors.

We have seen earlier that there is a natural correspondence between commutative monotone complete C^*-algebras and complete Boolean algebras. So it is not surprising that we begin with Boolean algebras.

4.1 Boolean Algebra Preliminaries

Most readers will have come across Boolean algebras. This short section outlines the few topics we need. Experts in this area should skip to the next section.

4.1.1 Basics

An elegant, lucid account of Boolean algebras, ample for our purposes, is given by Halmos [57] but a vast literature exists, see [12, 95]. Complete Boolean algebras are of importance in mathematical logic. As we mentioned earlier each complete

© Springer-Verlag London 2015   63
K. Saitô, J.D.M. Wright, *Monotone Complete C*-algebras and Generic Dynamics*,
Springer Monographs in Mathematics, DOI 10.1007/978-1-4471-6775-4_4

Boolean algebra can be identified with the lattice of projections in a *commutative monotone complete* C^*-algebra.

Given a non-empty set X, a non-empty collection of subsets, $\mathcal{F}$, is a *field* of sets if (i) $A \in \mathcal{F}$ implies that its compliment $A' \in \mathcal{F}$ and (ii) A and B in $\mathcal{F}$ implies $A \cup B$ and $A \cap B$ are in $\mathcal{F}$. These properties imply that $\varnothing$ and X are also in $\mathcal{F}$. Every field of sets is an example of a Boolean algebra and the formal definition of a Boolean algebra (which we shall give shortly) is an abstraction of the idea of a field of sets. (In fact there are many equivalent lists of axioms for Boolean algebras.)

For $E \subset X$ and $F \subset X$ the *symmetric difference* $E \triangle F$ is defined to be $(E \cap F') \cup (E' \cap F)$. It turns out that this operation is associative (see below) and that we can define a ring structure on $\mathcal{F}$ by putting:

$$0 = \varnothing \text{ and } 1 = X$$

$$E + F = E \triangle F$$

$$EF = E \cap F.$$

This ring is commutative, has a unit, and, for each E, $E^2 = E$. Any unital ring where $e^2 = e$ for each element is said to be a *Boolean ring*. In a Boolean ring, by expanding the identity $(a + a)^2 = a + a$, we find that $a + a = 0$. Every Boolean ring is commutative. To see this we use

$$a + b = (a + b)^2 = a^2 + ab + ba + b^2 = a + b + ab + ba.$$

When X is a set with only one element then the collection of all subsets has only two elements. So the corresponding Boolean ring contains just 0 and 1. The two element Boolean ring is a field, in the algebraic sense, that is, each non-zero element (there is only 1) has a multiplicative inverse. This use of "field" in two completely different senses is unfortunate but should not cause confusion. Following the notation of Halmos, we shall sometimes denote the two element Boolean ring by 2.

Then 2^X is the set of all functions from X to 2. When we add χ_E and χ_F pointwise (where we regard them as taking their values in 2) we get $\chi_E + \chi_F = \chi_{E \triangle F}$. (Hence "$\triangle$" is associative.) Similarly $\chi_E \chi_F = \chi_{E \cap F}$.

A Boolean algebra is a set B with two (distinct) distinguished elements 0 and 1, a binary operation $\vee$, a binary operation $\wedge$, and a unary operation $'$ with the following properties. For a, b, c in B

(i) $a \wedge (b \vee c) = (a \wedge b) \vee (a \wedge c)$ and $a \vee (b \wedge c) = (a \vee b) \wedge (a \vee c)$ (distributivity)
(ii) the operation $\vee$ is commutative and associative
(iii) the operation $\wedge$ is commutative and associative
(iv) $a \wedge a = a$ and $a \vee a = a$
(v) $a \wedge 0 = 0$ and $a \vee 0 = a$
(vi) $a'' = a$
(vii) $0' = 1$

(viii) $(a \vee b)' = a' \wedge b'$

  (ix) $a \wedge a' = 0$.

From these axioms we can deduce, for example $1' = 0'' = 0$. We also have $(a \wedge b)' = (a' \vee b')'' = a' \vee b'$ and $a \vee a' = (a' \wedge a'')' = 0' = 1$.

The above axioms are clearly satisfied by a field of subsets of a set X, when $\vee$ is interpreted as union, $\wedge$ is interpreted as intersection and $'$ as complimentation.

Given a Boolean algebra B we can give it the structure of a Boolean ring by defining $x + y = (x \wedge y') \vee (x' \wedge y)$ and $xy = x \wedge y$. This Boolean ring has, appropriately, 0 as its zero element and 1 as its unit. Conversely, given any Boolean ring we can give it the structure of a Boolean algebra by defining

$$x \wedge y = xy \text{ and } x \vee y = x + y + xy \text{ and } x' = 1 + x.$$

This correspondence between Boolean algebras and Boolean rings is bijective i.e. if we start with a Boolean algebra, construct the corresponding Boolean ring and then construct a Boolean algebra from this ring, we recover the original Boolean algebra.

Let B_1 and B_2 be Boolean algebras. A map $H : B_1 \mapsto B_2$ is a *Boolean homomorphism* if it preserves the Boolean operations. That is, for all x, y in B_1

$$H(x \vee y) = Hx \vee Hy$$

$$H(x \wedge y) = Hx \wedge Hy$$

$$Hx' = (Hx)'.$$

It can be deduced that $H(0) = 0$ and $H(1) = 1$. Furthermore, when these Boolean algebras are given their canonical Boolean ring structures, $H : B_1 \mapsto B_2$ is a ring homomorphism (with $H(1) = 1$) precisely when H is a Boolean homomorphism.

Let J be a subset of B. It is a (ring) ideal of B precisely when it is the kernel of a (ring) homomorphism. But this is equivalent to J being the kernel of a Boolean homomorphism. So we call J a Boolean ideal if (i) $0 \in J$, (ii) $a \in J$ and $b \in J$ implies $a \vee b \in J$ and (iii) $a \in J$ and $b \in B$ implies $a \wedge b \in J$.

There is a third way of looking at Boolean algebras. We can partially order a Boolean algebra B, by defining $p \leq q$ to mean $p \wedge q = p$. With this partial ordering, each Boolean algebra becomes a distributive lattice where a and b have least upper bound $a \vee b$ and greatest lower bound $a \wedge b$.

We define a Boolean algebra B to be *σ-complete* if each countable subset of B has a least upper bound. It is *complete* if every subset has a least upper bound.

4.1.2 Regular Open Sets

Let X be a (non-empty) topological space. Throughout this book we consider only Hausdorff spaces and so we shall suppose X to be Hausdorff even though this assumption is not strictly necessary here.

We recall that an open subset of X is *regular* if it is the interior of its closure.

As usual we denote the compliment of P in X by P'. It is convenient, in this section, to denote the closure of P by P^- and the interior of P by P^o. Then $P^o = P'^{-'}$. So Q is a regular open set precisely when $Q = Q^{-o} = Q^{-'-'}$. We now define $P^\perp$ to be $P^{-'}$. So P is a regular open set if, and only if, $P = P^{\perp\perp}$.

Lemma 4.1.1

(i) *Let R be any subset of X. Then R^{-o} is a regular open set. When R is open then R^{-o} is the smallest regular open set which contains R.*

(ii) *Let P and Q be regular open sets. Then $P \cap Q$ is a regular open set.*

(iii) *Let $\{P_\lambda : \lambda \in \Lambda\}$ be a (non-empty) collection of regular open sets. Let $O = \cup_{\lambda \in \Lambda} P_\lambda$. Then O^{-o} is the smallest regular open set containing each P_λ.*

Proof

(i) Let V be the open set R^{-o}. Then $V^- \subset R^-$. So $V^{-o} \subset R^{-o} = V$. But, since V is an open subset of V^-, $V \subset V^{-o}$. Thus V is a regular open set. When R is open then $R \subset R^{-o} = V$.

Let W be any regular open set with $R \subset W$. Then $R^- \subset W^-$ and so $V = R^{-o} \subset W^{-0} = W$.

(ii) Clearly $(P \cap Q)^- \subset P^-$. So $(P \cap Q)^{-o} \subset P^{-o} = P$ (because P is a regular open set). So $(P \cap Q)^{-o} \subset P \cap Q$. But, by (i), $P \cap Q \subset (P \cap Q)^{-o}$. So $P \cap Q = (P \cap Q)^{-o}$.

(iii) By applying (i) to O we get (iii).   □

Let $RegX$ be the set of regular open subsets of X. By Lemma 4.1.1, $RegX$ is a complete Boolean algebra. But, in general, it is not a field of sets. One reason is the union of two regular open sets need not be regular. For example, when X is the set of real numbers with its usual topology, the intervals $(0, 1)$ and $(1, 2)$ are regular open but their union is not. But if we put $0 = \varnothing$, $1 = X$, $P \wedge Q = P \cap Q$ and $P \vee Q = (P \cup Q)^{-o}$ then, if we use $P \mapsto P^\perp$ as unary operation, we obtain a complete Boolean algebra.

With no additional conditions on X, whenever O is an open subset, then $O^- \setminus O$ is a closed set with empty interior. We recall that a set is *nowhere dense* if its closure has empty interior. A set is *meagre* if it is the union of countably many nowhere dense subsets. It follows that the union of countably many meagre sets is, again, the union of countably many nowhere dense sets. Hence it is meagre.

Also any subset of a meagre subset is meagre. So the meagre subsets of X form a Boolean σ-ideal of $\mathcal{P}(X)$. But this ideal is only of interest if it is not too big. For example, if X consists of all the rationals in $\mathbb{R}$, with the topology induced by $\mathbb{R}$, then every subset of X is meagre. For this reason we shall now focus our attention on Baire spaces.

We recall that X is a Baire space if its only meagre open set is the empty set. Equivalently, the intersection of a sequence of dense open subsets of X is a dense subset of X.

Let X be a Baire space. Then $E \subset X$ is said to have the *Baire Property* if there exists an open set U such that $E \triangle U$ is meagre. Let $BP(X)$ be the collection of all subsets of X with the Baire Property. Clearly every open set is in $BP(X)$. It is straightforward to show that $BP(X)$ is a σ-field of subsets of X. Hence it contains the Borel sets of X, because they are the smallest σ-field of subsets which contains the open sets.

Lemma 4.1.2 *Let X be a Baire space. Then for each $E \in BP(X)$ there is a unique regular open set U such that $E \triangle U$ is meagre.*

Proof Let J be the ideal of meagre sets in $BP(X)$. This is both a Boolean ideal and a ring ideal when we regard $BP(X)$ as a Boolean ring. Then $\triangle$ is addition in this ring.

By the definition of the Baire Property, there exists an open set O such that $E \triangle O \in J$. Now let $U = O^{-o}$. Then, by (i) in the preceding lemma, U is a regular open set. Also $O \subset O^{-o} \subset O^-$. Since $O^- \backslash O$ is nowhere dense, $O \triangle U \in J$. So $E \triangle U = (E \triangle O) \triangle (O \triangle U)$ which is in J.

Now let V be any regular open set such that $E \triangle V \in J$. Then $V \triangle U = (V \triangle E) \triangle (E \triangle U)$ which is in J.

Then the open set $(U \cup V) \backslash (U \cap V)^-$ is contained in $(U \cup V) \backslash (U \cap V) = U \triangle V$ which is in J.

Because X is a Baire space, any open set which is meagre must be empty. So

$$U \cup V \subset (U \cap V)^-.$$

Since $U \cup V$ is open, $U \cup V \subset (U \cap V)^{-o}$.

Since $U \cap V$ is a regular open set, it now follows that $U \cup V \subset U \cap V$. Hence $U \subset V$ and $V \subset U$. So $U = V$. □

Let $Bor(X)$ be the smallest σ-field of subsets of X which contains all the open sets. Then $Bor(X)$ is the field of Borel subsets of X. When X is a Baire space, for each Borel set E let rE be the unique regular open set such that $E \triangle rE$ is meagre.

Theorem 4.1.3 (Birkhoff-Ulam) *Let X be a Baire space. Let r be the map defined above. Then r is a Boolean σ-homomorphism of $Bor(X)$ onto $RegX$. Furthermore the kernel of this homomorphism is the ideal of meagre Borel sets.*

Corollary 4.1.4 *Let E and F be Borel sets whose intersection is meagre. Then rE and rF are disjoint.*

Proof We have $rE \cap rF = rE \wedge rF = r(E \cap F) = \varnothing$. □

4.1.3 Structure Space

Let $\{0, 1\}$ be equipped with the discrete topology. Then, for any index set I, the product space $\{0, 1\}^I$ is compact Hausdorff. From the definition of the product topology, it has a base of clopen subsets. So it is zero-dimensional. (For compact Hausdorff spaces, zero-dimensional and totally disconnected are synonymous.) Let X be a closed subset of $\{0, 1\}^I$. Then X is compact and totally disconnected. Let $\mathcal{K}$ be the collection of clopen subsets of X. Then $\mathcal{K}$ is a Boolean algebra, in fact a field of subsets of X. Given $x \in X$ we can define $h_x : \mathcal{K} \mapsto \{0, 1\}$ by $h_x(K)$ is 1 if $x \in K$ and 0 if $x \notin K$. Then h_x may be regarded as a Boolean homomorphism of $\mathcal{K}$ onto the two element Boolean algebra, or equivalently, as a (unital) ring homomorphism of $\mathcal{K}$ onto the field with two elements. So each point of X corresponds to a homomorphism onto 2.

The Stone Representation Theorem, see below, implies that every Boolean algebra is isomorphic to the algebra of clopen subsets of some compact totally disconnected space.

The only Boolean ring without any non-trivial ideals is the (algebraic) field 2.

Lemma 4.1.5 *Let B be a Boolean algebra. Let $p \in B$ with $p \neq 0$ and $p \neq 1$. Then there is a homomorphism $h : B \mapsto 2$ such that $h(p) = 1$ and $h(p') = 0$.*

Proof Consider the ideal $(1 - p)B$. By a Zorn's Lemma argument, it is contained in a maximal ideal M. Let h be the quotient homomorphism of B onto B/M.

Then B/M is a Boolean ring and a field. So we can identify it with 2.

Since $1 - p \in M$, we have $1 - h(p) = h(1 - p) = 0$. $\qquad\qquad\square$

Theorem 4.1.6 (Stone) *Let B be a Boolean algebra. Let S_B be the set of all homomorphisms of B onto 2. Then S_B is a closed subset of 2^B and thus a totally disconnected compact Hausdorff space. For each $b \in B$ let $H(b) = \{h \in S_B : h(b) = 1\}$. Then H is an isomorphism of B onto the Boolean algebra of all clopen subset of S_B.*

Corollary 4.1.7 *When X is a totally disconnected, compact Hausdorff space and $\mathcal{K}$ is the Boolean algebra of clopen subsets of X, then X is homeomorphic to $S_{\mathcal{K}}$.*

Corollary 4.1.8 *Let κ be an infinite cardinal. Let X be a totally disconnected compact Hausdorff space. Let κ clopen subsets form a base for the topology of X. Then X is homeomorphic to a subspace of 2^κ.*

Proof Let $\mathcal{K}$ be the Boolean algebra of all clopen subsets of X. Then, by definition of the Stone structure space of a Boolean algebra, $S_{\mathcal{K}}$ is a closed subset of $2^{\mathcal{K}}$.

Hence X is homeomorphic to a closed subspace of $2^{\mathcal{K}}$. By compactness, each clopen subset of X is the union of finitely many base elements. So $\#\mathcal{K} = \kappa$. Hence $2^{\mathcal{K}}$ is homeomorphic to 2^κ. $\qquad\qquad\square$

We know that there is a duality between the category of compact Hausdorff spaces (with continuous maps as morphisms) and the category of commutative unital C^*-algebras (with $*$-homomorphisms as morphisms). Stone's theorem implies a similar duality between totally disconnected, compact Hausdorff spaces and Boolean algebras.

In particular, the Stone structure space of a Boolean algebra is extremally disconnected precisely when the Boolean algebra is complete.

4.2 Commutative Algebras: General Constructions

Let us recall that if a monotone complete C^*-algebra has a separating family of normal states then it is a von Neumann algebra. If it has *no* normal states, then it is said to be *wild*. In this chapter we give some constructions with commutative algebras which will lead to huge numbers of examples of wild commutative monotone complete C^*-algebras. We shall make use of the classification semigroup $\mathcal{W}$ and spectroid invariants discussed in Chap. 3. In subsequent chapters we shall use group actions on wild commutative algebras to exhibit huge numbers of wild factors.

4.2.1 Measurable Functions

Let Y be a non-empty set. Let us recall the following notation. Let $Bnd(Y)$ be the algebra of all bounded complex valued functions on Y, where multiplication, addition, involution etc. are determined pointwise. When this algebra is given the norm

$$\|f\| = \sup\{|f(t)| : t \in Y\}$$

it becomes a commutative, unital C^*-algebra; in fact, a von Neumann algebra. We use $Bnd(Y)_{sa}$ to denote the self-adjoint part of $Bnd(Y)$. Then $Bnd(Y)_{sa}$ is the (real) commutative algebra of all bounded real-valued functions in $Bnd(Y)$.

We saw, in Chap. 2, that $Bnd(Y)_{sa}$ is a vector lattice as well as a commutative algebra.

Let S be a σ-algebra of subsets of Y, that is, (i) $Y \in S$, (ii) $A \in S$ implies $Y \backslash A \in S$, (iii) if $A_n \in S$ for $n = 1, 2 \ldots$ then $\cup_1^\infty A_n \in S$. We call the pair (Y, S) a measurable space. We recall that a real valued function $f : Y \mapsto \mathbb{R}$ is *S-measurable* if $\{y \in Y : f(y) > t\} \in S$ for each $t \in \mathbb{R}$. A complex valued function F on Y is measurable if $F = f + ig$ where f and g are real valued measurable functions. By standard measure theoretic arguments (see, for example, [127]) sums and products of measurable functions are measurable. Also, if (f_n) is a sequence of measurable functions which converges pointwise to f then f is measurable. Let $B^\infty(Y, S)$ be the algebra of all bounded, complex valued, S-measurable functions. (We shall abbreviate this to $B^\infty(Y)$ when it is clear which σ-algebra is being used.) It is a closed $*$-subalgebra of $Bnd(Y)$.

Let $\mathcal{J}$ be a (Boolean) σ-ideal of $\mathcal{S}$, that is, (i) $\varnothing \in \mathcal{J}$ and $Y \notin \mathcal{J}$. (ii) $A \in \mathcal{J}$ and $B \in \mathcal{S}$ implies $B \cap A \in \mathcal{J}$ (iii) $A_n \in \mathcal{J}$ for $n = 1, 2 \ldots$ implies $\cup_{n=1}^{\infty} A_n \in \mathcal{J}$.

Let J be the set of all $f \in B^{\infty}(Y)$ such that $\{y \in Y : f(y) \neq 0\} \in \mathcal{J}$. Then it is straightforward to show that J is a (C^*-algebra) σ-ideal of $B^{\infty}(Y)$. So, see Chap. 2, $B^{\infty}(Y)/J$ is monotone σ-complete. But sometimes this quotient algebra is monotone complete. For a familiar example, let Y be the interval $[0, 1]$ and $\mathcal{S}$ the collection of Lebesgue measurable subsets. When $\mathcal{J}$ is the set of Lebesgue null sets then $B^{\infty}(Y)/J$ is L^{∞} which is monotone complete. We shall find many other examples.

Just as in classical measure theory, we shall say that a property holds *almost everywhere* (with respect to $\mathcal{J}$) if the subset of Y for which it does not hold is in $\mathcal{J}$. We shall abbreviate this to a.e.$(\mathcal{J})$ or, when there is no risk of ambiguity, write "a.e.".

Let $q : B^{\infty}(Y) \mapsto B^{\infty}(Y)/J$ be the quotient homomorphism. Then, see Chap. 2, q is a σ-homomorphism.

Lemma 4.2.1 *Let f and g be bounded, real-valued measurable functions on Y. Then $q(f) \geq 0$ if, and only if, $f \geq 0$ almost everywhere. Furthermore, $q(f) \geq q(g)$ if, and only if, $f \geq g$ almost everywhere.*

Proof

(i) Let $f = f^+ - f^-$ where f^+ and f^- are positive and $f^+ f^- = 0$.

   Then $f \geq 0$ a.e. precisely when $f^- = 0$ a.e. But this is equivalent to $q(f^-) = 0$.

   This is equivalent to $q(f) = q(f^+)$. This implies $q(f) \geq 0$.

   Conversely, $q(f) \geq 0$ implies, on multiplying by $q(f^-)$, that $q(f^-) = 0$.

(ii) We have $q(f) \geq q(g)$ is equivalent to $q(f - g) \geq 0$ which is equivalent to $f - g \geq 0$ a.e. $\square$

We recall that a measurable function s is *simple* if it takes only finitely many values. Equivalently, $s = \sum_1^n c_n \chi_{E_n}$ where $\{E_1, \ldots, E_n\}$ is a disjoint collection of measurable sets. The simple functions in $B^{\infty}(Y)$ are closed under addition and multiplication by scalars. In other words, they form a vector subspace of $B^{\infty}(Y)$. The following elementary lemma tells us this vector space is norm dense in $B^{\infty}(Y)$.

Lemma 4.2.2 *Let $f \in B^{\infty}(Y)_{sa}$. Then there exists a monotone increasing sequence of simple (measurable) functions which converges uniformly to f.*

Proof It suffices to show this when $0 \leq f \leq 1$ that is, when the range of f is in the interval $[0, 1]$.

Let $A = f^{-1}[0, \frac{1}{2})$ and $B = f^{-1}[\frac{1}{2}, 1]$. We define $s_1 = 0\chi_A + \frac{1}{2}\chi_B$ then $||f - s_1|| \leq \frac{1}{2}$.

We split A into the disjoint union $C \cup D$ where $C = f^{-1}[0, \frac{1}{4})$ and $D = f^{-1}[\frac{1}{4}, \frac{1}{2})$.

We split B into the disjoint union $E \cup F$ where $E = f^{-1}[\frac{1}{2}, \frac{1}{2} + \frac{1}{4})$ and $F = f^{-1}[\frac{1}{2} + \frac{1}{4}, 1]$.

We define $s_2 = 0\chi_C + \frac{1}{4}\chi_D + \frac{1}{2}\chi_E + \frac{3}{4}\chi_F$. Then

$$s_1 \le s_2 \le f$$

and $\|f - s_2\| \le \frac{1}{2^2}$.

By repeating this process, splitting the range of f into smaller and smaller disjoint pieces, we construct a monotone increasing sequence of simple functions (s_n) such that $\|f - s_n\| \le \frac{1}{2^n}$.   □

4.2.2 Baire Spaces and Category

We defined a *Baire space* to be a (Hausdorff) topological space T such that whenever (O_n) is a sequence of dense open sets then $\bigcap_{n=1}^{\infty} O_n$ is dense in T. Equivalently, the only meagre open subset of T is the empty set. Let us recall that a G_δ-set is the intersection of countably many open sets. When T is a Baire space and G is a *dense* G_δ-subset of T, then G is a Baire space in the relative topology. Since each closed subspace of a compact space is compact, every G_δ-subset of a compact space is a Baire space; similarly for complete metric spaces.

As before, all topological spaces considered are required to be Hausdorff.

We discussed semicontinuous, real-valued functions in Chap. 2 (see Lemma 2.3.3). On any topological space X, $f : X \mapsto \mathbb{R}$ is lower semicontinuous if, for each real number t, the set $\{x \in X : f(x) > t\}$ is open.

Let Y be a completely regular space and $f : Y \mapsto \mathbb{R}$ a bounded function. Let $U(f)$ be the set of all bounded, real-valued continuous functions, $a \in C(Y)_{sa}$, such that $a \ge f$. Then the *upper envelope* of f is defined to be

$$\hat{f}(y) = \inf\{a(y) : a \in U(f)\}.$$

Then $\hat{f}$ is a bounded upper semicontinuous function. (Apply Lemma 2.3.3 with $f_\lambda = -a_\lambda$.)

We can define the *lower envelope* of f to be the pointwise supremum of the continuous functions below f or, equivalently,

$$\underline{f} = -(\widehat{-f}).$$

Proposition 4.2.3 *Let Y be a completely regular space and $f : Y \mapsto \mathbb{R}$ a bounded lower semicontinuous function. Then*

$$\{y \in Y : \hat{f}(y) > f(y)\}$$

is a meagre set.

Proof Since $-\hat{f}$ is lower semicontinuous, so also, is $-\hat{f} + f$. Thus $\hat{f} - f$ is upper semicontinuous. So for each n, $\{y \in Y : \hat{f}(y) - f(y) \geq \frac{1}{n}\}$ is closed. Let O_n be the interior of this set.

Suppose, for some n, that O_n is not empty. Fix $y_0 \in O_n$. Since Y is completely regular, there exists a continuous function $h : Y \mapsto [0, \frac{1}{n}]$ such that $h(y_0) = \frac{1}{n}$ and h takes the constant value 0 on the closed set $Y \backslash O_n$.

For $y \in O_n$, we have $\hat{f}(y) - f(y) \geq \frac{1}{n} \geq h(y)$. For $y \in Y \backslash O_n$, we have $\hat{f}(y) - f(y) \geq 0 = h(y)$. So $\hat{f} \geq f + h \geq f$.

Let $a \in U(f)$. Then $a \geq \hat{f} \geq f + h$. So $a - h \in U(f)$. Hence $a(y_0) - h(y_0) \geq \hat{f}(y_0)$. Hence $\hat{f}(y_0) - \frac{1}{n} \geq \hat{f}(y_0)$. This is a contradiction so each O_n is empty. Hence $\{y \in Y : \hat{f}(y) > f(y)\}$ is the union of a sequence of closed nowhere dense sets. So it is a meagre set.   □

Corollary 4.2.4 *Let Y be a completely regular space. Let $u : Y \mapsto \mathbb{R}$ be a bounded upper semicontinuous function and let $\underline{u}$ be its lower envelope. Then $\underline{u}$ is lower semicontinuous and $\{y \in Y : u(y) > \underline{u}(y)\}$ is meagre.*

Proof In the proposition, put $u = -f$.   □

Lemma 4.2.5 *Let u be a bounded, upper semicontinuous function on a Baire space Y. Let $\{y : u(y) < 0\}$ be a meagre set. Then $u \geq 0$.*

Proof Because u is upper semicontinuous, $\{y \in Y : u(y) < 0\}$ is open. But $\{y \in Y : u(y) < 0\}$ is meagre. In a Baire space a meagre open set must be empty. So $u \geq 0$.   □

Lemma 4.2.6 *Let X be a completely regular space. Let $D = \{a_\lambda : \lambda \in \Lambda\}$ be a downward directed set in $C_b(X)_{sa}$, where $a_\lambda \geq 0$ for each λ. Let*

$$u(x) = inf\{a_\lambda(x) : \lambda \in \Lambda\}$$

for each x in X. If D has a greatest lower bound, 0, in $C_b(X)_{sa}$ then

$$M = \{x \in X : u(x) > 0\}$$

is meagre. Conversely, if X is a Baire space and M is meagre then D has a greatest lower bound, 0, in $C_b(X)_{sa}$.

Proof First we observe that u is positive and upper semicontinuous. If D has a greatest lower bound, a, in $C_b(X)_{sa}$ then $a = \underline{u}$, the lower envelope of u. Hence, if D has infimum 0 in $C_b(X)_{sa}$ then, by Corollary 4.2.4, M is meagre.

Conversely, let us suppose that M is meagre. Let $c \in C_b(X)_{sa}$ such that c is a lower bound for D. Then $c \leq u$. So $\{x \in X : c(x) > 0\}$ is a meagre open set. Because X is now required to be a Baire space, this set is empty. So $c \leq 0$.   □

In the following, Y is a (Hausdorff) completely regular Baire space and $Bor(Y)$ is the σ-field of Borel subsets of Y, that is, the smallest σ-field of subsets of Y which contains the open sets. Then every lower semicontinuous function is Borel measurable. Let $\mathfrak{M}_g$ be the set of all meagre Borel sets in Y; this is a Boolean σ-ideal of the Borel sets. For the rest of this section, $B(Y)$ is the algebra of bounded, Borel measurable functions; that is, we abbreviate $B^\infty(Y, Bor(Y))$ by $B(Y)$. Let $M(Y)$ be the set of all f in $B(Y)$ such that $\{y \in Y : f(y) \neq 0\}$ is meagre. Then $M(Y)$ is a σ-ideal and so the quotient map

$$q : B(Y) \mapsto B(Y)/M(Y)$$

is a σ-homomorphism.

For each Borel set E there is a unique regular open set rE such that the symmetric difference, $(E \backslash rE) \cup (rE \backslash E)$, is meagre. (See Sect. 4.1.)

Lemma 4.2.7 *Let g be a real-valued, bounded Borel measurable function on a completely regular Baire space Y such that $\alpha 1 \leq g \leq \beta 1$ for real numbers α and β. Then there exists a lower semicontinuous function f such that $\{y \in Y : f(y) \neq g(y)\}$ is meagre and f satisfies $\alpha 1 \leq f \leq \beta 1$. Furthermore, there exists an upper semicontinuous function u such that $u = g$ a.e.$(\mathfrak{M}_g)$ and u satisfies $\alpha 1 \leq u \leq \beta 1$.*

Proof By adding an appropriate constant and then multiplying by a suitable constant, it suffices to prove the result when $0 \leq g \leq 1$.

By Lemma 4.2.2, there exists (s_n), a monotone increasing sequence of positive, Borel measurable, simple functions, such that $s_n \to g$ uniformly.

For each n there exist pairwise disjoint Borel sets $\{E_j^n : j = 1, 2, \ldots, k(n)\}$ and non-negative constants $(c_j^n)(j = 1, 2, \ldots, k(n))$ such that

$$s_n = c_1^n \chi_{E_1^n} + \ldots + c_{k(n)}^n \chi_{E_{k(n)}^n}.$$

Then, see Sect. 4.1, $\{rE_j^n : j = 1, 2, \ldots, k(n)\}$ is a set of pairwise disjoint, regular open sets. Now let

$$f_n = c_1^n \chi_{rE_1^n} + \ldots + c_{k(n)}^n \chi_{rE_{k(n)}^n}.$$

Then f_n is a lower semicontinuous function bounded above by $max\{c_j^n : j = 1, 2, \ldots, k(n)\}$. Since each E_j^n only differs from rE_j^n on a meagre set, $s_n = f_n$ a.e.$(\mathfrak{M}_g)$.

Let $f(y) = sup f_n(y)$. Then f is lower semicontinuous, $0 \leq f \leq 1$ and $g = f$ a.e.$(\mathfrak{M}_g)$.

Put $u = \hat{f}$, the upper envelope of f. $\square$

Let us recall that a topological space is extremally disconnected when the closure of each open set is an open set. In such a space, each regular open set is a clopen set.

Corollary 4.2.8 *Let Y be an extremally disconnected, completely regular Baire space. Let g be a bounded Borel measurable function on Y. Then there is a bounded continuous function f in $C_b(Y)$ such that $g = f$ a.e.$(\mathfrak{M}_g)$.*

Proof By taking real and imaginary parts, it suffices to prove this when g is real valued. We use the same notation as in the proof of the lemma. But now, because each regular open set is a clopen set, each f_n is a bounded continuous function.

For any m and n, we have $|f_m(y) - f_n(y)| = |s_m(y) - s_n(y)| \leq ||s_m - s_n||$ for all y outside a meagre subset of Y. Because of continuity, the set

$$\{y \in Y : |f_m(y) - f_n(y)| > ||s_m - s_n||\}$$

is open. But it is also meagre and so must be empty. It follows that (f_n) is a Cauchy sequence which converges uniformly to f. So f is a (bounded) continuous function.   □

Theorem 4.2.9 *Let Y be a completely regular, Baire space. Then the commutative algebra $B(Y)/M(Y)$ is monotone complete.*

Proof Let $L = \{q(a_\lambda) : \lambda \in \Lambda\}$ be a set of self-adjoint elements which is bounded in norm. It suffices to consider the situation where $0 \leq q(a_\lambda) \leq 1$ for each λ. Then, by using Lemma 4.2.7, we can suppose that each a_λ is lower semicontinuous and that $0 \leq a_\lambda \leq 1$ for each $\lambda \in \Lambda$.

Let $a(y) = sup\{a_\lambda(y) : \lambda \in \Lambda\}$. Then a is lower semicontinuous. Clearly $q(a)$ is an upper bound for L. Let $q(f)$ be any other upper bound for L.

For any λ, we have $q(f) \geq q(a_\lambda)$. So $f \geq a_\lambda$ a.e.$(\mathfrak{M}_g)$. Now let $\hat{f}$ be the upper semicontinuous envelope of f.

Then $\hat{f} - a_\lambda$ is upper semicontinuous and $\hat{f} - a_\lambda \geq 0$ a.e.$(\mathfrak{M}_g)$. So, by Lemma 4.2.5, $\hat{f} \geq a_\lambda$. So $\hat{f} \geq a$.

Since $q(\hat{f}) = q(f)$ it follows that $q(f) \geq q(a)$. Thus $q(a)$ is the least upper bound of L.   □

Consider the quotient map q from $B(Y)$ onto $B(Y)/M(Y)$. Let q_0 be the restriction of q to $C_b(Y)$. Because Y is a Baire space, the only continuous function in $M(Y)$ is the constantly zero function. (See Lemma 4.2.5.) So q_0 is an injective map from $C_b(Y)$ into $B(Y)/M(Y)$. But an injective $*$-homomorphism from a C^*-algebra into another C^*-algebra is an isometry, so q_0 is an isometric embedding from $C_b(Y)$ into $B(Y)/M(Y)$. So we may identify $C_b(Y)$ with its image in $B(Y)/M(Y)$. We shall show that, with this identification, $C_b(Y)$ is a regular subalgebra of $B(Y)/M(Y)$.

Let Ω be the spectrum of $B(Y)/M(Y)$, that is, Ω is the compact Hausdorff space such that $C(\Omega)$ is isomorphic to $B(Y)/M(Y)$. Since this algebra is monotone complete, Ω must be extremally disconnected.

Corollary 4.2.10 *Let E be an extremally disconnected, completely regular, Baire space. Then $C_b(E)$ is monotone complete.*

Proof By Corollary 4.2.8, for each g in $B(E)$ there exists $f \in C_b(E)$ such that $q(g) = q(f)$. So q_0 is an isometric $*$-isomorphism from $C_b(E)$ onto $B(E)/M(E)$.   □

The definition of a regular subalgebra was introduced in Sect. 2.1. The intuitive idea is that $B(Y)/M(Y)$ may be thought of as a "Dedekind cut" completion of $C_b(Y)$, analogous to the way the rational numbers embed in the reals.

Corollary 4.2.11 *Let Y be a completely regular Baire space. The algebra $C_b(Y)$ is a regular subalgebra of $B(Y)/M(Y) \simeq C(\Omega)$.*

Proof Each self-adjoint element of $B(Y)/M(Y)$ is $q(u)$ for some bounded upper semicontinuous u.

Let $L = \{a \in C_b(Y)_{sa} : q(a) \leq q(u)\}$.

If a is in L then $u - a \geq 0$ a.e. So, by Lemma 4.2.5, $u \geq a$.

So $\underline{u}(y) = sup\{a(y) : a \in L\}$.

Arguing as in the proof of the theorem, $q(\underline{u})$ is the least upper bound of $\{q(a) : a \in L\}$.

By Corollary 4.2.4, $q(u) = q(\underline{u})$. $\qquad\square$

Lemma 4.2.12 *Let T be any (Hausdorff) completely regular space. Let $RegT$ be the complete Boolean algebra of regular open subsets of T. Then T has an isolated point if, and only if $RegT$ has an atom.*

Proof Suppose s is an isolated point of T. Then the closed set $\{s\}$ is also open. Hence it is in $Reg(T)$. Clearly it is an atom.

Let U be a regular open set. Suppose there exist distinct points y_1 and y_2 in U. Because Y is completely regular, each of these points has an open neighbourhood, such that the closures of these neighbourhoods are disjoint. So U contains two disjoint, non-empty, regular open sets. So U is not an atom.

So U is an atom of $Reg(T)$ if, and only if, it contains exactly one point. $\qquad\square$

Proposition 4.2.13 *Let Y be a completely regular Baire space. The Boolean algebra of projections in $B(Y)/M(Y) \simeq C(\Omega)$ is isomorphic to the Boolean algebra of regular open subsets of Y.*

Proof Let e be a projection in $B(Y)/M(Y)$. Then there is a real-valued Borel function f such that $e = q(f)$ and $0 \leq f \leq 1$. Then $q(f^n) = e$. Also the sequence (f^n) is monotone decreasing and converges pointwise to a projection in $B(Y)$. Since q is a σ-homomorphism it follows that $e = q(p)$ for some projection p in $B(Y)$. Thus $e = q(\chi_P)$ where P is a Borel subset of Y. It is straightforward to verify that q induces a Boolean homomorphism $q^{\#}$ from $Bor(Y)$ onto $Proj(C(\Omega))$. The kernel of $q^{\#}$ is the Boolean ideal of all meagre Borel sets, $\mathfrak{M}_g$. So $Proj(C(\Omega))$ is isomorphic to $Bor(Y)/\mathfrak{M}_g$ which is isomorphic to $RegY$, the complete Boolean algebra of regular open subsets of Y. In particular, for each projection $e \in Proj(C(\Omega))$ there exists a unique regular open set $O \subset Y$, such that $q(\chi_O) = e$. $\qquad\square$

Corollary 4.2.14 *The space Ω has no isolated points if, and only if, Y has no isolated points.*

Proof We have $Reg(\Omega) \simeq Proj(C(\Omega)) \simeq Reg(Y)$. So $Reg(\Omega)$ has no atoms if, and only if, $Reg(Y)$ has no atoms. So Ω has no isolated points if, and only if, Y has no isolated points. $\qquad\square$

4.2.3   Normal States and Wild Algebras

We are familiar with the notion of a normal state on a monotone complete C^*-algebra. But we can generalise this in a natural way to general C^*-algebras.

Let W be a unital C^*-algebra. Let D be any norm-bounded downward directed set in W_{sa}. If D has a greatest lower bound, d, in W_{sa}, then we write $\bigwedge D = d$.

Let ϕ be a positive linear functional on W. Then ϕ is said to be *normal* if, whenever $\bigwedge D = 0$, then $inf\{\phi(x) : x \in D\} = 0$.

Lemma 4.2.15 *Let W be a regular subalgebra of a monotone complete C^*-algebra M. Let ϕ be a positive normal functional on M then the restriction of ϕ to W is also normal.*

Proof Let D be downward directed in W and have infimum 0 in W. Then, because W is a regular subalgebra of M, Proposition 2.1.35 tells us that the infimum of D in M is also 0. So $inf\{\phi(x) : x \in D\} = 0$.   □

Proposition 4.2.16 *Let X be a compact Hausdorff space and let ϕ be a positive normal functional on $C(X)$. Let μ be the regular Borel measure on X which corresponds to ϕ. Then $\mu E = 0$ whenever E is a meagre Borel subset of X. Also, for any Borel set B,*

$$\mu B = \mu(clB).$$

Proof Let F be a closed, nowhere dense subset of X. Let $D = \{f \in C(X)_{sa} : \chi_F \leq f \leq 1\}$. Then, because X is completely regular and χ_F is upper semicontinuous, D is downward directed with pointwise infimum χ_F. It now follows from Lemma 4.2.6 that $\bigwedge D = 0$.

Because ϕ is normal, $inf\{\int f d\mu : f \in D\} = 0$.

But $\mu F = \int \chi_F d\mu \leq \int f d\mu$ for each f in D. So $\mu F = 0$.

It now follows from countable additivity that each meagre Borel set has measure 0.

Let B be any Borel set. Choose $\varepsilon > 0$. Because μ is a regular Borel measure, there is an open set O such that $B \subset O$ and $\mu O < \mu B + \varepsilon$.

Then $\mu(clB) \leq \mu(clO)$. But $clO \backslash O$ is a closed nowhere dense set. So $\mu(clO) = \mu O < \mu B + \varepsilon$.

It follows that $\mu(clB) \leq \mu B$. Since $B \subset clB$ the result follows.   □

Theorem 4.2.17 *Let X be a compact (Hausdorff) space. Let X have a dense meagre subset. Then the monotone complete C^*-algebra $B(X)/M(X) \simeq C(\Omega)$ is wild.*

Proof Every meagre set is contained in the union of a sequence of closed nowhere dense sets. So every meagre subset of X is contained in a meagre Borel set. Hence X has a dense meagre Borel set.

Let ϕ be a normal, positive linear functional on $C(\Omega)$. Since $C(X)$ is a regular subalgebra of $C(\Omega)$, by applying Proposition 2.1.35 we see that the restriction of ϕ to $C(X)$ is normal. Let μ be the corresponding regular Borel measure on X. Then, by Proposition 4.2.3, $\mu(X) = 0$. So $\phi(1) = 0$ which, by positivity, implies $\phi = 0$. So $C(\Omega)$ has no normal states. $\square$

Corollary 4.2.18 *Let X be a compact (Hausdorff) space with no isolated points. Let X be separable. Then $B(X)/M(X)$ is wild.*

Proof Since there are no isolated points, each one-point set in X is nowhere dense. So a countable dense set is a dense meagre set. $\square$

It follows from Corollary 4.2.18 that $B([0, 1])/M([0, 1])$ is wild.

It turns out that whenever Y is homeomorphic to a complete, separable metric space with no isolated points (i.e. is a perfect Polish space) then $B(Y)/M(Y)$ is isomorphic to $B([0, 1])/M([0, 1])$. (This follows from the observation that each perfect Polish space contains a dense G_δ-set which is homeomorphic to $\mathbb{N}^\mathbb{N}$, the space of irrationals. See Chapter 3 Section 36 in [97]. So the Boolean algebras $RegY$ and $Reg\mathbb{N}^\mathbb{N}$ are isomorphic.)

We call the algebra $B(Y)/M(Y)$ the *Dixmier algebra*.

Corollary 4.2.19 *Let μ be Lebesgue measure on $(-\pi, \pi)$. Let S_L be the spectrum of $L^\infty(-\pi, \pi)$. Then S_L is not separable.*

Proof There are no minimal projections, because each set of strictly positive Lebesgue measure can be split into two disjoint sets of strictly positive measure. So S_L has no isolated points. So if S_L were separable the algebra would be wild. But the map $f \mapsto \frac{1}{2\pi} \int f d\mu$ gives a normal state. $\square$

4.2.4 Separable Spaces

Our aim is to construct large numbers of commutative examples of wild monotone complete C^*-algebras. We know from Corollary 4.2.18 that a wild algebra can be obtained from any separable compact space which has no isolated points. So we begin by considering separable spaces more carefully.

Lemma 4.2.20 *Let X be a compact Hausdorff space. Then X is separable if, and only if, there is a (unital) $*$-isomorphism of $C(X)$ into ℓ^∞.*

Proof Suppose $\{x_n : n = 1, 2, \ldots\}$ is a dense subset of X. For each f in $C(X)$ let $\pi(f)$ be the sequence $(f(x_n))$. Then π is an isometric $*$-isomorphism of $C(X)$ into ℓ^∞.

Conversely, let π be an injective $*$-homomorphism of $C(X)$ into ℓ^∞. (Since these are C^*-algebras it follows that π is an isometric $*$-isomorphism onto a closed subalgebra.) Let $\beta\mathbb{N}$ be the Stone-Czech compactification of $\mathbb{N}$, the natural numbers equipped with the discrete topology; this can be identified with the spectrum of ℓ^∞. Then we may regard π as an injective $*$-homomorphism of $C(X)$ into $C(\beta\mathbb{N})$. So, by the Gelfand-Naimark duality between compact Hausdorff spaces and commutative unital C^*-algebras, there is a continuous surjective map σ from $\beta\mathbb{N}$ onto X. But $\mathbb{N}$ is a countable dense subset of $\beta\mathbb{N}$. So $\sigma[\mathbb{N}]$ is a countable subset of X.

Let U be any non-empty open subset of X. Then, because σ is surjective, $\sigma^{-1}[U]$ is a non-empty open set. So, for some $n \in \mathbb{N}$, $n \in \sigma^{-1}[U]$.

Thus $\sigma(n) \in U$. So $\sigma[\mathbb{N}]$ is a dense subset of X.   □

Corollary 4.2.21 *Let Y be a separable (Hausdorff) completely regular space. Then $C_b(Y)$ has a faithful state and is a subalgebra of ℓ^∞.*

Proof Since Y is dense in its Stone-Czech compactification, βY is separable.

By Lemma 4.2.20, $C(\beta Y) \simeq C_b(Y)$ can be regarded as a unital subalgebra of ℓ^∞. For $(x_n) \in \ell^\infty$ let $\lambda((x_n)) = \sum_{n=1}^\infty \frac{1}{2^n} x_n$. Then λ is a faithful state on ℓ^∞. So its restriction to $C_b(Y)$ is a faithful state on $C_b(Y)$.   □

Proposition 4.2.22 *Let Y be a separable, completely regular, Baire space. Let Ω be the spectrum of $B(Y)/M(Y)$. Then the extremally disconnected space Ω is separable and $C(\Omega)$ has a faithful state.*

Proof By Corollary 4.2.21 there exists a $*$-isomorphism π from $C_b(Y)$ into ℓ^∞. It now follows from Proposition 2.3.11 that π can be extended to a homomorphism $\hat{\pi}$ from $B(Y)/M(Y)$ into ℓ^∞.

We shall show that $\hat{\pi}$ is injective. Let $\hat{\pi}(z) = 0$. Now let $b = zz^*$. Then $\hat{\pi}(b) = 0$. Let $a \leq b$ and $a \in C_b(Y)_{sa}$. Then $\pi(a) \leq 0$. So $a \leq 0$.

Since $C_b(Y)$ is a regular subalgebra of $B(Y)/M(Y)$, b is the least upper bound of $\{a \in C_b(Y)_{sa} : a \leq b\}$. But 0 is an upper bound for this set. So $b \leq 0$. Hence $b = 0$. So $z = 0$.

So there is an isomorphism from $C(\Omega) \simeq B(Y)/M(Y)$ into ℓ^∞. So, by Lemma 4.2.20, Ω is separable. The existence of a faithful state now follows from Corollary 4.2.21.   □

Corollary 4.2.23 *Let Y be a separable, completely regular (Hausdorff) Baire space. Let f be any positive bounded lower semicontinuous function on Y. Then there exists (a_n), a monotone increasing sequence of continuous, positive real-valued functions, such that $\|a_n\| \leq \|f\|$ and a lower semicontinuous function h such that $f = h$ a.e.($\mathfrak{M}_g$) and*

$$h(x) = sup\{a_n(x) : n = 1, 2 \ldots\}.$$

Proof Let $D = \{a \in C_b(Y)_{sa} : 0 \leq a \leq f\}$. Then $q(f)$ is the least upper bound of D in $B(Y)/M(Y)$. Since this algebra has a faithful state, we can use Theorem 2.1.14 to find a monotone increasing sequence in D, (a_n), such that this sequence has supremum $q(f)$.

Now let $h(x) = sup\{a_n(x) : n = 1, 2 \ldots\}$ for each x. Then $q(h)$ is the least upper bound of (a_n). Thus $q(h) = q(f)$. So $h = f$ (a.e.).   □

4.2.5 Baire Measurability

The name of Baire is associated with several different mathematical topics. This can cause some confusion.

In any topological space an F_σ-set is one which is the union of countably many closed sets. Let X be a completely regular (Hausdorff) Baire space. When $f : X \mapsto \mathbb{R}$ is continuous then the open set $f^{-1}(t, \infty)$ is the union of the closed sets $f^{-1}[t + \frac{1}{n}, \infty)$. So $f^{-1}(t, \infty)$ is an open F_σ set. When X is normal, in particular, when X is compact, every open F_σ-subset arises in this way from a continuous f. But there exist completely regular spaces which are not normal.

So we define $Baire(X)$ to be the smallest σ-field of sets for which every bounded continuous, real-valued, function on X is measurable. We shall abbreviate $B^\infty(X, Baire(X))$ by $B^\infty(X)$. Then $B^\infty(X)$ is the σ-envelope of $C_b(X)$ in $B(X)$.

Then $B^\infty(X)$ is the smallest σ-subalgebra of $B(X)$ which (i) contains $C_b(X)$ and (ii) is closed under pointwise sequential limits of bounded sequences. Then $B^\infty(X)$ is the algebra of bounded, Baire measurable functions on X.

Let $M_0(X) = B^\infty(X) \cap M(X)$. Then $M_0(X)$ is a σ-ideal of $B^\infty(X)$. Clearly we may regard $B^\infty(X)/M_0(X)$ as a subalgebra of $B(X)/M(X)$ and $C_b(X)$ as a subalgebra of $B^\infty(X)/M_0(X)$. By Corollary 4.2.11, $C_b(X)$ is a regular subalgebra of $B(X)/M(X)$. In other words, each self-adjoint element of $B(X)/M(X)$ is the supremum of the elements of $C_b(X)$ which lie below it. So, also, $C_b(X)$ is a regular subalgebra of $B^\infty(X)/M_0(X)$. From this it follows that $B^\infty(X)/M_0(X)$ may be identified with the *regular σ-completion* of $C_b(X)$. (See Definition 2.1.37.) In Chap. 5 we shall show that every C^*-algebra has a regular σ-completion.

As an immediate consequence of Corollary 4.2.23 we have:

Corollary 4.2.24 *Let Y be a separable, (Hausdorff) completely regular, Baire space. Then $B(Y)/M(Y) = B^\infty(Y)/M_0(Y)$.*

Proof By Corollary 4.2.23 and Lemma 4.2.7, each real valued, bounded Borel function coincides with a Baire measurable function, outside a meagre set. □

4.2.6 Cantor Product Spaces

Let $\{0, 1\}$ be given the discrete topology. Then for any index set I, the space $\{0, 1\}^I$ is a compact Hausdorff, totally disconnected space. When I is infinite then (see Sect. 4.1) the collection of clopen subsets of $\{0, 1\}^I$ has cardinal #I. Following the notation of Halmos, we shall usually abbreviate $\{0, 1\}^I$ by 2^I.

We shall call $2^\mathbb{N}$ the *Cantor space* because it is homeomorphic to Cantor's "middle third" space. So it is metrisable and separable. It follows from the above, that the collection of clopen subsets of $2^\mathbb{N}$ is countable. Each element of this space is the characteristic function of a subset of $\mathbb{N}$. So

$$\#2^\mathbb{N} = \#\mathcal{P}(\mathbb{N}) = \#\mathbb{R}.$$

We shall call $2^{\mathbb{R}}$ the *Big Cantor space*. This space turns out to be separable but not metrisable. (A separable metric space has at most $\#\mathbb{R}$ elements. But $\#2^{\mathbb{R}} = \#\mathcal{P}(\mathbb{R}) > \#\mathbb{R}$.)

Lemma 4.2.25 *The Big Cantor space is separable.*

Proof Separability follows from a more general theorem that says the product of continuum many copies of a separable space is separable. See [78] or [167]. But the following simple argument is all that is needed here.

Let $T = 2^{\mathbb{N}}$ be the usual Cantor space. Because $\#T = \#\mathbb{R}$, the Big Cantor space can be identified with 2^{T}. Let

$$D = \{\chi_K : K \subset T \text{ is clopen}\}.$$

Then D is a countable set.

Let $\{s_1, \ldots, s_n\}$ and $\{t_1, \ldots, t_m\}$ be disjoint finite subsets of T. We can find a clopen K such that $\{s_1, \ldots, s_n\} \subset K$ and $\{t_1, \ldots, t_m\}$ is disjoint from K. Then χ_K is in the open set

$$\{x \in 2^{T} : x(s_r) = 1 \text{ for } r = 1, \ldots, n \text{ and } x(t_k) = 0 \text{ for } k = 1, \ldots, m\}.$$

Since opens sets like this give a base for the topology of 2^{T} it follows that the countable set D is dense in the Big Cantor Space.   □

Proposition 4.2.26 *Let X be a separable, totally disconnected, compact Hausdorff space. Then X is homeomorphic to a closed subspace of $2^{\mathbb{R}}$.*

Proof By Corollary 4.1.8, if X has a base of at most $\#\mathbb{R}$ clopen sets, then it is homeomorphic to a closed subspace of $2^{\mathbb{R}}$.

Let C be a countable dense subset of X. For each clopen set $K \subset X$, $K \cap C$ is a dense subset of K. So $K \mapsto K \cap C$ is an injection of the collection of all clopen subsets of X into the collection of all subsets of a countable set. So X has at most $\#\mathbb{R}$ clopen sets. So X is homeomorphic to a closed subspace of $2^{\mathbb{R}}$.   □

4.3 Constructing and Classifying Wild Commutative Algebras

We shall give explicit constructions of large numbers of wild commutative monotone complete C^*-algebras. The examples given here are all subalgebras of ℓ^{∞}. We use spectroids to show that these examples take $2^{\mathbb{R}}$ distinct values in the classification semigroup $\mathcal{W}$.

Our strategy is to apply the results of Sect. 4.2 in the following way:

Let C be any infinite, countable subset of $2^{\mathbb{R}}$. Let X be the closure of C in $2^{\mathbb{R}}$. Then X is compact Hausdorff, totally disconnected and separable. If X has an isolated point then, since C is dense in X, this point must be in C. So, unless C has an isolated point in the relative topology, the commutative monotone complete C^*-algebra $B^\infty(X)/M(X) \simeq C(\hat{X})$ is wild. By Proposition 4.2.22, $\hat{X}$ is separable, so $C(\hat{X})$ is a subalgebra of ℓ^∞.

By varying the countable set C in a carefully controlled way we find $2^\mathbb{R}$ wild commutative algebras, each of which takes different values in the classification semigroup $\mathcal{W}$. We go on to show that $2^\mathbb{R}$ of these algebras can be represented as quotients of $B^\infty(2^\mathbb{N})$ by "exotic" σ-ideals. So we may use exotic spaces and ideals of meagre subsets or, the Cantor metric space and exotic σ-ideals.

In a later chapter we shall consider group actions on commutative algebras constructed here and use them to find $2^\mathbb{R}$ small wild factors; each taking different values in $\mathcal{W}$.

We begin by recalling some elementary topological results which will be needed later. In contrast to the notation in Sect. 4.1, we shall denote the closure of a set S by clS and its interior by $intS$.

Lemma 4.3.1 *Let K be a (Hausdorff) topological space and let X be a dense subset of K. Let U be any open subset of K. Then $clU = cl(X \cap U)$. In particular, when U is a clopen subset of K, then we have $U = cl(U \cap X)$.*

Proof Let $y \in clU$. Let V be an open subset of K such that $y \in V$. Then $V \cap U$ is not empty.

Since X is dense in K, $(V \cap U) \cap X \neq \varnothing$, which implies that $V \cap (U \cap X) \neq \varnothing$. So, we have $y \in cl(U \cap X)$; that is, $clU \subset cl(U \cap X)$. The reverse inclusion is trivial and it follows that $clU = cl(U \cap X)$. □

Corollary 4.3.2 *Let $L \cap X = M \cap F \cap X$, where L, M are clopen subsets of K and F is a closed subset of K. Then $L = M \cap intF$.*

Proof Applying the preceding lemma, $L = cl(L \cap X) \subset cl(F \cap X) \subset F$. So $L \subset intF$.

Again, by the lemma, $L \cap X \subset M \cap X$ implies $L \subset M$. Thus $L \subset M \cap intF$.

But $(M \cap intF) \cap X \subset M \cap F \cap X = L \cap X$. By applying the lemma again, $cl(M \cap intF) \subset L$. Hence $L = M \cap intF$. □

As before, we use the notation $F(S)$ to denote the collection of all finite subsets (including the empty set) of a set S.

To carry out the strategy outlined above we shall need some technicalities. None of them are difficult but they are mildly intricate. For a first reading, they may be skimmed over.

We shall start with the following:

Definition 4.3.3 A pair $(T, \mathbf{O})$ is said to be *feasible* if it satisfies the following conditions:

(i) T is a set of cardinality $c = 2^{\aleph_0}$; $\mathbf{O} = (O_n)$ $(n = 1, 2, \dots)$ is an infinite sequence of non-empty subsets of T, with $O_m \neq O_n$ whenever $m \neq n$.

(ii) Let M be a finite subset of T and $t \in T \backslash M$. For each natural number m there exists $n > m$ such that $t \in O_n$ and $O_n \cap M = \emptyset$. In other words $\{n \in \mathbb{N} : t \in O_n \text{ and } O_n \cap M = \emptyset\}$ is an infinite set.

An example satisfying these conditions can be obtained by putting $T = 2^{\mathbb{N}}$, the Cantor space, and letting $\mathbf{O}$ be an enumeration (without repetitions) of the (countable) collection of all non-empty clopen subsets of $2^{\mathbb{N}}$.

For the rest of this section $(T, \mathbf{O})$ will be a fixed but arbitrary feasible pair.

Definition 4.3.4 Let $(T, \mathbf{O})$ be a feasible pair and let R be a subset of T. Then R is said to be *admissible* if

(i) R is a subset of T, with $\#R = \#(T \backslash R) = c = 2^{\aleph_0}$.
(ii) O_n is not a subset of R for any natural number n.

Return to the example where T is the Cantor space and $\mathbf{O}$ an enumeration of the non-empty clopen subsets. Then, whenever $R \subset 2^{\mathbb{N}}$ is nowhere dense and of cardinality c, then R is admissible.

For example, let $R = \{y \in 2^{\mathbb{N}} : y(2n) = 1 \text{ for all } n \in \mathbb{N}\}$. Then clearly R is a closed subset which has empty interior and $\#R = \#(T \backslash R) = c$.

Lemma 4.3.5 *Let $(T, \mathbf{O})$ be any feasible pair and let R be an admissible subset of T. Then there are 2^c subsets of R which are admissible.*

Proof Let $S \subset R$ where $\#S = c$. Then S is admissible. $\square$

Since $F(\mathbb{N}) \times F(T)$ has cardinality c, we can identify the Big Cantor space with $2^{F(\mathbb{N}) \times F(T)}$. For each $k \in F(\mathbb{N})$, let $f_k \in 2^{F(\mathbb{N}) \times F(T)}$ be the characteristic function of the set

$$\{(l, L) : L \in F(T \backslash R), l \subset k \text{ and } O_n \cap L = \emptyset \text{ whenever } n \in k \text{ and } n \notin l\}.$$

Let X_R be the countable set $\{f_k : k \in F(\mathbb{N})\}$. Let K_R be the closure of X_R in the Big Cantor space. Then K_R is a (separable) compact Hausdorff totally disconnected space with respect to the relative topology induced by the product topology of the Big Cantor space. Let $A_R = B^{\infty}(K_R)/M_0(K_R)$, the regular σ-completion of $C(K_R)$. By the discussion in the previous section, A_R is monotone complete and is a (unital) C^*-subalgebra of ℓ^{∞}. Furthermore, the only way it could fail to be wild, is if one of the points in X_R were an isolated point in K_R. We shall show that this does not happen; so that the algebra must be wild.

The projections in A_R form a complete Boolean algebra which satisfies the countable chain condition (because it embeds in ℓ^{∞} which supports a faithful state) and which is Boolean isomorphic to $Reg(K_R)$, the Boolean algebra of regular open subsets of K_R.

For each $(k, K) \in F(\mathbb{N}) \times F(T)$ let $E_{(k,K)} = \{x \in K_R : x(k, K) = 1\}$. The definition of the product topology of the Big Cantor space implies that $E_{(k,K)}$ and its compliment $E^c_{(k,K)}$ are clopen subsets of K_R. It also follows from the definition of the product topology that finite intersections of such clopen sets form a base for the topology of K_R. Hence their intersections with X_R give a base for the relative topology of X_R. We shall see that, in fact, $\{E_{(k,K)} \cap X_R : k \in F(\mathbb{N}), K \in F(T \backslash R)\}$ is a base for the topology of X_R. To establish this we first need to prove some preliminary technical results. We shall then show that $Reg(K_R)$ is generated as a Boolean σ-algebra by the countable set $\{E_{(\{n\}, \varnothing)} : n \in \mathbb{N}\}$. Using this we shall find a natural representation of A_R in the form $B^\infty(2^\mathbb{N})/J_R$ where J_R is a σ-ideal of the bounded Baire measurable functions on the Cantor space with $C(2^\mathbb{N}) \cap J_R = \{0\}$.

Lemma 4.3.6 *We have $E_{(k,K)} = \varnothing$ unless $K \subset T \backslash R$.*

Proof Suppose K is not a subset of $T \backslash R$. Then, for any $h \in F(\mathbb{N})$, it follows from the definition of f_h, that $f_h(k, K) = 0$. So $X_R \subset E^c_{(k,K)}$. Then by Lemma 4.3.1, $K_R = clX_R = E^c_{(k,K)}$. Thus $E_{(k,K)} = \varnothing$. □

Lemma 4.3.7 *Let $x \in X_R$ and $(k, K) \in F(\mathbb{N}) \times F(T)$. Let $x \in E^c_{(k,K)}$. Then there exists $(l, L) \in F(\mathbb{N}) \times F(T \backslash R)$ such that $x \in E_{(l,L)} \subset E^c_{(k,K)}$.*

Proof First suppose that $K \cap R \neq \varnothing$. Then by the preceding lemma $K_R = E^c_{(k,K)}$.
For any $h \in F(\mathbb{N}), f_h \in E_{(h,\varnothing)}$. But $E_{(h,\varnothing)} \subset K_R = E^c_{(k,K)}$.
So we may now assume that $K \cap R = \varnothing$.
Let $x = f_h$. Then $f_h \in E^c_{(k,K)} \iff not(k \subset h \ \&(\forall n(n \in h \backslash k \implies O_n \cap K = \varnothing))) \iff k \backslash h \neq \varnothing$ or $\exists n_1 \in h \backslash k$ such that $O_{n_1} \cap K \neq \varnothing$.

(1) First let us deal with the situation where $k \backslash h \neq \varnothing$. Then there exists $n_0 \in k \backslash h$. Since R is admissible, we can find $t_0 \in T \backslash R$ such that $t_0 \in O_{n_0}$. Then it is straight forward to verify that $f_h \in E_{(h,\{t_0\})}$. It remains to show that $E_{(h,\{t_0\})} \cap E_{(k,K)} = \varnothing$. Suppose that this is false. Then we can find $f_g \in E_{(h,\{t_0\})} \cap E_{(k,K)}$. So $h \subset g$ and $k \subset g$. Thus $n_0 \in g \backslash h$. Then $f_g \in E_{(h,\{t_0\})}$ implies $t_0 \notin O_{n_0}$. But this is a contradiction. So $f_h \in E_{(h,\{t_0\})} \subset E^c_{(k,K)}$.

(2) Now consider the case where $\exists n_1 \in h \backslash k$ such that $O_{n_1} \cap K \neq \varnothing$. Consider $E_{(\{n_1\},\varnothing)}$. It is clear that f_h is an element of this set. We now wish to show that $E_{(\{n_1\},\varnothing)} \cap E_{(k,K)} = \varnothing$. Suppose this is false. Then we can find $f_g \in E_{(\{n_1\},\varnothing)} \cap E_{(k,K)}$. Then $n_1 \in g \backslash k$. So $O_{n_1} \cap K = \varnothing$. This is a contradiction. So $f_h \in E_{(\{n_1\},\varnothing)} \subset E^c_{(k,K)}$. □

Lemma 4.3.8 *Let (l, L) and (k, K) be any elements of $F(\mathbb{N}) \times F(T \backslash R)$ such that $E_{(l,L)} \cap E_{(k,K)} \neq \varnothing$. Then $E_{(l,L)} \cap E_{(k,K)} = E_{(l \cup k, L \cup K)}$.*

Proof Since $E_{(l,L)} \cap E_{(k,K)}$ is not empty and X_R is dense in K_R, $E_{(l,L)} \cap E_{(k,K)} \cap X_R$ is not empty. Let $f_h \in E_{(l,L)} \cap E_{(k,K)} \cap X_R$. Then $l \subset h$ and $k \subset h$. So $l \cup k \subset h$.
Also (i) $O_n \cap L = \varnothing$ for all $n \in h \backslash l$ and $O_n \cap K = \varnothing$ for all $n \in h \backslash k$.
Since $k \backslash l \subset h \backslash l$ and $l \backslash k \subset h \backslash k$ we have
(ii) $O_n \cap L = \varnothing$ for all $n \in (l \cup k) \backslash l$ and $O_n \cap K = \varnothing$ for all $n \in (l \cup k) \backslash k$.
From (i) we have $O_n \cap (L \cup K) = \varnothing$ whenever $n \in h \backslash (l \cup k)$. So $f_h \in E_{(l \cup k, L \cup K)}$.
Thus $(E_{(l,L)} \cap E_{(k,K)} \cap X_R) \subset E_{(l \cup k, L \cup K)} \cap X_R$. Hence $E_{(l,L)} \cap E_{(k,K)} \subset E_{(l \cup k, L \cup K)}$.
By the above, $E_{(l \cup k, L \cup K)}$ is not empty. So $E_{(l \cup k, L \cup K)} \cap X_R$ is not empty. Let $f_g \in E_{(l \cup k, L \cup K)}$. Then $l \cup k \subset g$. Also, for all $n \in g \backslash (l \cup k)$, we have $O_n \cap (L \cup K) = \varnothing$.

By (ii) we also have $O_n \cap L = \varnothing$ for $n \in (l \cup k)\backslash l$. Hence $f_g \in E_{(l,L)}$. Similarly $f_g \in E_{(k,K)}$.

It now follows that $E_{(l\cup k, L\cup K)} \cap X_R$ is a subset of $E_{(l,L)} \cap E_{(k,K)}$. Taking closures and applying Lemma 4.3.1, gives $E_{(l,L)} \cap E_{(k,K)} = E_{(l\cup k, L\cup K)}$. $\square$

Lemma 4.3.9 *Let U be an open subset of K_R and $x \in U \cap X_R$. Then there exists $(k, K) \in F(\mathbb{N}) \times F(T\backslash R)$ such that $x \in E_{(k,K)} \subset U$.*

Proof It follows from the definition of the product topology on the Big Cantor space and Lemma 4.3.7, that $x \in \bigcap_{j=1}^{q} E_{(h(j),H(j))} \subset U$ where $(h(j), H(j)) \in F(\mathbb{N}) \times F(T\backslash R)$ for $j = 1, 2, \ldots, q$.

Let $k = \bigcup_{j=1}^{q} h(j)$ and let $K = \bigcup_{j=1}^{q} H(j)$. Then, by repeated applications of Lemma 4.3.8, we have $x \in E_{(k,K)} \subset U$. $\square$

Corollary 4.3.10 *Let U be a non-empty regular open subset of K_R. Then there exists a sequence $(k(j), K(j))(j = 1, 2, \ldots)$ in $F(\mathbb{N}) \times F(T\backslash R)$ such that, in the complete Boolean algebra of regular open subsets of K_R,*
$$U = \bigvee_{j=1} E_{(k(j),K(j))}.$$

Proof Since $X_R \cap U$ is a countable set it can be enumerated by $(x_j)(j = 1, 2, \ldots)$. By Lemma 4.3.9, for each j, we can find $(k(j), K(j))$ in $F(\mathbb{N}) \times F(T\backslash R)$ such that $x_j \in E_{(k(j),K(j))} \subset U$. So $X_R \cap U \subset \bigcup_{j=1} E_{(k(j),K(j))} \subset U$. On taking closures, we get
$$clU \subset cl \bigcup_{j=1}^{\infty} E_{(k(j),K(j))} \subset clU.$$

Because U is a regular open set, $U = int(clU)$. We recall that the supremum. of a sequence of regular open sets in $Reg(K_R)$ is formed by taking the closure of their union, and then taking the interior of that set. So $U = \bigvee_{j=1} E_{(k(j),K(j))}$. $\square$

The following technical lemma will not be needed in this section, but it seems natural to prove it here.

Lemma 4.3.11 *Let (l, L) and (k, K) be in $F(\mathbb{N}) \times F(T\backslash R)$. Let $E_{(l,L)} = E_{(k,K)}$. Then $l = k$ and $L = K$.*

Proof For $h \in F(\mathbb{N}), f_h \in E_{(l,L)} \iff l \subset h$, and whenever $n \in h\backslash l$ then $O_n \cap L = \varnothing$.

It follows that $f_l \in E_{(l,L)}$ and so $f_l \in E_{(k,K)}$. Thus $k \subset l$. Similarly we can show that $l \subset k$. So $l = k$.

Suppose that L is not a subset of K. Then there exists $t \in L\backslash K$. Then, by feasibility, there exists m such that $m \notin k, t \in O_m$ and $O_m \cap K = \varnothing$. Let $h = k \cup \{m\}$. Hence $f_h \in E_{(k,K)}$. So then $f_h \in E_{(l,L)}$. So $O_m \cap L = \varnothing$. But $t \in O_m \cap L$. This contradiction shows that $L \subset K$. Similarly $K \subset L$. $\square$

Corollary 4.3.12 *Let* (l, L) *and* (k, K) *be in* $F(N) \times F(T\backslash R)$. *If* $E_{(l,L)} \subset E_{(k,K)}$ *then* $k \subset l$ *and* $K \subset L$. *Conversely, if* $k \subset l$ *and* $K \subset L$ *then either* $E_{(l,L)} \cap E_{(k,K)} = \emptyset$ *or* $E_{(l,L)} \subset E_{(k,K)}$.

Proof First suppose $E_{(l,L)} \subset E_{(k,K)}$. Then $E_{(l,L)} = E_{(l,L)} \cap E_{(k,K)}$. Since $f_l(l, L) = 1$ this intersection is not empty. So, by Lemma 4.3.8, $E_{(l,L)} = E_{(l\cup k, L\cup K)}$. By Lemma 4.3.11, $l = l \cup k$ and $L = L \cup K$. i.e. $k \subset l$ and $K \subset L$.

Conversely, let $k \subset l$ and $K \subset L$. By Lemma 4.3.8, either $E_{(l,L)} \cap E_{(k,K)} = \emptyset$ or $E_{(l,L)} \cap E_{(k,K)} = E_{(l\cup k, L\cup K)} = E_{(l,L)}$. $\qquad\square$

Proposition 4.3.13 *The algebra* $A_R = B^\infty(K_R)/M_0(K_R) = C(\widehat{K_R})$ *is wild and non-atomic.*

Proof It follows from the work of the preceding Section that it suffices to show that none of the elements of X_R is an isolated point in K_R.

Suppose this is false and f_g is an isolated point. Then, by Lemma 4.3.9, for some $k \in F(\mathbb{N})$ and $K \in F(T\backslash R)$, $E_{(k,K)} = \{f_g\}$.

Since K is a finite set and $T\backslash R$ is infinite, we can find $t_0 \in (T\backslash R)\backslash K$. It now follows from the definition of feasibility that we can find $n_0 \notin g$ such that $t_0 \in O_{n_0}$ and $O_{n_0} \cap K = \emptyset$. Let $h = g \cup \{n_0\}$. Then $k \subset h$ and, for $n \in h\backslash k$, $O_n \cap K = \emptyset$. So $f_h \in E_{(k,K)}$. Hence $f_h = f_g$. But $f_g(\{n_0\} \cup k, K) = 0$ whereas $f_h(\{n_0\} \cup k, K) = 1$. This is a contradiction. So the proposition is proved. $\qquad\square$

Lemma 4.3.14 *For each* $k \in F(\mathbb{N})$ *and* $t \in T\backslash R$ *we have*

$$E_{(k,\{t\})} = \bigcap_{n\in k} E_{(\{n\},\emptyset)} \cap int\left(\bigcap_{n\notin k, t\in O_n} (K_R\backslash E_{(\{n\},\emptyset)}) \right).$$

Proof By Corollary 4.3.2 it suffices to prove that

$$X_R \cap E_{(k,\{t\})} = X_R \cap \bigcap_{n\in k} E_{(\{n\},\emptyset)} \cap \bigcap_{n\notin k, t\in O_n} (K_R\backslash E_{(\{n\},\emptyset)}). \qquad (\#)$$

Let $f_g \in X_R \cap E_{(k,\{t\})}$. So $f_g(k, \{t\}) = 1$. Thus (a) $k \subset g$ and (b) for every $n \in g\backslash k$ we have $t \notin O_n$. So, by (a), $f_g(\{n\}, \emptyset) = 1$ for each $n \in k$. Thus $f_g \in E_{(\{n\},\emptyset)}$ for every $n \in k$.

Now consider $n \notin k$. If $n \in g$ then $n \in g\backslash k$. So $O_n \cap \{t\} = \emptyset$. Hence if $n \notin k$ and $t \in O_n$ then $n \notin g$. So $f_g(\{n\}, \emptyset) = 0$. Thus $f_g \in K_R\backslash E_{(\{n\},\emptyset)}$. It now follows that f_g is an element of the right hand side of (#).

Conversely, let us take f_h to be an element of the right hand side of (#). Then $f_h(\{n\}, \emptyset) = 1$ for each $n \in k$. So $k \subset h$. Now fix $n \in h\backslash k$.

Then $f_h(\{n\}, \emptyset) = 1$. If $t \in O_n$ then $f_h \in (X_R\backslash E_{(\{n\},\emptyset)})$ which would imply $f_h(\{n\}, \emptyset) = 0$. Hence $t \notin O_n$. It follows that $f_h(k, \{t\}) = 1$. So the equality (#) is established. $\qquad\square$

Proposition 4.3.15 *Let $Reg(K_R)$ be the complete Boolean algebra of regular open subsets of K_R; it is the smallest σ-complete subalgebra of itself which contains the countable set $\{E_{(\{n\},\varnothing)} : n = 1, 2, \ldots\}$.*

Proof Let S be the σ-subalgebra of $Reg(K_R)$ generated by $\{E_{(\{n\},\varnothing)} : n = 1, 2, \ldots\}$. Fix $k \in F(\mathbb{N})\backslash\{\varnothing\}$ and consider $E_{(k,\varnothing)}$. Then $f_k \in E_{(\{n\},\varnothing)}$ for each $n \in k$. So, by Lemma 4.3.8, $E_{(k,\varnothing)} = \bigcap_{n \in k} E_{(\{n\},\varnothing)}$. Hence $E_{(k,\varnothing)} \in S$. We observe that $E_{(\varnothing,\varnothing)} \cap X_R = \{f_g : g \in F(\mathbb{N})\} = X_R$. So $E_{(\varnothing,\varnothing)} = K_R$.

We now consider $E_{(k,K)}$ where $K \neq \varnothing$. If $E_{(k,K)} \neq \varnothing$ then $K \subset T\backslash R$. So $K = \{t_1, t_2, \ldots, t_n\}$ where each t_j is in $T\backslash R$. By Lemma 4.3.14, $E_{(k,\{t\})} \in S$ for each $k \in F(\mathbb{N})$ and $t \in T\backslash R$. Also $f_k \in E_{(k,\{t\})}$. It now follows from Lemma 4.3.8 that $E_{(k,K)} \in S$. We can now apply Corollary 4.3.10 to deduce that $Reg(K_R) \subset S$. □

Corollary 4.3.16 *Let q_R be the canonical quotient homomorphism of $B^\infty(K_R)$ onto $B^\infty(K_R)/M_0(K_R)$. Let B be a Boolean σ-subalgebra of the Baire subsets of K_R such that $E_{(\{n\},\varnothing)} \in B$ for each $n \in \mathbb{N}$. Then $\{q_R(\chi_S) : S \in B\}$ is the set of all projections in $B^\infty(K_R)/M_0(K_R)$.*

Proof This follows from Proposition 4.3.15 and the observation that the map $O \mapsto q_R(\chi_O)$ is a Boolean isomorphism from $Reg(K_R)$ onto the Boolean algebra of all projections in $B^\infty(K_R)/M_0(K_R)$. □

Definition 4.3.17 Let

$$\mathbf{N} : T \longmapsto \mathcal{P}(\mathbb{N})$$

be the map defined by

$$\mathbf{N}(t) = \{n \in \mathbb{N} : t \in O_n\}$$

for each $t \in T$.

We remark that the definition of feasibility implies that $\mathbf{N}(t)$ is an infinite set for every $t \in T$. Feasibility also implies that it is an injective map. Its definition is independent of any choice of R.

Proposition 4.3.18 *For each $t \in T$ let $C(t)$ be the closed set defined by*

$$C(t) = \bigcap_{n \in \mathbf{N}(t)} (K_R \backslash E_{(\{n\},\varnothing)}).$$

Then $C(t)$ has empty interior if, and only if, $t \in R$.

Proof We recall that $f_g(\{n\}, \varnothing) = 0$ if, and only if, $n \notin g$. So $f_g \in C(t)$ if, and only if, $g \cap \mathbf{N}(t) = \varnothing$.

First assume that $t \in T\backslash R$. Then $f_h \in E_{(\varnothing,\{t\})} \iff (n \in h \text{ implies } t \notin O_n) \iff h \cap \mathbf{N}(t) = \varnothing \iff f_h \in C(t)$. Thus $f_\varnothing \in E_{(\varnothing,\{t\})} \cap X_R = C(t) \cap X_R$.

By Corollary 4.3.2, given a closed set C in K_R and a clopen E in K_R, if $X_R \cap C = X_R \cap E$, then $E = intC$. So the clopen set $E_{(\varnothing, \{t\})} = intC(t)$. This set is non-empty because $f_\varnothing$ is an element.

Conversely, let us assume that $C(t)$ has non-empty interior. So there exists $(k, K) \in F(\mathbb{N}) \times F(T \backslash R)$ such that $\varnothing \neq E_{(k,K)} \subset C(t)$. First suppose that $t \notin K$. Since $(T, \mathbf{O})$ is feasible we can find $n \notin k$ such that $t \in O_n$ and $O_n \cap K = \varnothing$. Let $h = \{n\} \cup k$. Then it follows from this that $f_h \in E_{(k,K)}$. So $f_h \in C(t)$. Thus $h \cap \mathbf{N}(t) = \varnothing$. In particular, $t \notin O_n$. This contradiction shows that we must have $t \in K \subset T \backslash R$.

So $C(t)$ has non-empty interior if, and only if, $t \in T \backslash R$. The required result follows. $\qquad\qquad\square$

We shall now see how to represent $A_R = B^\infty(K_R)/M_0(K_R) = C(\hat{X})$ as a quotient of the algebra of Baire functions on the (classical) Cantor space. The key fact which makes this possible is the existence of a countable set of generators.

Let Γ be a map from the Big Cantor space onto the small Cantor space, defined as follows:

For $x \in 2^{F(\mathbb{N}) \times F(T)}$ let $\Gamma(x)(n) = x((\{n\}, \varnothing))$ for $n = 1, 2, \ldots$. Then Γ is a map from the Big Cantor space into $2^{\mathbb{N}}$, the classical Cantor space.

Let $J = \{(\{n\}, \varnothing) : n \in \mathbb{N}\}$. Then, trivially, we may identify $2^{\mathbb{N}}$ with 2^J. So Γ may be regarded as a restriction map and, by the definition of the topology for product spaces, it is continuous.

Let $\Sigma = \{y \in 2^{\mathbb{N}} : y(n) = 0$ for all but finitely many $n\}$. Then Σ is a countable dense subset of $2^{\mathbb{N}}$ such that $\Gamma[X_R] = \Sigma$. Let Γ_R be the restriction of Γ to K_R. Then Γ_R is a continuous map from the compact Hausdorff space K_R onto a compact Hausdorff space. Since X_R is dense in K_R, it follows that $\Gamma_R[K_R] = 2^{\mathbb{N}}$. This map Γ_R is never an open map, see the remarks at the end of this section.

Let $E_n = \{y \in 2^{\mathbb{N}} : y(n) = 1\}$. Then $\Gamma_R^{-1}[E_n] = E_{(\{n\}, \varnothing)}$ for $n = 1, 2, \ldots$.

Since Γ_R is continuous, it follows that whenever $f \in B^\infty(2^{\mathbb{N}})$ then $f \circ \Gamma_R$ is in $B^\infty(K_R)$. We define a σ-homomorphism γ_R from $B^\infty(2^{\mathbb{N}})$ to $B^\infty(K_R)$ by $\gamma_R(f) = f \circ \Gamma_R$. As in Corollary 4.3.16, we let q_R be the canonical quotient homomorphism of $B^\infty(K_R)$ onto $B^\infty(K_R)/M_0(K_R)$.

Definition 4.3.19 Let $I_R = \{f \in B^\infty(2^{\mathbb{N}}) : f \circ \Gamma_R \in M_0(K_R)\}$.

Theorem 4.3.20 *We have I_R is a σ-ideal of $B^\infty(2^{\mathbb{N}})$ and $B^\infty(2^{\mathbb{N}})/I_R$ is isomorphic to $B^\infty(K_R)/M_0(K_R) \approx C(\widehat{K_R}) = A_R$. Also $I_R \cap C(2^{\mathbb{N}}) = \{0\}$.*

Proof The mapping $q_R \circ \gamma_R$ is a σ-homomorphism, so its kernel is a σ-ideal.

But $q_R \circ \gamma_R(f) = 0 \iff \gamma_R(f) \in M_0(K_R) \iff f \circ \Gamma_R \in M_0(K_R) \iff f \in I_R$. Thus I_R is a σ-ideal and $B^\infty(2^{\mathbb{N}})/I_R$ is isomorphic to

$$q_R \circ \gamma_R[B^\infty(2^{\mathbb{N}})] \subset B^\infty(K_R)/M_0(K_R).$$

We observe that γ_R maps the characteristic function of E_n to the characteristic function of $E_{(\{n\}, \varnothing)}$. It now follows from Corollary 4.3.16 that the range of $q_R \circ \gamma_R$

contains every projection in $B^\infty(K_R)/M_0(K_R)$. Since the range of $q_R \circ \gamma_R$ is a closed subalgebra of $B^\infty(K_R)/M_0(K_R)$ it must coincide with $B^\infty(K_R)/M_0(K_R)$.

Finally, if $f \in I_R \cap C(2^{\mathbb{N}})$, then $f \circ \Gamma_R$ is a continuous function in $M_0(K_R)$. So $\{y \in K_R : f \circ \Gamma_R(y) \neq 0\}$ is an open meagre set. So, by the Baire Category Theorem for compact Hausdorff spaces, this set is empty i.e. $f = 0$. This completes the proof. $\square$

We denote the natural quotient homomorphism from $B^\infty(2^{\mathbb{N}})$ to $B^\infty(2^{\mathbb{N}})/I_R$ by π_R.

To avoid "subscripts of subscripts" we shall denote the characteristic function of a set S by $\chi(S)$.

We now let $D_n = 2^{\mathbb{N}} \backslash E_n$. So γ_R maps $\chi(D_n)$ to $\chi(E^c_{(\{n\},\varnothing)})$. For each $k \in F(\mathbb{N}) \backslash \{\varnothing\}$ let $d_k = \chi(\bigcap_{n \in k} E^c_{(\{n\},\varnothing)})$. Then γ_R maps $\chi(\bigcap_{n \in k} D_n)$ to d_k.

Because γ_R is a σ-homomorphism, it maps $\chi(\bigcap_{n \in N(t)} D_n)$ to $\chi(C(t))$. Then, by Proposition 4.3.18, $\chi(C(t)) \in M_0(K_R)$ if and only if $t \in R$. So $\chi(\bigcap_{n \in N(t)} D_n) \in I_R$ if and only if $t \in R$.

Proposition 4.3.21 *When $R \neq S$ then $I_R \neq I_S$.*

Proof Without loss of generality, we may suppose that there exists $t \in R \backslash S$. Then $\chi(\bigcap_{n \in N(t)} D_n) \in I_R$ but $\chi(\bigcap_{n \in N(t)} D_n) \notin I_S$. So $I_R \neq I_S$. $\square$

Corollary 4.3.22 *There are 2^c distinct ideals I_R.*

Proof By Lemma 4.3.5 there are 2^c admissible sets. $\square$

Remark 4.3.23 When $R \neq S$ then $I_R \neq I_S$. But it does not necessarily follow that $B^\infty(2^{\mathbb{N}})/I_R$ is not isomorphic to $B^\infty(2^{\mathbb{N}})/I_S$. To show that there are 2^c algebras A_R which are not equivalent and hence, in particular, not isomorphic, we make use of the machinery of representing functions and spectroids, modulo $(T, \mathbf{N})$, where $\mathbf{N}$ is the map defined in Definition 4.3.17. We define a particular representing function for $B^\infty(2^{\mathbb{N}})/I_R$ by defining $f_R(k) = \pi_R(\chi(\bigcap_{n \in k} D_n))$ for $k \neq \varnothing$, and putting $f_R(\varnothing) = 1$.

Then we have:

Lemma 4.3.24 *For each admissible R the function f_R is a representing function for $B^\infty(2^{\mathbb{N}})/I_R$. Then R is represented by f_R, modulo $(T, \mathbf{N})$. In other words, $R \in \partial_{(T,\mathbf{N})}(B^\infty(2^{\mathbb{N}})/I_R)$.*

Proof First we note that $\bigcap_{n \in k} D_n$ is a non-empty clopen set for each finite set k. So $\chi(\bigcap_{n \in k} D_n)$ is a non-zero continuous function. Hence it is not in I_R. It is clear that f_R is downward directed. Now consider $\bigwedge_{k \in F(\mathbb{N})} f_R(k) = \pi_R(\chi(\bigcap_{n=1}^{\infty} D_n))$. But

$\bigcap_{n=1}^{\infty} D_n$ is a single point set. So $\pi_R(\chi(\bigcap_{n=1}^{\infty} D_n))$ is zero or an atomic projection. Since $B^{\infty}(2^{\mathbb{N}})/I_R \approx A_R$ which has no atoms, $\bigwedge_{k \in F(\mathbb{N})} f_R(k) = 0$. Thus f_R is a representing function.

We now calculate $R_{(T,\mathbb{N})}(f_R) = \{t \in T : \bigwedge_{k \in F(\mathbb{N}(t))} f_R(k) = 0\}$. We have $\bigwedge_{k \in F(\mathbb{N}(t))} f_R(k) = 0$ precisely when $\chi(\bigcap_{n \in \mathbb{N}(t)} D_n) \in I_R$.

In the notation of Proposition 4.3.18, $\gamma_R(\chi(\bigcap_{n \in \mathbb{N}(t)} D_n)) = \chi(C(t))$. So, by applying Proposition 4.3.18, we see that $\chi(C(t)) \in I_R$ precisely when $t \in R$. Thus $R = R_{(T,\mathbb{N})}(f_R)$.

So R is in $\partial_{(T,\mathbb{N})}(B^{\infty}(2^{\mathbb{N}})/I_R) = \partial_{(T,\mathbb{N})}(A_R)$. $\qquad\qquad\square$

Corollary 4.3.25 *Let $\mathcal{R}$ be the collection of all admissible subsets of T. Then there exists $\mathcal{R}_0 \subset \mathcal{R}$, with $\#\mathcal{R}_0 = 2^c$ and such that $\{A_R : R \in \mathcal{R}_0\}$ is a set for which A_R is not equivalent to A_S unless $R = S$. Also the spectroid of A_R, modulo $(T, \mathbb{N})$, is not equal to the spectroid of A_S, modulo $(T, \mathbb{N})$, when $R \neq S$.*

Proof By Lemma 4.3.5 $\#\mathcal{R} = 2^c$. From Lemma 4.3.24, for each $R \in \mathcal{R}, R \in \partial_{(T,\mathbb{N})}(A_R)$. The result now follows from Theorem 3.3.13 and Corollary 3.3.14. $\qquad\square$

There are a number of interesting side issues which we have omitted. For example, it can be proved that none of the algebras constructed above is isomorphic to the Dixmier algebra. If, for a given R, the continuous map Γ_R were open then this would imply that $B^{\infty}(2^{\mathbb{N}})/I_R$ is isomorphic to the Dixmier algebra. So Γ_R is never open.

In the construction of commutative monotone complete algebras described in this section we have assumed that $\#(T\backslash R)$, the cardinality of the compliment of the admissible set used, is c. If we replaced this assumption by requiring $\#(T\backslash R) = \aleph_0$, then the constructions would still work but it turns out that the algebras obtained would all be isomorphic to the Dixmier algebra.

It can be shown that the spectroid of any small C^*-algebra has cardinality c. In particular the spectroid of $\mathbb{C}$ has cardinality c. This is not too surprising, since the spectroid of $\mathbb{C}$ equals the spectroid of each small von Neumann algebra.

This section is taken from our paper [144]. The compact separable spaces discussed here are modifications of spaces constructed by Hamana [66]. Our indebtedness to his work is clear.

Chapter 5
Convexity and Representations

In this chapter we apply methods from convexity to representations of C^*-algebras. We construct regular σ-completions and show that the regular σ-completion of a small C^*-algebra is its regular completion.

We discuss small C^*-algebras and completely positive maps.

5.1 Function Representations of Non-commutative Algebras

Unless we specifically state otherwise, the C^*-algebras considered in this chapter will be assumed to have a unit element. For most results this is not essential but it permits a simpler and more straightforward exposition.

We shall need a few tools from the theory of compact convex sets and of partially ordered Banach spaces. We shall remind the reader of some of the basics. For more complete information an excellent source is [2]. See also [123]. We shall also need to recall some standard results from C^*-algebra theory.

When a C^*-algebra is commutative it is isometrically isomorphic to $C(\Omega)$, the C^*-algebra of all complex valued continuous functions on its spectrum Ω. See [50, 89, 150]. But for non-commutative algebras the Gelfand-Naimark Theorem does not apply.

Given a C^*-algebra A, its self-adjoint part, A_{sa}, is a partially ordered Banach space (over the real numbers). Let $a \in A_{sa}$. Then, by spectral theory,

$$||a|| = \inf\{\lambda : -\lambda 1 \le a \le \lambda 1\}.$$

This implies that 1 is an *order unit* for the partially ordered Banach space A_{sa}. It also shows that for any self-adjoint a, its norm $||a||$ coincides with the order-unit norm of a.

© Springer-Verlag London 2015

K. Saitô, J.D.M. Wright, *Monotone Complete C*-algebras and Generic Dynamics*,
Springer Monographs in Mathematics, DOI 10.1007/978-1-4471-6775-4_5

Let K be the state space of A, that is, K is the set of states of A with the $\sigma(A^*, A)$-topology. Then K is a compact convex subset of the dual space A^*, when A^* is equipped with the $\sigma(A^*, A)$-topology. Equivalently, K is a compact convex subset of the (real) dual space A_{sa}^* (of A_{sa}), when A_{sa}^* is equipped with the $\sigma(A_{sa}^*, A_{sa})$-topology. By a theorem of Kadison (see page 70 [121] and Theorem II.1.8 Page 74 [2]) the real, order unit normed, Banach space A_{sa} is isometrically order isomorphic to $A(K)$. Here $A(K)$ is the Banach space of all continuous, real valued, affine functions on K, equipped with the supremum norm.

This can be generalised from C^*-algebras. Let V be a partially ordered vector space which is Archimedean, possesses a distinguished order unit u and is a Banach space with respect to the order-unit norm. We call V an *order unit Banach space*. Let Q be the convex set of all $\phi \in V^*$ such that $\phi \geq 0$ and $\phi(u) = 1$. Then Q is compact in the $\sigma(V^*, V)$-topology. We call Q the *state space* of V. By Kadison's theorem, V is isometrically order isomorphic to $A(Q)$. For details, see [2]. See also [82].

When A is *commutative* with spectrum Ω then Ω can be identified with the extreme points of the state space of A.

Let A be a unital C^*-algebra. We recall some standard results. (See Section 12 [34], also Theorem 3.10.3 and Proposition 3.7.8 [121] and Section 2 of Chapter 3 [161]). For each state ϕ of A there is a corresponding Gelfand-Naimark-Segal representation. The direct sum of all these representations is the *universal representation* (π_u, H_u). We shall identify A with its image in $L(H_u)$, that is, with $\pi_u[A]$. The von Neumann envelope of A in $L(H_u)$ is the double commutant $\pi_u[A]''$. By the Sherman-Takeda Theorem, this von Neumann envelope can be identified with A^{**}, the second (Banach space) dual of A. See [87, 151, 158]. We shall make this identification from now onward. The weak operator topology on $\pi_u(A)''$ induces the $\sigma(A^{**}, A^*)$-topology on A^{**}. (Observe that every positive functional $\phi \in A^*$ can be identified with a vector state on $\pi_u(A)''$.) Similarly, A''_{sa} (which is $(\pi_u(A)'')_{sa}$) can be identified with the second dual of the *real* Banach space A_{sa}. Also the $\sigma(A_{sa}^{**}, A_{sa}^*)$-topology corresponds to the weak operator topology.

We can extend the Kadison representation as follows. Let $A''[K]$ be the space of all bounded (real valued) affine functions on K, equipped with the supremum norm. Then, see Lemma III.6.7 page 161 of [161], the Kadison isomorphism of A_{sa} onto $A(K)$, can be extended to an isometric isomorphism of A_{sa}^{**} onto $A''[K]$. In this isomorphism, $a \in \pi_u[A]''_{sa}$ is mapped to $\tilde{a}$, where

$$\phi(a) = \tilde{a}(\phi) \text{ for each } \phi \in K.$$

Furthermore a (norm bounded) net a_λ in $\pi_u[A]''_{sa}$ converges to a in the weak operator topology if, and only if, $(\widetilde{a_\lambda})$ is a (norm bounded) net which converges pointwise to $\tilde{a}$.

We shall frequently find it convenient to write "a" instead of "$\tilde{a}$". This minor abuse helps us to avoid some cumbersome notation ("hats on tildes").

Incidentally, for any Hilbert space H, if a sequence in $L(H)$ converges in the weak operator topology then it is bounded in norm. This is just a consequence of the Uniform Boundedness Theorem, see Exercise 5.3.2.

We shall discuss the Pedersen-Baire envelope of A, (see Definition 2.2.23), and the Brown-Borel envelope of A in Sect. 5.3. These are both C^*-subalgebras of A'' and both of them contain A as a subalgebra. We shall make use of them when proving representation theorems for monotone σ-complete C^*-algebras and monotone complete C^*-algebras.

We shall see, in Sect. 5.6, that for every monotone complete C^*-algebra A there exists a *commutative* monotone complete C^*-algebra C and a faithful, normal, positive linear map $\Gamma : A \mapsto C$. So in the notation of Chap. 3,

$$A \precsim C.$$

5.2 Compact Convex Sets: Preliminaries

When we apply convexity theory to C^*-algebras, the compact convex sets we use, will usually be the state spaces of the algebras. But many of the basic geometric results can be proved much more generally with no extra effort. So in this section K will be a compact convex subset of a Hausdorff, locally convex (real) vector space $\mathcal{V}$. In our applications, in later sections, K will usually be the state space of a C^*-algebra A and $\mathcal{V}$ will be the space A_{sa}^*, equipped with the $\sigma(A_{sa}^*, A_{sa})$-topology.

We shall denote the set of extreme points of K by ∂K. We recall that, by the Krein-Milman Theorem, the convex hull of ∂K is dense in K.

For any convex $X \subset \mathcal{V}$, a real valued function $f : X \mapsto \mathbb{R}$ is *convex* if,

$$f(\lambda x + (1 - \lambda)y) \le \lambda f(x) + (1 - \lambda)f(y)$$

for $\lambda \in [0, 1]$ and for any x and y in X. It is *concave* if $-f$ is convex. Clearly f is affine when it is both convex and concave. We use $A(X)$ to denote the vector space of all continuous (real-valued) affine functions on X.

All functions on K considered in this section are real valued.

Exercise 5.2.1 Let $f \in A(\mathcal{V})$. Show that $f - f(0)1$ is a continuous linear functional on $\mathcal{V}$.

Hint: When $f(0) = 0, f(\frac{1}{2}x) = \frac{1}{2}f(x)$. So

$$\frac{1}{2}f(x + y) = f\left(\frac{1}{2}x + \frac{1}{2}y\right) = \frac{1}{2}f(x) + \frac{1}{2}f(y).$$

In the following let $A(K; \mathcal{V}) = \{f|_K : f \in A(\mathcal{V})\}$.

Lemma 5.2.2 *Let f be a lower semicontinuous convex function on K. Then, for each $x \in K$,*

$$f(x) = \sup\{a(x) : a \in A(K; \mathcal{V}) \text{ and } a < f\};$$

where $a < f$ signifies $a(x) < f(x)$ for each $x \in K$.

Proof See Proposition 1.1.2 [2] or Lemma III.6.1 [161]. By multiplying by -1 we get a corresponding result for upper semicontinuous concave functions.   ☐

Lemma 5.2.3 *Let a be a lower semicontinuous (real valued) affine function on K. Then there is an increasing net, (a_λ), in $A(K; \mathcal{V})$ such that*

$$a(x) = \lim a_\lambda(x).$$

The set $\{f \in A(K; \mathcal{V}) : f < a\}$ is upward directed.

Proof See Corollary 1.1.4 [2] or Lemma III.6.2 [161]. By multiplying by -1 we get a corresponding result for upper semicontinuous affine functions.

Both the preceding lemmas were obtained in [102].   ☐

Corollary 5.2.4 *Let a be a continuous (real valued) affine function on K. Then there is an increasing sequence in $A(K; \mathcal{V})$ which converges to a in the uniform norm. In particular, $A(K; \mathcal{V})$ is dense in $A(K)$.*

Proof By Lemma 5.2.3 and Dini's Theorem, there is a increasing net in $A(K; \mathcal{V})$ which converges uniformly to a. That means $\|a - a_\lambda\| \to 0$. It follows that we may choose a monotone increasing sequence in $A(K; \mathcal{V})$ which converges in norm to a.
          ☐

Exercise 5.2.5 Suppose that the compact space K is a Polish space, that is, homeomorphic to a complete separable metric space. Show that ∂K is a G_δ-set.

Hint: Take a complete metric d on K. Then consider the sets

$$F_n = \{x \in K : x = \frac{1}{2}(y + z) \text{ for some } y \text{ and some } z \text{ in } K \text{ where } d(y, z) \geq 1/n\}.$$

Show that each F_n is closed. Show also that if a point in K is not extreme then it is in $\cup_n F_n$.

Each G_δ-subset of a Polish space is a Baire space. So Exercise 5.2.5 implies that when K is Polish then ∂K is a Baire space. A key theorem of Choquet says that for *every* compact convex K, ∂K is a Baire space in the relative topology induced by K.

Lemma 5.2.6 (Choquet) *Let K be a compact convex subset of V and let $e \in \partial K$. Then a fundamental system of open neighbourhoods of e is given by the family of sets of the form*

$$\{x \in \partial K : a(x) > 0\}$$

where $a \in A(K)$ and $a(e) > 0$.

Proof See 25.13 in [26] or page 355 [34]. For a generalisation see [37].   □

Theorem 5.2.7 (Choquet) *Let K be a compact convex subset of V. Then ∂K is a Baire space in the relative topology.*

Proof See 27.9 in [26] or B14 of [34].   □

Corollary 5.2.8 *Let A be a unital C^*-algebra and let K be its state space. Then ∂K, the set of all pure states of A, is a Baire space.*

The following theorem is based on [170].

Theorem 5.2.9 *Let $\mathcal{F}$ be a bounded subset of $A(K)$ such that there exists g, a greatest lower bound in $A(K)$ for $\mathcal{F}$. Let*

$$\bar{g}(x) = \inf\{f(x) : f \in \mathcal{F}\}$$

for each $x \in K$. Then $\bar{g}$ is a bounded, upper semicontinuous concave function on K and the set

$$\{x \in \partial K : \bar{g}(x) > g(x)\}$$

is a meagre subset of ∂K in the relative topology. Furthermore, if $\mathcal{F}$ is downward directed then $\bar{g}$ is affine.

Proof Clearly $\bar{g}$ is upper semicontinuous and concave. If, in addition, $\mathcal{F}$ is downward directed then $\bar{g}$ is the pointwise limit of a directed set of affine functions, and so affine.

Returning to the general situation, let

$$M_n = \{x \in K : \bar{g}(x) - g(x) \geq 1/n\}$$
$$= \bigcap_{f \in \mathcal{F}} \{x \in K : f(x) - g(x) \geq 1/n\}.$$

Then M_n is the intersection of a family of closed convex sets. So it is a closed convex subset of K.

Suppose that there exists an open subset U of K such that

$$M_n \cap \partial K \supset U \cap \partial K \neq \varnothing.$$

Choose $e \in U \cap \partial K$. By Lemma 5.2.6, there exists $h \in A(K)$ such that

$$O = \{x \in K : h(x) > 0\}$$

is an open neighbourhood of e and $O \cap \partial K \subset U \cap \partial K$.

Since we can multiply h by $1/n\|h\|$ we can assume, without loss of generality, that $h(x) \leq 1/n$ for all $x \in K$. For $x \in \partial K \backslash O$ we have

$$h(x) \leq 0 \leq \overline{g}(x) - g(x).$$

For $x \in \partial K \cap O$ we have

$$h(x) \leq 1/n \leq \overline{g}(x) - g(x)$$

because $O \cap \partial K \subset M_n$.

So for all $x \in \partial K$ and each $f \in \mathcal{F}$

$$0 \leq f(x) - g(x) - h(x).$$

When an affine continuous function is positive on ∂K then it is positive on the closed convex hull of ∂K. So, by the Krein-Milman Theorem,

$$g + h \leq f$$

for all $f \in \mathcal{F}$. Hence $g + h \leq g$. So $h \leq 0$. But this contradicts the existence of e. So $M_n \cap \partial K$ is a closed nowhere dense set in the relative topology of ∂K.

Since the union of a sequence of nowhere dense sets is meagre, the theorem follows.   □

Corollary 5.2.10 *Let $\mathcal{U}$ be a bounded subset of $A(K)$ such that there exists g, a least upper bound in $A(K)$ for $\mathcal{U}$. Let*

$$\underline{g}(x) = \sup\{f(x) : f \in \mathcal{U}\}$$

for each $x \in K$. Then $\underline{g}$ is a bounded, lower semicontinuous convex function on K and the set

$$\{x \in \partial K : \underline{g}(x) < g(x)\}$$

is a meagre subset of ∂K in the relative topology. Furthermore, if $\mathcal{U}$ is upward directed then g is affine.

Proof Let $\mathcal{F} = \{-h : h \in \mathcal{U}\}$ and apply Theorem 5.2.9.   □

Definition 5.2.11 The *Baire envelope*, $A^\infty(K)$, of $A(K)$ is the smallest space of bounded, Baire measurable functions which contains $A(K)$ and which is closed under pointwise limits of uniformly bounded, monotone increasing sequences.

Definition 5.2.12 The *Borel envelope*, $A^b(K)$, of $A(K)$ is the smallest space of bounded, Borel measurable functions which contains all bounded, lower semi-continuous affine functions on K and which is closed under pointwise limits of uniformly bounded, monotone increasing sequences.

Remark 5.2.13 We use $A''(K)$ to denote the space of all bounded (real valued) affine functions on K. Suppose K is the state space of a C^*-algebra A. Then $A''(K)$ can be identified with A''_{sa}, the self-adjoint part of A''. In turn this can be identified with A^{**}_{sa}, the second dual of the real Banach space A_{sa}. Also, by Kadison, A_{sa} is identified with $A(K)$. See Page 161 of [161]. Also, as will be discussed later, both $A^\infty(K)$ and $A^b(K)$ are the self-adjoint parts of C^*-algebras when K is the state space of a C^*-algebra.

Definition 5.2.14 Let f and g be bounded functions on K. Then $f \equiv g$ if $\{x \in \partial K : f(x) \neq g(x)\}$ is meagre. When $f \equiv g$ we shall say $f = g$ *almost everywhere* and also write this as $f = g$ a.e. More generally

$$f \geq g \text{ a.e.}$$

when $\{x \in \partial K : f(x) < g(x)\}$ is meagre.

Exercise 5.2.15 Let (f_n) and (g_n) be bounded sequences of bounded functions on K.

Let $f(x) = \sup\{f_n(x) : n = 1, 2 \ldots\}$ and $g(x) = \sup\{g_n(x) : n = 1, 2 \ldots\}$. If $f_n \geq g_n$ a.e. for each n show that $f \geq g$ a.e.

Hint: The union of countably many meagre sets is meagre.

Show also that $\inf f_n \geq \inf g_n$ a.e.

Let $\overline{\lim} f_n(x) = \inf_{m \geq 1}(\sup_{n \geq m} f_n(x))$. Show that $\overline{\lim} f_n \geq \overline{\lim} g_n$ a.e. Deduce that if $f_n = g_n$ a.e. then $\overline{\lim} f_n = \overline{\lim} g_n$ a.e.

We observed in Chapter 2, see [44], that even when $C(S)$ is monotone complete, it may fail to have a locally convex Hausdorff topology in which bounded, monotone sequences converge to their order limits. The following representation theorems [170] are useful when dealing with spaces which exhibit this "bad behaviour". Just as for C^*-algebras, $A(K)$ is monotone complete if each upper bounded, upward directed subset has a least upper bounded in $A(K)$. Monotone σ-complete is defined similarly.

Theorem 5.2.16 *Let $A(K)$ be monotone complete. Then corresponding to each $f \in A^b(K)$ there is a unique $kf \in A(K)$ such that $kf \equiv f$. Furthermore, k is a positive linear, idempotent map of $A^b(K)$ onto $A(K)$ with the following properties.*

(1) *Let (f_n) be a monotone increasing, upper bounded sequence in $A^b(K)$ with pointwise limit $\sup f_n$, and with $\bigvee_{n=1}^\infty k(f_n)$ as the least upper bound of (kf_n) in $A(K)$. Then $k(\sup f_n) = \bigvee_{n=1}^\infty k(f_n)$.*

(2) *Let $\mathcal{F}$ be an upward directed, upper bounded subset of $A(K)$ with pointwise supremum f. Let $\bigvee \mathcal{F}$ be the least upper bound of $\mathcal{F}$ in $A(K)$. Then*

$$k\underline{f} = \bigvee \mathcal{F}.$$

Proof Suppose that $f \equiv g_1$ and $f \equiv g_2$ where g_1 and g_2 are in $A(K)$. Then

$$\{x \in \partial K : g_1(x) \neq g_2(x)\}$$

is a meagre, open subset of ∂K. So, by Theorem 5.2.7, this set is empty. So when kf exists it is unique. If $f \geq 0$ then

$$\{x \in \partial K : kf(x) < 0\}$$

is an open, meagre subset of ∂K and hence empty. So, by Krein-Milman, $kf \geq 0$.

Let

$$W = \{f \in A^b(K) : \exists kf \in A(K) \text{ such that } f \equiv kf\}.$$

When f and g are in W then $f + g \equiv kf + kg$. Since $kf + kg$ is in $A(K)$, we have $k(f + g) = kf + kg$. So W is a vector subspace of $A^b(K)$ containing $A(K)$. Also k is a positive linear map of W onto $A(K)$.

Let $\mathcal{F}$ be an upper bounded, upward directed subset of $A(K)$. Let f be its pointwise supremum then, by Corollary 5.2.10, f is a lower semicontinuous function with $\underline{f} \equiv \bigvee \mathcal{F}$. So

$$k\underline{f} = \bigvee \mathcal{F}.$$

By Lemma 5.2.3, each bounded lower semicontinuous affine function on K is the pointwise supremum of an upward directed set in $A(K)$. Hence W contains all the bounded lower semicontinuous affine functions on K.

Let (f_n) be a monotone increasing sequence in W bounded above by a constant m. Since k is a positive linear map, (kf_n) is a monotone increasing sequence in $A(K)$ bounded above by m.

So $\lim_{n \to \infty} kf_n(x) = \sup kf_n(x)$ exists for all $x \in K$. By Corollary 5.2.10, $\sup kf_n \equiv \bigvee_{n=1}^{\infty} kf_n$.

By Exercise 5.2.15

$$\sup f_n \equiv \sup kf_n.$$

So $k(\sup f_n) = k(\sup kf_n) = \bigvee_{n=1}^{\infty} kf_n$. So $\sup f_n \in W$. Hence W contains $A^b(X)$. $\quad\square$

Theorem 5.2.17 *Let $A(K)$ be monotone σ-complete. Then corresponding to each $f \in A^{\infty}(K)$ there is a unique $kf \in A(K)$ such that $kf \equiv f$. Furthermore, k is a positive linear, idempotent map of $A^{\infty}(K)$ onto $A(K)$. Let (f_n) be a monotone* *increasing, upper bounded sequence in $A^\infty(K)$ with pointwise limit $\sup f_n$, and with $\bigvee_{n=1}^\infty k(f_n)$ as the least upper bound of (kf_n) in $A(K)$. Then*

$$k(\sup f_n) = \bigvee_{n=1}^\infty k(f_n).$$

Proof The proof is a simplified version of the proof of Theorem 5.2.16. ☐

Theorems 5.2.16 and 5.2.17 can be applied to obtain extension theorems for $A(K)$-valued maps, see Proposition 5.4.7.

The following application of Choquet's Lemma gives a well-known result which will be needed later.

Proposition 5.2.18 *Let H be a separable Hilbert space. Let $L(H)$ be the algebra of bounded operators on H. Let K be its state space. Then ∂K is separable.*

Proof Since H is a separable metric space, $\{\xi \in H : ||\xi|| = 1\}$ has a countable dense subset $\{\xi_n : n = 1, 2, \ldots\}$. For each n, let y_n be the normal pure state defined by $y_n(a) = < a\xi_n, \xi_n >$.

Let $e \in \partial K$ and let U be any open neighbourhood of e in ∂K. By Lemma 5.2.6, there exists a self-adjoint $a \in L(H)$ such that

$$e \in \{y \in \partial K : y(a) > 0\} \subset U.$$

If $y_n \notin \{y \in \partial K : y(a) > 0\}$ for any n then $< a\xi_n, \xi_n > \le 0$ for all n. But this implies $a \le 0$, which is impossible. So ∂K has a countable dense subset.

It follows from the Krein-Milman Theorem that K is also separable. ☐

Proposition 5.2.19 *Let A be a unital C^*-algebra with a separable state space. Then A possesses a faithful state. If A is monotone σ-complete then it is monotone complete.*

Proof Let $\{x_n : n = 1, 2, \ldots\}$ be a dense set of states. Let $y = \sum_{n \ge 1} \frac{1}{2^n} x_n$. Then y is a faithful state. If A is monotone σ-complete, then by Theorem 2.1.14, it is monotone complete. ☐

5.3 Envelopes of C^*-Algebras

Let B be a closed $*$-subalgebra of $L(H)$ for some Hilbert space H. Then there are several interesting candidates for larger C^*-subalgebras of $L(H)$ which contain B. The most familiar is the closure of B in the weak operator topology, the von Neumann envelope of B in $L(H)$.

Another natural construction is due to Davies [32]. When Y is a subset of $L(H)$, it is said to be *sequentially closed* in the weak operator topology if, whenever (T_n) is a sequence in Y which converges in the weak operator topology to T then $T \in Y$. Given any $Y \subset L(H)$ let $\Sigma(Y)$ be the intersection of all subsets of $L(H)$ which contain Y and are sequentially closed in the weak operator topology of $L(H)$. Clearly $\Sigma(Y)$ is sequentially closed in the weak operator topology and is the smallest such set containing Y.

Lemma 5.3.1 *Let B be a closed $*$-subalgebra of $L(H)$ for some Hilbert space H. Then $\Sigma(B)$ is a C^*-algebra and is sequentially closed in the weak operator topology of $L(H)$. We do not require B to have a unit element.*

Proof Take complex numbers λ and μ. Take $a \in B$. Now let W be the set of all $y \in \Sigma(B)$ such that

$$(\lambda a + \mu y) \in \Sigma(B), y^* \in \Sigma(B) \text{ and } ay \in \Sigma(B).$$

Then W contains B and is sequentially closed in the weak operator topology. So $W = \Sigma(B)$.

Now let V be the set of all $x \in L(H)$ such that, for every $b \in \Sigma(B)$,

$$(\lambda x + \mu y) \in \Sigma(B) \text{ and } xb \in \Sigma(B).$$

Then V contains B and is sequentially closed in the weak operator topology. So $V = \Sigma(B)$. Since a norm convergent sequence in $L(H)$ is also convergent in the weak operator topology, $\Sigma(B)$ is a C^*-subalgebra of $L(H)$. $\square$

Exercise 5.3.2 Let $\{T_\lambda : \lambda \in \Lambda\}$ be a subset of $L(H)$ such that, for each ξ and η in H,

$$\sup\{|<T_\lambda \xi, \eta>| : \lambda \in \Lambda\} < \infty.$$

Prove that $\{\|T_\lambda\| : \lambda \in \Lambda\}$ is bounded.

Hint: Fix η and use the Uniform Boundedness Theorem to show that $\{\|T_\lambda^* \eta\| : \lambda \in \Lambda\}$ is bounded.

As before, let A be a unital C^*-algebra, with universal representation (π_u, H_u). We identify A with its embedding, $\pi_u[A]$, in $L(H_u)$. Its second dual is identified with $\pi_u[A]''$, its von Neumann envelope in $L(H_u)$. Its state space is the compact convex set K.

Definition 5.3.3 The *Davies-Baire envelope* of A is the smallest subset of $L(H_u)$ which contains A and is sequentially closed in the weak operator topology of $L(H_u)$. We denote it by A^Σ. By Lemma 5.3.1, it is a C^*-algebra. When A''_{sa} is identified with $A''(K)$ then A^Σ_{sa} is identified with $A^\Sigma(K)$; where $A^\Sigma(K)$ is the smallest (real) subspace of $A''(K)$ which (i) contains $A(K)$ and (ii) is closed under pointwise limits of sequences (see Exercise 5.3.2).

Let H be any Hilbert space. Let (T_n) be a norm bounded, monotone increasing sequence in $L(H)_{sa}$. Then this sequence converges in the weak operator topology and also in the strong operator topology. Denote its limit by $\lim T_n$. Then $\lim T_n$ is the least upper bound of the sequence in $L(H)_{sa}$.

We defined the *Pedersen-Baire envelope* of A, A^∞, in Definition 2.2.23. Then A^∞_{sa} is the smallest subspace of $L(H_u)_{sa}$ which contains A_{sa} and is closed under strong limits of bounded monotone increasing sequences. Comparison with Definition 5.2.11, shows that A^∞_{sa} corresponds to $A^\infty(K)$. It is clear that A^∞ is a σ-subspace of $A^\Sigma \subset L(H_u)$. The fact that it is a σ-subalgebra of A^Σ and $L(H_u)$ follows from Theorem 4.5.4 in [121]. Using Pedersen's methods and those of Kadison [85] we shall obtain, in Theorem 5.3.7, a result which, for unital algebras, is more general.

Remark 5.3.4

(i) Let B be a non-unital C^*-algebra and let $A = B^1$. Since B is a (maximal closed) two-sided ideal of A, we can identify B'' with a weakly closed two-sided ideal of A'', where C'' is the universal enveloping von Neumann algebra of a C^*-algebra of C.

Let B^∞_{sa} be the smallest σ-closed subset of A''. Since B'' is σ-closed in A'', it follows that $B^\infty_{sa} \subset B''$. That is, B^∞_{sa} is the smallest of all such in B''_{sa}.

By Pedersen's theorem, $B^\infty = B^\infty_{sa} + iB^\infty_{sa}$ is a C^*-subalgebra of B'' and so of A''. We note here that even when B is non-unital, his argument works to show that B^∞ is a C^*-subalgebra of B''. That is, the Pedersen-Baire envelope of B is obtained by taking the monotone σ-closure in A'' and so we have

$$B \subset B^\infty \subset B'' \subset A''.$$

Hence it follows that

$$B \subset A \subset B^\infty + \mathbb{C}1 \subset A^\infty \subset A''.$$

Since $B^\infty + \mathbb{C}1$ is monotone σ-closed in A'', it follows that $A^\infty = B^\infty + \mathbb{C}1$. Here 1 is the unit element of A'', where $B'' = A''z$ for a central projection z of A''.

(ii) Let A be a (unital) C^*-algebra and let A^∞ be the Pedersen-Baire envelope of A. Let $\{\pi_a, H_a\}$ be the reduced atomic representation of A (see page 103, 4.3.7 [121]). Let $\widetilde{\pi}_a$ be the unique normal $*$-homomorphism, from A'' into $L(H_a)$, which extends π_a to A''. Then $\widetilde{\pi}_a|_{A^\infty}$ is *injective* (see page 114, Corollary 4.5.13 [121]).

(iii) Let $A = C(\Omega)$ be the commutative C^*-algebra of all complex valued continuous functions on a compact Hausdorff space Ω. Then $A^\infty \cong B^\infty(\Omega)$, where $B^\infty(\Omega)$ is the monotone σ-complete C^*-algebra of all bounded Baire measurable functions on Ω. To see this, let π_a be the reduced atomic representation of A. Note that $\partial K = \Omega$, $\pi_a(x) = \{x(\omega)\}_{\omega \in \Omega}$ ($x \in A$) and $H_a = \ell^2(\Omega)$. We have a unique normal extension $\widetilde{\pi}_a$ of π_a, as a unital σ-normal $*$-homomorphism $\widetilde{\pi}_a : A^{**} \mapsto Bnd(\Omega) \subset L(\ell^2(\Omega))$. Moreover, we have $\widetilde{\pi}_a(A^\infty) = B^\infty(\Omega)$. Since $\widetilde{\pi}_a|_{A^\infty}$ is injective, it follows that $A^\infty \cong B^\infty(\Omega)$ via the map $\widetilde{\pi}_a|_{A^\infty}$.

5.3.1   Open Problem

It is clear that the Pedersen-Baire envelope of A is contained in the Davies-Baire envelope of A, that is

$$A^\infty \subset A^\Sigma.$$

Do they coincide? Or is there an A where the Davies-Baire envelope is strictly larger than the Pedersen-Baire envelope? This problem has been open for nearly 50 years. Pedersen succeeded in showing that the answer is positive for some classes of A, see [121]. But, in general, this is a mystery. We shall refer to this question later.

Many of the results we prove using A^∞ could easily be extended to results using A^Σ. But there seems little point in doing so, unless we can find A where $A^\infty \neq A^\Sigma$.

5.3.2   Monotone Envelopes and Homomorphisms

We recall the following. Let C be a self-adjoint (complex) subspace of a C^*-algebra. Then $C = C_{sa} + iC_{sa}$ where C_{sa} is the real vector space of self-adjoint elements of C. This uses the elementary observation $z = \frac{1}{2}(z + z^*) + i\frac{1}{2i}(z - z^*)$.

Definition 5.3.5 Let B be a monotone complete C^*-algebra. Let A be a unital C^*-subalgebra of B. Let M be the smallest monotone closed subset of B_{sa} containing A_{sa}. We recall from Sect. 2.1, that M is monotone closed if it satisfies the following: whenever (x_α) is an increasing net of elements of M with $x_\alpha \uparrow x$ in B_{sa} for some $x \in B_{sa}$, $x \in M$ and whenever (y_γ) is a decreasing net of elements in M with $y_\gamma \downarrow y$ in B_{sa} for some $y \in B_{sa}$, $y \in M$. Then, we shall call $M + iM$ the *monotone closure of A in B* and we denote it by $M(A)$. If $M(A) = B$, then we say that B is *monotone generated by A* or A monotone generates B. (Arguing as in Exercise 2.1.29, we find that M is a (real) vector subspace of B_{sa}.)

Definition 5.3.6 Let C be a monotone σ-complete C^*-algebra and let B be a unital C^*-subalgebra of C. Let M_σ be the smallest monotone σ-closed subset of C containing A_{sa}. As in Sect. 2.1, M_σ is monotone σ-closed if it satisfies the following: whenever (x_n) is an increasing sequence of elements of M_σ with $x_n \uparrow x$ in C_{sa} for some $x \in C_{sa}$, then $x \in M_\sigma$ and whenever (y_n) is a decreasing sequence of elements of M_σ with $y_n \downarrow y$ in C_{sa} for some $y \in C_{sa}$, then $y \in M_\sigma$. Then, we call $M_\sigma + iM_\sigma$ the monotone σ-closure of A and denote it by $M_\sigma(A)$. If $M_\sigma(A) = C$ then C is said to be *(monotone) σ-generated* by A. Equivalently, A is said to σ-generate C. (By Exercise 2.1.29, we find that M_σ is a (real) vector subspace of B_{sa}.)

The following theorem (and its sequential analogue) is a useful tool which the reader should know about. But its proof may be postponed to a later reading. Although the proof may seem intricate at first sight, it is based on repeated applications of the same basic strategy used in Exercise 2.1.29.

Let us recall that a function Φ from one C^*-algebra into another is a *-*map* if $\Phi(z^*) = \Phi(z)^*$for all z. Clearly a *-map takes self-adjoint elements to self-adjoint elements.

Theorem 5.3.7 *Let B_1 and B_2 be monotone complete C^*-algebras and let A be a unital C^*-subalgebra of B_1. Let $\Phi : B_1 \mapsto B_2$ be a* unital, normal, positive linear *-map. Suppose that $\Phi|_A$ is a *-homomorphism from A into B_2. Then, $M(A)$ is a monotone closed unital C^*-subalgebra of B_1 and the restricted map $\Phi|_{M(A)}$ is a normal, unital *-homomorphism from the monotone complete C^*-algebra $M(A)$ into B_2.*

Proof See the proof of Theorem 4.5.4 Page 110 in [121] and see also [85]. Because of our more complicated situation, we give a detailed argument.

Arguments used in Exercises 2.1.28 and 2.1.29 can also be used to show that M is the smallest monotone closed subspace of $(B_1)_{sa}$ which contains A_{sa}. Hence $M(A)$ is the smallest monotone closed *-subspace of B_1 that contains A. Since Φ is a positive linear unital map from B_1 to B_2 it is bounded in norm on B_1 and hence on M.

The 1st step: We wish to show that $M(A)$ is norm closed in B_1. Let (x_n) be a sequence in M such that $\|x_n - x\| \longrightarrow 0$ for an element $x \in (B_1)_{sa}$. By taking a subsequence, we may assume that $\|x_n - x\| \leq 1/2^{n+2}$ $(n = 1, 2, \cdots)$. So $\|x_{n+1} - x_n\| \leq 1/2^{n+1}$. From this it follows that $(x_n - 2^{-n}1)$ is a monotone increasing sequence in M which converges in norm to x. So by Lemma 2.1.7, $(x_n - 2^{-n}1) \uparrow x$ in $(B_1)_{sa}$. Since M is monotone closed, it follows that $x \in M$. So $M(A)$ is norm closed.

The 2nd step: Take any $x \in M$. We shall show that $x^n \in M$ and $\Phi(x^n) = \Phi(x)^n$ for all positive integers n. Let W_1 be the set

$$\{a \in M : a^n \in M \text{ and } \Phi(a^n) = \Phi(a)^n \text{ for all positive integers } n\}.$$

Clearly W_1 contains A_{sa}. We observe that, for any real number t, and any $a \in W_1$ we have $ta \in W_1$. This follows from M being a vector space and the linearity of $\Phi|_M$.

We shall show that W_1 is monotone closed. It will then follow that $W_1 = M$. Since $a \in W_1$ if, and only if, $-a \in W_1$, we need only consider *increasing* nets. Also, see the remarks following Corollary 2.1.23, we only need to consider nets which are bounded in norm. Take any norm bounded increasing net (x_γ) in W_1. Then $x_\gamma \uparrow x$ in $(B_1)_{sa}$ for some $x \in (B_1)_{sa}$. Since M is monotone closed, $x \in M$. We wish to show that $x \in W_1$.

Since W_1 is stable under multiplication by positive scalars we only need to consider nets where $\|x_\gamma\| \leq 1$. So, we also have that $-1 \leq x \leq 1$, that is, $\|x\| \leq 1$. For every $t \in \mathbb{R}$ with $0 \leq t < 1$, using norm convergence, we have the identity

$$(1 - tx_\gamma)^{-1} = 1 + \sum_{k \geq 1} t^k (x_\gamma)^k.$$

Since each $t^k(x_\gamma)^k$ is in M and M is a norm closed subspace of a Banach space, $(1 - tx_\gamma)^{-1} \in M$ for all γ and t with $0 \leq t < 1$.

Let us recall [121] that, in any C^*-algebra, if $0 \leq c \leq d$ and c is invertible, then d is also invertible and $d^{-1} \leq c^{-1}$.

Let (a_λ) be a norm bounded decreasing net of positive elements in $(B_1)_{sa}$ such that $a_\lambda \downarrow a$ for some $a \in (B_1)_{sa}$. Suppose that a is invertible. Then each a_λ is invertible and (a_λ^{-1}) is a norm bounded increasing net of positive elements such that $a_\lambda^{-1} \uparrow a^{-1}$.

So $(1 - tx_\gamma)^{-1} \uparrow (1 - tx)^{-1}$ in $(B_1)_{sa}$. Since M is monotone closed, it follows that $(1 - tx)^{-1} \in M$ for all t with $0 \leq t < 1$. Since Φ is a *positive* linear map, we have $\|\Phi(x_\gamma)\| \leq 1$ and $\|\Phi(x)\| \leq 1$ for each γ. Since Φ is a bounded linear operator

$$\Phi\left((1 - tx_\gamma)^{-1}\right) = 1 + \sum_{k \geq 1} t^k \Phi((x_\gamma)^k).$$

Because each $x_\gamma \in W_1$ we have

$$\Phi\left((1 - tx_\gamma)^{-1}\right) = 1 + \sum_{k \geq 1} t^k (\Phi(x_\gamma))^k = (1 - t\Phi(x_\gamma))^{-1}.$$

Since Φ is normal, we have $\Phi(x_\gamma) \uparrow \Phi(x)$ in $(B_2)_{sa}$. Arguing as above $(1 - t\Phi(x_\gamma))^{-1} \uparrow (1 - t\Phi(x))^{-1}$ in $(B_2)_{sa}$.

Because Φ is normal and $(1 - tx_\gamma)^{-1} \uparrow (1 - tx)^{-1}$ in $(B_1)_{sa}$, it follows that $\Phi\left((1 - tx_\gamma)^{-1}\right) \uparrow \Phi\left((1 - tx)^{-1}\right)$ in $(B_2)_{sa}$. So, we have

$$(1 - t\Phi(x))^{-1} = \Phi\left((1 - tx)^{-1}\right) \text{ for all } t \in \mathbb{R} \text{ with } 0 \leq t < 1.$$

The fact that $\frac{1}{t^2}((1 - tx)^{-1} - (1 + tx)) \in M$ and

$$\left\| \frac{1}{t^2}((1 - tx)^{-1} - (1 + tx)) - x^2 \right\| \longrightarrow 0 \ (t \to +0)$$

tells us that $x^2 \in M$. Since Φ is linear and norm continuous, it follows that

$$\Phi\left(\frac{1}{t^2}\left((1 - tx)^{-1} - (1 + tx)\right)\right) = \frac{1}{t^2}\left[(1 - t\Phi(x))^{-1} - (1 + t\Phi(x))\right].$$

By letting $t \to 0$ through positive values we find $\Phi(x^2) = (\Phi(x))^2$.

We observe that, for each natural number k,

$$\left\| \frac{1}{t^k}\left[(1 - tx)^{-1} - (1 + tx + \cdots + (tx)^{k-1})\right] - x^k \right\| \longrightarrow 0 \ (t \to +0).$$

By induction on k we find $x^k \in M$ and $\Phi(x^k) = \Phi(x)^k$ for each k. So $x \in W_1$. Thus W_1 is monotone closed. This implies that $W_1 = (B_1)_{sa}$. Hence we have $x^n \in M$ and $\Phi(x^n) = (\Phi(x))^n$ for all $n \in \mathbb{N}$ and for every $x \in M$.

The 3rd step: We shall show that $M(A)$ is a $*$-subalgebra and Φ is a multiplicative map.

To do this, first of all, note the following consequences of the 2nd Step.

Take any x and y in $(B_1)_{sa}$. Let us recall that the Jordan product $x \circ y$ is defined to be $\frac{1}{2}(xy + yx)$.

We have the following algebraic identities

$$2(x \circ y) = xy + yx = (x + y)^2 - x^2 - y^2 \tag{#}$$

$$2xyx = (xy + yx)x + x(xy + yx) - (yx^2 + x^2y) \tag{##}$$

$$= 4(x \circ y) \circ x - 2y \circ x^2$$

$$(x + i1)^* y(x + i1) = xyx + i(xy - yx) + y. \tag{###}$$

Hence for every pair x and y in M, we have, by the 2nd step, $xy + yx = (x + y)^2 - x^2 - y^2 \in M$ and

$$\Phi(xy + yx) = \Phi((x + y)^2) - \Phi(x^2) - \Phi(y^2)$$

$$= (\Phi(x) + \Phi(y))^2 - \Phi(x)^2 - \Phi(y)^2$$

$$= \Phi(x)\Phi(y) + \Phi(y)\Phi(x).$$

So M is closed under the Jordan product and, for each x and y in M

$$\Phi(x \circ y) = \Phi(x) \circ \Phi(y).$$

Using this in (##) we see that xyx is in M and

$$\Phi(xyx) = 2(\Phi(x) \circ \Phi(y)) \circ \Phi(x) - \Phi(y) \circ \Phi(x)^2$$

$$= \Phi(x)\Phi(y)\Phi(x).$$

We shall show that, for every pair x and y in M,

$$i(xy - yx) \in M \text{ and } i(\Phi(x)\Phi(y) - \Phi(y)\Phi(x)) = \Phi(i(xy - yx)).$$

To do this, first fix $x \in A_{sa}$. Now let W_2 be the set of all $y \in M$ such that $(x + i1)^* y(x + i1) \in M$ and

$$\Phi((x + i1)^* y(x + i1)) = (\Phi(x) + i1)^* \Phi(y)(\Phi(x) + i1).$$

Clearly A_{sa} is a subset of W_2. Since the maps $y \longmapsto z^*yz$ and Φ are normal and order preserving for any $z \in B_1$, it follows that W_2 is monotone closed in B_1. Hence $W_2 = M$.

So, for all $x \in A_{sa}$ and $y \in M$ we have $(x + i1)^*y(x + i1) \in M$ and

$$\Phi((x + i1)^*y(x + i1)) = (\Phi(x) + i1)^*\Phi(y)(\Phi(x) + i1).$$

So, by (###) $i(xy - yx) \in M$. By also applying (###) in $\Phi[M]$ and M we find

$$\Phi(i(xy - yx)) = i(\Phi(x)\Phi(y) - \Phi(y)\Phi(x)).$$

By multiplying this equation by -1, then interchanging x and y, we find that, for all $x \in M$ and all $y \in A_{sa}$, $i(xy - yx) \in M$ and

$$i\Phi(xy - yx) = i(\Phi(x)\Phi(y) - \Phi(y)\Phi(x)).$$

Now let W_3 be the set of all $y \in M$ such that, for every $x \in M$, $(x+i1)^*y(x+i1) \in M$ and

$$\Phi((x + i1)^*y(x + i1)) = (\Phi(x) + i1)^*\Phi(y)(\Phi(x) + i1).$$

From the preceding paragraph, $A_{sa} \subset W_3$. Arguing as in the W_2 case above, W_3 is monotone closed. So $W_3 = M$. Using (###) again, we have $i(xy - yx) \in M$ and

$$i\Phi(xy - yx) = i(\Phi(x)\Phi(y) - \Phi(y)\Phi(x))$$

for all x and y in M. We already have $xy + yx \in M$ and $\Phi(xy + yx) = \Phi(x)\Phi(y) + \Phi(y)\Phi(x)$. Since Φ is linear on $M(A)$ it follows that $xy \in M(A)$ and

$$\Phi(xy) = \Phi(x)\Phi(y).$$

From this it follows that Φ is a $*$-homomorphism from $M(A)$ to B_2. $\square$

By a minor modification of the above proof we can obtain the following "sequential" analogue of the above theorem.

Corollary 5.3.8 *Let B_1 and B_2 be unital, monotone σ-complete C^*-algebras. Let A be a unital C^*-subalgebra of B_1. Let M_σ be the smallest monotone σ-closed subset containing A_{sa}. Then $M_\sigma(A) = M_\sigma + iM_\sigma$ is a unital monotone σ-closed $*$-subalgebra of B_1. Let $\Phi : B_1 \mapsto B_2$ be a unital, σ-normal, positive linear $*$-map such that $\Phi|_A$ is a $*$-homomorphism from A into B_2. Then Φ is a $*$-homomorphism from $M_\sigma(A)$ into B_2.*

Fix a Hilbert space H and let $B_1 = B_2 = L(H)$. Let Φ be the identity map from B_1 to B_2. Then Corollary 5.3.8 gives Pedersen's theorem for A unital, see [121].

By a more careful argument, the validity of Pedersen's result can be established for non-unital algebras; see [121].

5.3.3 Borel Envelopes

We have discussed how A_{sa} can be identified with $A(K)$ and A_{sa}^{**} with $A''(K)$, the bounded affine functions on K. We saw that A_{sa}^{∞}, the self adjoint part of the Pedersen-Baire envelope of A, can be identified with $A^{\infty}(K)$. We also observed that when $A = C(\Omega)$ then A^{∞} can be identified with the algebra of bounded Baire measurable functions on Ω. We seek a non-commutative generalisation of the algebra of bounded Borel measurable functions on Ω. It is easy to identify a plausible candidate but it has been very hard to show that it is a C^*-algebra. This has only recently been accomplished by Brown [19]. We will expound his proof here.

First we define A_{sa}^{m} to be the set of those elements of A''_{sa} which are limits, in the weak operator topology, of increasing nets in A_{sa}. See 3.11.6 in [121] and page 157 of [161]. Then each element of A_{sa}^{m} corresponds to the pointwise limit of an increasing net in A_{sa}. Hence it corresponds to a bounded lower semicontinuous affine function on K. Conversely, by Lemma 5.2.3, each bounded lower semicontinuous affine function on K corresponds to an element of A_{sa}^{m}.

Let $V = A_{sa}^{m} - A_{sa}^{m}$ and let V_0 be the norm closure of V. Then, by page 166, Proposition 6.14 of [161], V_0 is a norm closed Jordan subalgebra of A_{sa}. Let A_{sa}^{b} be the smallest subset of A''_{sa} that contains $V = A_{sa}^{m} - A_{sa}^{m}$ and is a monotone σ-closed subset of A''_{sa}.

We now define the Brown-Borel envelope of A.

Definition 5.3.9 Let $A^{b} = A_{sa}^{b} + iA_{sa}^{b}$ where A_{sa}^{b} is the smallest subset of A''_{sa} that contains $V = A_{sa}^{m} - A_{sa}^{m}$ and is a monotone σ-closed subset of A''_{sa}. Then A^{b} is the Brown-Borel envelope of A. When we identify $A''(K)$ with A''_{sa} we find that A_{sa}^{b} corresponds to $A^{b}(K)$, see Definition 5.2.12.

Lemma 5.3.10 The set A_{sa}^{b} is a norm closed subspace of A''_{sa}.

Proof Since V is a real vector space, it follows from Exercise 2.1.28 that A_{sa}^{b} is a monotone σ-closed subspace of A''_{sa}. By applying the argument used in Step 1 of the proof of Theorem 5.3.7 we find that A_{sa}^{b} is a norm closed subspace of A''_{sa}. □

Lemma 5.3.11 The set A_{sa}^{b} is a norm closed Jordan subalgebra of A''_{sa}.

Proof By Lemma 5.3.10, A_{sa}^{b} contains the norm closure of V. This is V_0 which is a norm closed Jordan subalgebra of A''_{sa}. We shall show $x^n \in A_{sa}^{b}$ for all $x \in A_{sa}^{b}$. Let $W = \{x \in A_{sa}^{b} : x^n \in A_{sa}^{b} \text{ for all } n \in \mathbb{N}\}$. Since the norm closure of V is a Jordan subalgebra, $V_0 \subset W$. By imitating the argument used in the 2nd Step of the proof of Theorem 5.3.7, W is a monotone σ-closed subspace of A_{sa}^{b} that contains V.

Hence $W = A_{sa}^b$, which implies that $x^2 \in A_{sa}^b$ for all $x \in A_{sa}^b$ and so A_{sa}^b is a Jordan subalgebra of A''_{sa}.   □

Corollary 5.3.12 *The vector space A^b is a norm closed Jordan $*$-subalgebra of A''.*

Lemma 5.3.13 (Brown [19]) *A^b is a two-sided A-submodule of A''.*

Proof Take $t \in A$ and consider the map $\theta_t : A'' \mapsto A''$ defined by $\theta_t(s) = t^*st$ for $s \in A^{**}$. Clearly θ_t is a normal positive linear map and satisfies $\theta_t(A_{sa}^m) \subset A_{sa}^m \subset A_{sa}^b$, which means that $\theta_t(V) \subset V$. So, $\theta_t(A_{sa}^b) \subset A_{sa}^b$. Take any $s, t \in A$ and $x \in A''_{sa}$. Since

$$4t^*xs = \sum_{k=0}^3 i^k (s + i^k t)^* x (s + i^k t)$$

if $x \in A_{sa}^b$, then it follows that $t^*xs \in A^b$. Hence A^b is a two-sided A-submodule of A''.   □

Let us observe that under the canonical embedding of A into A'', the universal enveloping von Neumann algebra (page 122 of [161]) $M_2(A)''$ of the C^*-algebra $M_2(A)$, is $*$-isomorphic to $M_2(A'')$ in a canonical way. Indeed, note that $M_2(A) \cong A \otimes M_2(\mathbb{C})$ (see Sect. 2.4) via a canonical isomorphism Π and $A'' \otimes M_2(\mathbb{C})$ is the universal enveloping von Neumann algebra of $A \otimes M_2(\mathbb{C})$ (see page 203–204 of [161]). So, $A'' \otimes M_2(\mathbb{C})$, which is $*$-isomorphic to $M_2(A'')$, is also $*$-isomorphic to $M_2(A)''$ in a canonical way. Hence we may assume that $M_2(A)'' = M_2(A'')$. Let us consider the map $\phi : A'' \ni x \mapsto \begin{pmatrix} x & 0 \\ 0 & 0 \end{pmatrix} \in M_2(A'')$ and $\psi : M_2(A'') \ni \begin{pmatrix} a_{11} & a_{12} \\ a_{21} & a_{22} \end{pmatrix} \mapsto a_{11} \in A''$. Then ϕ and ψ are norm continuous, normal positive linear maps and so we have

$$\phi(A_{sa}^m) \subset (M_2(A))_{sa}^m \text{ and } \psi((M_2(A))_{sa}^m) \subset A_{sa}^m.$$

Lemma 5.3.14 (Brown [19]) *Take any $x \in A''$. Then $x \in A_{sa}^b$ if and only if $\begin{pmatrix} x & 0 \\ 0 & 0 \end{pmatrix} \in M_2(A)_{sa}^b$. Also, for any $y \in M_2(A'')(\equiv M_2(A)'')$, $y \in M_2(A)^b$ if, and only if, on putting $y = \begin{pmatrix} y_{11} & y_{12} \\ y_{21} & y_{22} \end{pmatrix}$, $y_{ij} \in A^b$ for every pair i and j with $1 \leq i, j \leq 2$.*

Proof Let W_1 be the set of all $y \in M_2(A'')_{sa}$ such that $\psi(y) \in A_{sa}^b$. Since ψ is normal and positive, W_1 is monotone σ-closed and contains $M(A)_{sa}^m$. Hence we have $W_1 \supset M_2(A)_{sa}^b$. Let W_2 be the set of all $x \in A''$ such that $\phi(x) \in M_2(A)_{sa}^b$. Since ϕ is normal and positive, by a similar argument, W_2 is monotone σ-closed and contains A_{sa}^m. Hence $W_2 \supset A_{sa}^b$. The first part of the statement follows.

Since $\sum_{1 \leq i,j \leq 2} \mathbf{e}_{i1} \begin{pmatrix} x_{ij} & 0 \\ 0 & 0 \end{pmatrix} \mathbf{e}_{1j} = x$ and $\mathbf{e}_{1i} x \mathbf{e}_{j1} = \begin{pmatrix} x_{ij} & 0 \\ 0 & 0 \end{pmatrix}$ for each i, j and x, the first statement and Lemma 5.3.13 can be applied to show Lemma 5.3.14.   □

Theorem 5.3.15 (Brown [19]) A^b is a C^*-subalgebra of A'' whose self-adjoint part is A^b_{sa}.

Proof Take any $x \in A^b_{sa} + iA^b_{sa}$. Then by Lemma 5.3.14, $y = \begin{pmatrix} 0 & x^* \\ x & 0 \end{pmatrix} \in M_2(A)^b_{sa}$. Since $M_2(A)^b$ is a Jordan subalgebra of $M_2(A'')_{sa}$, it follows that $y^2 \in M_2(A)^b_{sa}$, which implies that

$$M_2(A)^b_{sa} \ni y^2 = \begin{pmatrix} 0 & x^* \\ x & 0 \end{pmatrix} \begin{pmatrix} 0 & x^* \\ x & 0 \end{pmatrix} = \begin{pmatrix} x^*x & 0 \\ 0 & xx^* \end{pmatrix}.$$

Hence it follows that $x^*x \in A^b_{sa}$. Now take any pair x and y in A^b. Since $(x+y^*)^*(x+y^*) = x^*x + yy^* + (yx + x^*y^*) \in A^b_{sa}$, it follows that $yx + x^*y^* \in A^b_{sa}$. On the other hand, also $(x+iy^*)^*(x+iy^*) = x^*x + yy^* - i(yx - x^*y^*) \in A^b_{sa}$, which implies that $yx \in A^b$. Hence A^b is a C^*-subalgebra of A''.

So the Brown-Borel envelope of A is a C^*-algebra which is monotone σ-closed in A'' and contains A^m_{sa}. Its self-adjoint part is A^b_{sa}. So $(A^b)_{sa} = A^b_{sa}$. We re-iterate that when we make the usual identifications of A''_{sa} with $A''(K)$ then

$$A^b_{sa} = A^b(K). \qquad \square$$

5.4 Representation Theorems

Let C be a monotone σ-complete algebra. Let us recall, see just before Example 2.1.3, that a *σ-compatible topology* is a Hausdorff locally convex vector topology such that each bounded monotone increasing sequence converges to its least upper bound in C. It would be very convenient if there always existed a σ-compatible topology for C. But this is not true in general. However the representation theorems provide a substitute: The Pedersen-Baire envelope of C has a σ-compatible topology and C is the quotient of its Pedersen-Baire envelope by a σ-ideal. There exists a σ-homomorphism h from C^∞ onto C such that h is idempotent. Also, when (c_n) is a monotone increasing sequence in C whose least upper bound in C is c, then

$$h(\lim c_n) = c.$$

Here $\lim c_n$ is the limit of the sequence in C^∞, with respect to the topology induced on C^∞ by the strong operator topology of C''. Since the sequence is monotonic, its limit in the weak operator topology of C'' is the same.

When C is monotone complete we can say more: There is an idempotent, σ-homomorphism h from its Brown-Borel envelope onto C with the following property. Let (c_λ) be an increasing net in C with least upper bound c in C. As remarked in Chap. 2, we may suppose that this net is bounded below and, hence bounded in norm. Then, using the strong operator topology, $\lim c_\lambda$ exists in C^b and

$$h(\lim c_\lambda) = c.$$

In Sect. 5.2 we obtained representations of monotone (σ-) complete order unit spaces as quotients of their Baire (Borel) envelopes. By using "meagre ideals", we get corresponding representation theorems for monotone complete, and σ-complete C^*-algebras.

We revert to our earlier notation where A is a unital C^*-algebra with state space K. As before, we regard A as embedded in $L(H_u)$. We let A'' be the double commutant of A which is also identified with the second dual of A. We also identify A_{sa} with $A(K)$ and A''_{sa} with $A''(K)$, the vector space of all bounded, real-valued, affine functions on K.

Lemma 5.4.1 *Let ∂K be equipped with the relative topology induced by K. Let $\mathcal{M}$ be the set of all $z \in A''$ such that $\{\varphi \in \partial K : \varphi(z) \neq 0\}$ is a meagre subset of ∂K. Then $\mathcal{M}$ is a $*$-subspace of A'' which is sequentially closed in the weak operator topology of $L(H_u)$. For any unitary in A, $u\mathcal{M}u^* = \mathcal{M}$. Also $\mathcal{M} \cap A = \{0\}$.*

Proof To show that $\mathcal{M}$ is a $*$-subset of A'', we first note that $\varphi(z^*) = \overline{\varphi(z)}$ for all $\varphi \in K$ and $z \in A''$. Then, for each $\varphi \in \partial K$, $\varphi(z) \neq 0$ if, and only if, $\varphi(z^*) \neq 0$. So $z \in \mathcal{M}$ if, and only if, $z^* \in \mathcal{M}$. For any pair x and y in A'', we have

$$\{\varphi \in \partial K : \varphi(x + y) \neq 0\} \subset \{\varphi \in \partial K : \varphi(x) \neq 0\} \cup \{\varphi \in \partial K : \varphi(y) \neq 0\}.$$

So it is clear that $\mathcal{M}$ is a subspace of A''.

Let (z_n) be a sequence in $\mathcal{M}$ which converges in the weak operator topology to z. Then for each $\phi \in K$

$$\phi(z_n) \to \phi(z).$$

We have

$$\{\varphi \in \partial K : \varphi(z) \neq 0\} \subset \bigcup_{n=1}^{\infty} \{\varphi \in \partial K : \varphi(z_n) \neq 0\}.$$

Because the union of countably many meagre sets is meagre, it follows that $z \in \mathcal{M}$.

For each unitary $u \in A$ and any $\varphi \in K$, define a map from K into K by

$$T_u\varphi(x) = \varphi(uxu^*) \text{ for } x \in A.$$

Then $T_{u^{-1}}T_u$ is the identity operator because

$$(T_{u^*}T_u\phi)(x) = T_u\phi(u^*xu) = \phi(uu^*xuu^*) = \phi(x).$$

Interchanging u and u^* we see that $T_uT_{u^*}$ is also the identity. So T_u is a bijection of K onto K.

Let (φ_γ) be any net in K such that $\varphi_\gamma \longrightarrow \varphi$ with respect to the $\sigma(A^*, A)$-topology of K. Then

$$\varphi_\gamma(uxu^*) \longrightarrow \varphi(uxu^*) \text{ for each } x \in A.$$

Hence $T_u\varphi_\gamma \longrightarrow T_u\varphi$ with respect to $\sigma(A^*, A)$. So T_u is continuous. Since $(T_u)^{-1} = T_{u^*}$ it follows that T_u is a homeomorphism of K onto itself.

It is straightforward to check that T_u is an affine map of K onto K. From this it is easy to see that T_u is a homeomorphism of ∂K onto ∂K. In particular T_u maps meagre subsets of ∂K onto meagre subsets of ∂K.

For any $x \in A''$ there is, by the density theorem of Kaplansky, a bounded net (a_λ) in A which converges to x in the weak operator topology. So for any unitary u and any $\phi \in \partial K$,

$$T_u\phi(x) = \lim T_u\phi(a_\lambda) = \lim \phi(ua_\lambda u^*) = \phi(uxu^*).$$

It follows that $uxu^* \in \mathcal{M}$ if, and only if, $x \in \mathcal{M}$.

Finally, let $z \in \mathcal{M} \cap A$. Let $z = a + ib$ where a and b are self-adjoint. Then $a = \frac{1}{2}(z + z^*)$ is in $\mathcal{M} \cap A$. On identifying a with an affine continuous function on K, the set

$$\{\phi \in \partial K : a(\phi) \neq 0\}$$

is both open and meagre. Since ∂K is a Baire space, this set must be empty. So $a = 0$. A similar argument applies to b. So $z = 0$. □

Lemma 5.4.2 *Let* $\mathcal{M}^+ = \{z \in \mathcal{M} : z \geq 0\}$ *and let*

$$\mathcal{N} = \{z \in A'' : zz^* \in \mathcal{M}^+ \text{ and } z^*z \in \mathcal{M}^+\}.$$

Then $\mathcal{N}$ *is a* C^*-*algebra which is sequentially closed in the weak operator topology of* $L(H_u)$ *and* $\mathcal{N}^+ = \mathcal{M}^+$. *Furthermore, for any* $a \in A$, $a\mathcal{N} \subset \mathcal{N}$ *and* $\mathcal{N}a \subset \mathcal{N}$. *Also* $A \cap \mathcal{N} = \{0\}$.

Proof Since $\mathcal{M}^+$ is a norm closed hereditary subcone of $(A'')^+$, the set $L = \{z \in A'' : z^*z \in \mathcal{M}^+\}$ is a norm closed left ideal of A''. Let $\mathcal{N} = L \cap L^*$. Then $\mathcal{N}$ is a C^*-subalgebra of A''. (See page 15, Theorem 1.5.2 [121].) If $a \in \mathcal{N}$ and $a \geq 0$, then $a^{\frac{1}{2}} \in \mathcal{N}^+ \subset L$ and so $a \in \mathcal{M}^+$. Conversely if $b \in \mathcal{M}^+$, then $b^{\frac{1}{2}} \in L$ and $b^{\frac{1}{2}} \in L^*$, which implies $b \in \mathcal{N}$. Hence it follows that $\mathcal{N}^+ = \mathcal{M}^+$. Since $\mathcal{N}$ is the complex linear span of $\mathcal{N}^+$, $\mathcal{N} \subset \mathcal{M}$. So, by Lemma 5.4.1 $\mathcal{N} \cap A = \{0\}$.

Since $\mathcal{M}$ is sequentially closed in the weak operator topology, $\Sigma(\mathcal{N}) \subset \mathcal{M}$. By Lemma 5.3.1, $\Sigma(\mathcal{N})$ is a C^*-algebra. Then

$$\mathcal{N}^+ \subset \Sigma(\mathcal{N})^+ \subset \mathcal{M}^+.$$

It follows that $\Sigma(\mathcal{N}) = \mathcal{N}$. So $\mathcal{N}$ is sequentially closed in the weak operator topology of $L(H_u)$.

For any unitary $u \in A$ and $z \in \mathcal{N}$, $(zu)(zu)^* = zz^* \in \mathcal{M}^+$. Also

$$(zu)^*(zu) = u^*z^*zu \in u^*\mathcal{M}^+u.$$

By Lemma 5.4.1, $u^*\mathcal{M}^+u = \mathcal{M}^+$. So $zu \in \mathcal{N}$. Similarly $uz \in \mathcal{N}$. Since every element of A is the linear combination of at most four unitaries, for each $a \in A$ and each $z \in \mathcal{N}$ we have $za \in \mathcal{N}$ and $az \in \mathcal{N}$.   □

Lemma 5.4.3 *Let $W = \mathcal{N} + A$. Then W is a C^*-subalgebra of A'' such that $\mathcal{N}$ is a closed two-sided ideal of W. Let k be the natural projection of W onto A. Then k is a $*$-homomorphism whose kernel is the ideal $\mathcal{N}$ (of W). Let (c_λ) be a norm-bounded, upward directed net in A_{sa}. Let $\lim c_\lambda$ be the limit of this net in A'', with respect to the weak operator topology of $L(H_u)$. If (c_λ) has a least upper bound, c, in A, then,*

$$(c - \lim c_\lambda) \in \mathcal{N}^+$$

and $\lim c_\lambda \in W$. Also

$$k(\lim c_\lambda) = c.$$

Proof Let W be the C^*-subalgebra of A'' obtained by taking the norm closure of $A + \mathcal{N}$. By Lemma 5.4.2, $\mathcal{N}$ is a closed two-sided ideal of W and A is a C^*-subalgebra of W. By Corollary 1.5.8 in [121], $A + \mathcal{N}$ is a C^*-subalgebra of W. So $W = A + \mathcal{N}$. Clearly $k : W \mapsto A$ is a $*$-homomorphism with kernel $\mathcal{N}$.

By Corollary 5.2.10, $c - \lim c_\lambda \in \mathcal{M}^+$. But $\mathcal{M}^+ = \mathcal{N}^+$.

Also $\lim c_\lambda = c - (c - \lim c_\lambda)$. So $\lim c_\lambda \in W$. Since k has kernel $\mathcal{N}$, $k(\lim c_\lambda) = k(c) = c$.   □

Corollary 5.4.4 *Let A be monotone σ-complete. Then W is a σ-closed subalgebra of A'' and k is a σ-homomorphism.*

Proof Let (w_n) be any norm bounded increasing sequence in W_{sa}. Then $w_n = c_n + d_n$, where $c_n \in A_{sa}$ and $d_n \in \mathcal{N}$ for each n.

We have $k(w_n) = c_n$. So (c_n) is monotone increasing and norm bounded. So it has a least upper bound c in A. By Lemma 5.4.3, $\lim c_n$ is in W and $k(\lim c_n) = c$. Since (w_n) is monotone increasing and norm bounded, it converges in the weak operator topology to $\lim w_n \in A''$. Since

$$d_n = w_n - c_n$$

it follows that (d_n) is a sequence in $\mathcal{N}$ which converges in the weak operator topology to $\lim w_n - \lim c_n$. Since $\mathcal{N}$ is sequentially closed in the weak operator topology and $\lim c_n$ is in W, it follows that $\lim w_n$ is in W. So W is σ-closed in A''. Also, by Lemma 5.4.3,

$$k(\lim w_n) = k(\lim c_n) = c = \bigvee_{n \geq 1} c_n = \bigvee_{n \geq 1} k(w_n). \qquad \square$$

We now come to the representation theorem for monotone σ-complete C^*-algebras.

Theorem 5.4.5 *Let A be monotone σ-complete. There exists a σ-homomorphism q from A^∞ onto A such that $q(a) = a$ for each $a \in A$. Then $A^\infty \cap \mathcal{N}$ is a σ-ideal of A^∞ and is the kernel of q. So A is isomorphic to $A^\infty / (A^\infty \cap \mathcal{N})$.*

Proof The smallest σ-closed subspace of A''_{sa} which contains A_{sa} is A^∞_{sa}. So $A^\infty \subset W$. Let q be the restriction of k to A^∞. The result follows from Corollary 5.4.4. $\quad\square$

We recall that the algebra $\mathcal{N}$ is the complex linear span of $\mathcal{M}^+$. We shall see from the results of Sect. 5.6, that in Theorem 5.6.5, we may replace $A^\infty \cap \mathcal{N}$ by $A^\infty \cap \mathcal{M}$.

When specialised to *commutative* algebras, Theorem 5.4.5 corresponds to the Loomis-Sikorski theorem for Boolean σ-algebras [153].

By applying results of Birkhoff-Ulam for complete Boolean algebras, see Theorem 4.1.3, every commutative monotone complete C^*-algebra can be represented as follows. Let S be the spectrum of a commutative monotone complete C^*-algebra then $C(S)$ is isomorphic to the quotient of the algebra of bounded Borel measurable functions on S modulo the ideal of meagre Borel functions. This may be thought of as a special case, for commutative algebras, of the following representation theorem. See Theorem 4.2.9.

Theorem 5.4.6 *Let A be monotone complete. There exists a σ-homomorphism q from A^b onto A such that $q(a) = a$ for each $a \in A$. Then $A^b \cap \mathcal{N}$ is a σ-ideal of A^b and is the kernel of q. So A is isomorphic to $A^b / (A^b \cap \mathcal{N})$. Let (c_λ) be a norm bounded increasing net in A_{sa} with least upper bound c in A_{sa}. Let $\lim c_\lambda$ be its strong (and so weak) operator limit in A'' (and so is in A^b), then*

$$q(\lim c_\lambda) = c.$$

Furthermore, given $f \in A^b_{sa}$, $q(f) \leq 0$ if and only if $f \leq 0$ a.e.. So $q(f) = 0$ if, and only if, $f = 0$ a.e.

Proof Let (c_λ) be a norm-bounded, upward directed net in A_{sa}. Then $\lim c_\lambda$ is in A^b. By Lemma 5.4.3 $\lim c_\lambda$ is also in W. By definition, A^b_{sa} is the smallest σ-closed subspace of A''_{sa} which contains all x that correspond to lower semicontinuous affine functions on K. So $A^b \subset W$. Let q be the restriction of k to A^b.

For any $g \in A_{sa}^b$ we have $q(g - q(g)) = 0$. So $g - q(g)$ is in $\mathcal{N}$. So $g = q(g)$ a.e..
It now follows that if $f \in A_{sa}^b$ and $f \leq 0$ a.e. then $q(f) \leq 0$ a.e.. So

$$\{\phi \in \partial K : q(f)(\phi) > 0\}$$

is an open meagre subset of the Baire space ∂K. Hence this set is empty. Thus
$q(f) \leq 0$.

Conversely, let $q(f) \leq 0$. Since $f = q(f)$ a.e. it follows that $f \leq 0$ a.e.

Finally, $q(f) = 0$ if, and only if, $q(f) \leq 0$ and $q(-f) \leq 0$. So $q(f) = 0$ if, and
only if, $f = 0$ a.e. In other words, $A^b \cap \mathcal{N} = A^b \cap \mathcal{M}$.   □

We give a typical application of this Representation Theorem below. The
representation theorems obtained in Sect. 5.2 can be used in a similar way.

Proposition 5.4.7 *Let A be monotone complete and let D be a unital C^*-algebra.
Let $T : D \mapsto A$ be a positive $*$-linear map. Then T has a unique extension to a
positive, linear $\tilde{T} : D^b \mapsto A$ with the following properties. First, if (x_n) is a bounded,
monotone increasing sequence in D_{sa}^b then*

$$\tilde{T}(\lim x_n) = \bigvee_{n \geq 1} \tilde{T} x_n.$$

Secondly, if $(a_\lambda)_{\lambda \in \Lambda}$ is a bounded, upward directed net in D_{sa}, then

$$\tilde{T}(\lim a_\lambda) = \bigvee_{\lambda \in \Lambda} \tilde{T} a_\lambda.$$

Furthermore, if T is a $*$-homomorphism then so also is $\tilde{T}$.

Proof (Sketch) Given a bounded linear operator T from a Banach space X into a
Banach space Y, its second adjoint $T^{**} : X^{**} \mapsto Y^{**}$ is an extension of T. Also
T^{**} is a continuous map from X^{**}, equipped with the $\sigma(X^{**}, X^*)$ topology, to Y^{**},
equipped with the $\sigma(Y^{**}, Y^*)$ topology. Identify X with D and Y with A. Then T^{**}
is an extension of T which is weak-operator topology continuous on bounded balls
of D''. Let $W = \{x \in D'' : T^{**}x \in A^b\}$ and show $D^b \subset W$. Now let $\tilde{T}$ be qT^{**}
restricted to D^b. Finally, if $T : D \mapsto A$ is a $*$-homomorphism then $T^{**} : D'' \mapsto A''$
is a $*$-homomorphism. See Theorem 5.3.7.   □

Exercise 5.4.8 Show that if, in Proposition 5.4.7, A is only assumed to be monotone
σ-complete and D^b is replaced by D^∞ then we can obtain a sequential version of
Proposition 5.4.7 by applying Theorem 5.4.5.

Remark 5.4.9 If, in Exercise 5.4.8 or in Proposition 5.4.7, we put $D = C(X)$ we
may regard $\tilde{T}$ as an "A-valued integral". By using Theorem 5.2.16 and Proposition 5.4.7 we can replace A_{sa} by more general partially ordered vector spaces. See
[170].

The results of this section are mainly from [171] and [170] but with some
modifications of the original arguments. The recent achievement of Brown, see
Sect. 5.3, has allowed us to state Theorem 5.4.6 in a neater form than the original.

For any C^*-algebra A, it follows from Lemma 5.4.2 that $A^\infty \cap \mathcal{N}$ is a σ-ideal of A^∞. So the algebra $A^\infty/(A^\infty \cap \mathcal{N})$ is monotone σ-complete by Proposition 2.2.21. Since $A \cap \mathcal{N} = \{0\}$, $q|A$ is an injective $*$-homomorphism, and hence an isometric $*$-isomorphism of A into $A^\infty/(A^\infty \cap \mathcal{N})$. But a much stronger result is true. The quotient algebra is a regular σ-completion of A. We shall prove this in Sect. 5.6. In the final section of this chapter we shall show that when A is a small C^*-algebra then its regular σ-completion is also small. So it has a faithful state and thus is monotone complete. So when A is a small C^*-algebra its regular σ-completion is its regular completion. See the discussion at the end of Sect. 5.6.

5.4.1 Open Problem

We know $A^\infty \subset A^\Sigma$. This induces a natural embedding of $A^\infty/(A^\infty \cap \mathcal{N})$ into $A^\Sigma/(A^\Sigma \cap \mathcal{N})$. Are these algebras isomorphic? This could have a positive answer even if $A^\infty \neq A^\Sigma$.

For which A is $A^b/(A^b \cap \mathcal{N})$ the regular completion of A?

5.5 Compact Convex Sets: Semicontinuity and Approximations

We shall need some more results on convexity. As before K is a compact convex subset of a Hausdorff, locally convex topological vector space. Let Y be a subset of K, with $\partial K \subset Y \subset K$. For most applications $Y = K$ or $Y = \partial K$. All functions are real valued unless we specify otherwise.

Definition 5.5.1 Let f be a bounded real valued function on Y. For each $x \in K$ let

$$\underline{f}(x) = \sup\{a(x) : a \in A(K),\ a(\phi) \leq f(\phi) \text{ for every } \phi \in Y\}$$

and

$$\hat{f}(x) = \inf\{a(x) : a \in A(K),\ a(\phi) \geq f(\phi) \text{ for every } \phi \in Y\}.$$

Then it is clear that $\underline{f}$ is lower semicontinuous. It is straightforward to check that $\underline{f}$ is a convex function which is bounded below by $-\|f\|$. Since each $a \in A(K)$ attains its maximum on ∂K, it follows that $\underline{f}$ is bounded above by $\|f\|$. We note that $\underline{f}$ is defined on the whole of K. If $g : Y \mapsto \mathbb{R}$ is bounded and $f \leq g$ on Y, then it is easy to see that $\underline{f} \leq \underline{g}$ on K. Also $\hat{f} \leq \hat{g}$. (See Proposition 1.1.6 [2].)

Similarly, $\hat{f}$ is a concave, upper semicontinuous function with

$$-\|f\| \leq \hat{f} \leq \|f\|.$$

By Lemma 5.2.2, when $Y = K$ and f is lower semicontinuous and convex, then $f = \underline{f}$. Similarly, when f is a bounded, concave, upper semicontinuous function on K, by applying Lemma 5.2.2 to $-f$, we have $f = \hat{f}$.

Lemma 5.5.2 *Let u be a bounded upper semicontinuous concave function on K. Let $u \geq 0$ a.e., that is, $\{x \in \partial K : u(x) < 0\}$ is a meagre subset of ∂K. Then $u \geq 0$.*

Proof By upper semicontinuity, $\{x \in \partial K : u(x) < 0\}$ is open. By hypothesis it is also meagre. But any meagre open subset of a Baire space is empty. So u is non-negative on ∂K. Thus the closed convex set $\{x \in K : u(x) \geq 0\}$ contains ∂K. So, by the Krĕin-Milman theorem, $u(x) \geq 0$ for all $x \in K$, that is, $u \geq 0$.   □

We have the following:

Lemma 5.5.3 *Let u be a bounded upper semicontinuous function on K. Then $u(x) = \hat{u}(x)$ for each $x \in \partial K$.*

Proof This is a theorem of Hervé [73]. It is proved in Proposition 1.4.1 [2].   □

Corollary 5.5.4 *Let f be a bounded continuous, real-valued function on ∂K. Then*

$$f(x) = \hat{f}(x) = \underline{f}(x) \text{ for each } x \in \partial K.$$

Proof See Corollary 1.4.2 [2] or [125].   □

Lemma 5.5.3 has the following variant which starts with a bounded upper semicontinuous function defined only on the extreme points of K.

Lemma 5.5.5 *Let v be a bounded upper semicontinuous function on ∂K. Then v can be extended to a concave, upper semicontinuous, bounded function on K. In particular,*

$$v(x) = \hat{v}(x)$$

for each $x \in \partial K$. Furthermore, for all $x \in K$,

$$-\|v\| \leq \hat{v}(x) \leq \|v\|.$$

Proof Since K is compact Hausdorff, ∂K is completely regular. So, by Lemma 2.3.4, there is a bounded, downward directed net, $(f_\lambda : \lambda \in \Lambda)$, of bounded continuous functions on ∂K such that

$$\inf\{f_\lambda(x) : \lambda \in \Lambda\} = v(x)$$

for each $x \in \partial K$. We can assume, without loss of generality, that

$$-\|v\| \le \widehat{f_\lambda}(x) \le \|v\| \text{ for all } x \in K.$$

Let u be the pointwise infimum, on K, of $\{\widehat{f_\lambda} : \lambda \in \Lambda\}$. Then u is bounded, upper semicontinuous and concave. Also, by applying Corollary 5.5.4 to each f_λ, we see that $u(x) = v(x)$ for each $x \in \partial K$.

Since $f_\lambda \ge v$ on ∂K, we have $\widehat{f_\lambda} \ge \hat{v}$ on K. Hence $u \ge \hat{v}$. Thus

$$v(x) = \hat{v}(x) \text{ for each } x \in \partial K. \qquad \square$$

In general, ∂K need not be a Borel subset of K. But we have the following.

Corollary 5.5.6 *A bounded (real valued) function on ∂K is a Borel function if, and only if, it is the restriction of a bounded Borel function on K.*

Proof Let V be the set of all $F : K \mapsto \mathbb{R}$, where F is a bounded, Borel measurable function on K and $F|_{\partial K}$ is a Borel function on ∂K.

Then V is a real vector space and, for any F and G in V, FG is in V. Also, if u is a bounded, upper semicontinuous function on K then $u \in V$. Let (F_n) be a bounded, monotone increasing sequence in V with pointwise limit F. Then F is Borel on K and $F|_{\partial K}$ is Borel on ∂K. So $F \in V$. Hence V contains all the bounded (real-valued) Borel functions on K. So the restriction of a Borel function on K is a Borel function on ∂K. We now prove the converse. Let W be the set of all $f : \partial K \mapsto \mathbb{R}$ such that, f is Borel measurable and f can be extended to a bounded Borel measurable function on K. By Lemma 5.5.5, each bounded, upper semicontinuous function on ∂K is in W. Also W is a (real) subalgebra of $B(\partial K)$.

Now let (f_n) be a bounded, monotone increasing sequence in W, with $0 \le f_n \le 1$ for each n. This sequence converges pointwise to a Borel function f.

For each n there is a bounded Borel function F_n on K such that $F_n|_{\partial K} = f_n$. Since $F_n \wedge 1$ is a Borel function on K whose restriction to ∂K is f_n, we may assume that F_n is bounded above by 1. Similarly we may suppose it is bounded below by 0. Let

$$F(x) = \limsup_n F_n(x).$$

Then F is a bounded Borel function on K. But $F|_{\partial K} = f$. So $f \in W$. Hence W contains all the bounded Borel functions on ∂K. $\qquad \square$

Lemma 5.5.7 *Let u be a bounded concave upper semicontinuous function on K. Then $u = \underline{u}$ a.e., that is, $\{x \in \partial K : u(x) - \underline{u}(x) \ne 0\}$ is a meagre subset of ∂K.*

Proof First we observe that $u|_{\partial K}$ is upper semicontinuous. Since ∂K is a Baire space, we can apply Corollary 4.2.4 to $u|_{\partial K}$. So we find a bounded lower semicontinuous $w : \partial K \to \mathbb{R}$ such that $u|_{\partial K} \ge w$ and

$$\{x \in \partial K : u(x) - w(x) \ne 0\}$$

is a meagre subset of ∂K. Applying Lemma 5.5.5 to $-w$ we find that $-(\widehat{-w}) = \underline{w}$ is a bounded, lower semicontinuous convex function on K with

$$w(x) = \underline{w}(x) \text{ for all } x \in \partial K.$$

By Lemma 5.5.2, $u - \underline{w} \geq 0$. Hence $\underline{u} \geq \underline{w}$. Thus $u = \underline{u}$ a.e. $\square$

Proposition 5.5.8 *Let f be a (bounded) Borel function on K. Then there exists a bounded, concave, upper semicontinuous u such that $f = u$ a.e.*

Proof By Corollary 5.5.6, $f\,|_{\partial K}$ is a bounded Borel function on the completely regular space ∂K. By Lemma 4.2.7, there exists an upper semicontinuous v on ∂K, such that $v = f$ a.e. and v has the same upper bound and lower bounds as $f\,|_{\partial K}$. Then, using Lemma 5.5.5, $v = \hat{v}$ on ∂K. So, $\hat{v} = f$ a.e.. Put $u = \hat{v}$. $\square$

5.6 Applications of Convexity to Completions

We now return to the situation where A is a unital C*-algebra with state space K. As usual K is equipped with the compact topology induced by the $\sigma(A^*, A)$ topology.

Exercise 5.6.1 For each invertible $a \in A$ and each $\phi \in K$, let

$$T_a\phi(x) = \frac{\phi(axa^*)}{\phi(aa^*)}.$$

Show that (i) T_a maps K into K; (ii) $T_a : K \mapsto K$ is continuous; (iii) $T_aT_{a^{-1}}$ is the identity map on K; (iv) $T_a : \partial K \mapsto \partial K$ is a homeomorphism.

Hint: For (ii) look at Lemma 5.4.1. For (iv) demonstrate that T_a maps extreme points of K to extreme points.

Let A be a unital C^*-algebra and let A^∞ be the Pedersen-Baire envelope of A, see Definition 2.2.23. As before, we identify A^∞_{sa} with $A^\infty(K)$ and A''_{sa} with $A''(K)$. We recall that $\mathcal{M}$ is the set of all $z \in A''$ such that

$$\{\phi \in \partial K : \phi(z) \neq 0\}$$

is meagre. Let $\mathcal{N}$ be the complex linear span of $\mathcal{M}^+$. Then $\mathcal{N}$ is a *-subalgebra of A'' which is sequentially closed in the weak operator topology. See Lemma 5.4.2. We saw in Sect. 5.4 that $A^\infty \cap \mathcal{N}$ is a σ-ideal of A^∞.

Lemma 5.6.2 *Let $g \in A''(K)$ such that $g \leq 0$ a.e. Then $aga^* \leq 0$ a.e. when a is an invertible element of A. Also, for any positive $b \in A^+$, we have $bgb \leq 0$ a.e.*

Proof First let a be an invertible element of A. Let T_a be the homeomorphism from ∂K onto ∂K defined in Exercise 5.6.1. Since $\{\varphi \in \partial K : \varphi(g) > 0\}$ is a meagre subset of ∂K and $\{\varphi \in \partial K : \varphi(aga^*) > 0\} = T_{a^{-1}}\{\varphi \in \partial K : \varphi(g) > 0\}$. It follows that

$$aga^* \leq 0 \text{ a.e.}$$

Let $b \in A^+$ then $b + \frac{1}{n}$ is invertible for each n. Let $f_n = (b + \frac{1}{n})g(b + \frac{1}{n})$. Then $\{\phi \in \partial K : \phi(f_n) > 0\}$ is meagre for each n. Also (f_n) converges in norm to bgb. Hence

$$\{\phi \in \partial K : \phi(bgb) > 0\}$$

is contained in the meagre set

$$\bigcup_{n \geq 1}\{\phi \in \partial K : \phi(f_n) > 0\}.$$

So $bgb \leq 0$ a.e.   □

As before q is the quotient homomorphism of A^∞ onto $A^\infty/(A^\infty \cap \mathcal{N})$. By Proposition 2.2.21, $A^\infty/(A^\infty \cap \mathcal{N})$ is monotone σ-complete and q is σ-normal.

Lemma 5.6.3 Let $g \in A_{sa}^\infty$. Then $g \leq 0$ a.e. if, and only if, $q(g) \leq 0$.

Proof Let $g = g^+ - g^-$ where $g^+g^- = 0$. Let e be the range projection of g^+. So $e \in A_{sa}^\infty$.

Suppose $q(g) \leq 0$. Then

$$0 \leq q(g^+) = q(eg) = q(e)q(g)q(e) \leq 0.$$

So g^+ is in $A^\infty \cap \mathcal{N}$. Thus $g = -g^-$ a.e. So $g \leq 0$ a.e. Conversely, suppose $g \leq 0$ a.e. Let

$$V = \{x \in A_{sa}^\infty : x^2gx^2 \leq 0 \text{ a.e.}\}.$$

By Lemma 5.6.2, $A_{sa} \subset V$. Let (x_n) be a monotone increasing sequence in V which converges to x in A_{sa}^∞. Then $x_n \to x$ in the strong operator topology. So $x_n^2gx_n^2 \to x^2gx^2$ in the strong operator topology and hence in the weak operator topology. So, regarded as a sequence in $A''(K)$, it converges pointwise. It follows from Exercise 5.2.15 that $x^2gx^2 \leq 0$ a.e. So $x \in V$. Thus $A_{sa}^\infty \subset V$. In particular

$$g^+ = ege \leq 0 \text{ a.e.}$$

So $g^+ \in \mathcal{M}^+$. Thus g^+ is in $A^\infty \cap \mathcal{N}$. So

$$q(g) = q(g^+) - q(g^-) = -q(g^-) \leq 0.$$ □

Corollary 5.6.4 Let $g \in A^\infty$. Then $q(g) = 0$ if, and only if, $g \in \mathcal{M}$. That is $A^\infty \cap \mathcal{N} = A^\infty \cap \mathcal{M}$.

Proof Since q is a $*$-homomorphism it suffices to prove the result for g self-adjoint.

Then $q(g) = 0$ if, and only if, $q(g) \geq 0$ and $q(-g) \geq 0$. That is if, and only if, $g = 0$ a.e. $\square$

In the next theorem we show that A has a regular σ-completion.

Theorem 5.6.5 *The algebra* $A^\infty/(A^\infty \cap \mathcal{N})$ *is a regular σ-completion of A. If $\mathcal{D} \subset A_{sa}$ has an infimum d in A_{sa} then $q(d)$ is the infimum of $q[\mathcal{D}]$ in $A^\infty/(A^\infty \cap \mathcal{N})$.*

Proof As before, let q be the quotient homomorphism. We have seen that the restriction of q to A is an isometric $*$-isomorphism of A into the σ-algebra $A^\infty/(A^\infty \cap \mathcal{N})$. As usual, we identify A''_{sa} with $A''(K)$, the bounded real-valued affine functions on K.

Let W be the smallest monotone σ-closed subset of $q[A^\infty_{sa}]$ containing $q[A_{sa}]$. Let $V = \{a \in A^\infty_{sa} : q(a) \in W\}$. Then V is a monotone σ-closed subset of A^∞_{sa} containing A_{sa}. Hence $V = A^\infty_{sa}$. So $A^\infty/(A^\infty \cap \mathcal{N})$ is monotone σ-generated by $q[A]$.

Next, we shall show that $q[A]$ is a regular subalgebra of $A^\infty/(A^\infty \cap \mathcal{N})$. To show this, take any $f \in A^\infty[K]$. By Proposition 5.5.8 there exists a bounded, concave, upper semicontinuous u such that $u = f$ a.e.. Let

$$L = \{a \in A(K) : q(a) \leq q(f)\}.$$

Then, by Lemma 5.6.3, $a \in L$ if, and only if, $a \leq f$ a.e. This is equivalent to $a \leq u$ a.e. By applying Lemma 5.5.2 to $u - a$ we find that $a \in L$ if, and only if, $a \leq u$. This is equivalent, see Definition 5.5.1, to $a \leq \underline{u}$.

Now let $h \in A^\infty[K]$ be such that $q(h)$ is an upper bound for $\{q(a) : a \in L\}$. Let

$$L_h = \{a \in A(K) : q(a) \leq q(h)\}.$$

Then $L \subset L_h$.

By Proposition 5.5.8, there exists a bounded, concave, upper semicontinuous v such that $v = h$ a.e. Arguing as above, we find that $a \in L_h$ if, and only if, $a \leq v$. Since $L \subset L_h$, whenever $a \in L$, $a \leq v$. Hence $\underline{u} \leq v$.

By Lemma 5.5.7 $\underline{u} = u$ a.e. We know that $u = f$ a.e. and $v = h$ a.e. So $f \leq h$ a.e. Hence, by Lemma 5.6.3, $q(f) \leq q(h)$.

It now follows that $q(f)$ is the least upper bound of

$$\{q(a) : a \in A \text{ and } q(a) \leq q(f)\}.$$

It follows that $q[A]$ is a regular subalgebra of $A^\infty/(A^\infty \cap \mathcal{N})$.

So we have constructed a regular σ-completion of A.

Finally, the fact that q preserves existing infima follows from Proposition 2.1.35, because $q[A]$ is a regular subalgebra of $A^\infty/(A^\infty \cap \mathcal{N})$. $\square$

For the moment we drop the requirement that $A(K)$ be the self-adjoint part of a C*-algebra. So K is an arbitrary compact convex set in a Hausdorff locally convex vector space. To avoid cumbersome notation, in this chapter we shall use $B(X)$ to denote the bounded Borel measurable, *real-valued*, functions on a topological space

X. For F in $B(K)$ let $R(F) = F|_{\partial K}$. By Corollary 5.5.6 the map R is a surjection of $B(K)$ onto $B(\partial K)$.

Let $M(K)$ be the bounded Borel functions $f : K \mapsto \mathbb{R}$ such that $f = 0$ a.e., that is, $\{\phi \in \partial K : f(\phi) \neq 0\}$ is a meagre subset of ∂K. (This conflicts slightly with our earlier notation in Chap. 4 where $M(Y)$ is used to denote the set of all bounded Borel functions on Y, which vanish on a meagre subset of Y.) Then $B(K)/M(K)$ is clearly isomorphic to the quotient of $B(\partial K)$ by the ideal of bounded meagre Borel functions on ∂K. So, by Theorem 4.2.9 $B(K)/M(K)$ is monotone complete.

For $f \in B(K)$ let $[f]$ be the equivalence class in $B(K)/M(K)$ which contains f. So $[f] = f + M(K)$.

Let us observe that in [123] it is shown that every Archimedean partially ordered vector space has an (essentially unique) Dedekind vector lattice completion. The work of the following exercise shows that $B(K)/M(K)$ may be regarded as the Dedekind completion of $A(K)$.

Exercise 5.6.6 Let f, g be in $B(K)$ and let a, b be in $A(K)$. Prove the following:

(i) $[f] \geq 0$ if and only if, $f \geq 0$ a.e. that is $\{\phi \in \partial K : f(\phi) < 0\}$ is a meagre subset of ∂K. Hint: This is easier than Lemma 5.6.3 because $B(K)$ is a commutative algebra.

(ii) $[f] \geq [g]$ if, and only if, $f \geq g$ a.e.

(iii) $[a] \geq [b]$ if, and only if, $a \geq b$. Hint: Use Lemma 5.5.2.

(iv) Show that $a \mapsto [a]$ is a bipositive isometry of $A(K)$ into $B(K)/M(K)$.

(v) Let y be in $B(K)/M(K)$. By Proposition 5.5.8, $y = [u]$ for some bounded concave upper semicontinuous u. Let $L = \{a \in A(K) : [a] \leq y\}$. Show that $a \in L$ implies $a \leq u$. Hint: Use (iii) and Lemma 5.5.2.

(vi) Let z be an upper bound for $\{[a] : a \in L\}$. Show that $z \geq y$. Hint: Let $z = [w]$ where w is bounded concave and upper semicontinuous. Arguing as in (v), show that $w \geq u$.

(vii) Deduce that $[w]$ is the least upper bound of $\{[a] : a \in A_{sa}$ and $[a] \leq [w]\}$.

(viii) Let D be a (non-empty) subset of $A(K)$. Suppose that D has a least upper bound, b, in $A(K)$. Show that $[b]$ is the least upper bound, in $B(K)/M(K)$, of $\{[d] : d \in D\}$. Hint: Apply Proposition 2.1.35.

We now revert to assuming K is the state space of a (unital) C^*-algebra.

Proposition 5.6.7 *Let A be any unital C^*-algebra. Then there exists a faithful bipositive linear map Γ from A into a commutative monotone complete C^*-algebra C. If A is monotone complete, then Γ is normal. So, in the notation of Chap. 3, $A \precsim C$. If A is monotone σ-complete, then Γ is σ-normal.*

Proof It suffices to define Γ on $A_{sa} = A(K)$. We take for C the commutative algebra whose self-adjoint part is $B(K)/M(K)$. Using the notation of Exercise 5.6.6, let

$$\Gamma a = [a] \text{ for } a \in A(K).$$

Then Exercise 5.6.6 (iv) shows that Γ is faithful and bipositive. By Exercise 5.6.6 (viii) Γ is normal. $\square$

Lemma 5.6.8 *Let Z be a partially ordered (real) vector space; Y a regular subspace of Z; X a regular subspace of Y. Then X is a regular subspace of Z.*

Proof For any $z \in Z$, let

$$L_Y(z) = \{b \in Y : b \leq z\}$$
$$L_X(z) = X \cap L_Y(z)$$
$$U_Y(z) = \{b \in Y : z \leq b\}.$$

Now fix $x \in Z$. We wish to show x is the least upper bound of $L_X(x)$. Let $y \in Z$ be an upper bound of $L_X(x)$. Let $u \in U_Y(y)$. So $u \in Y$ is an upper bound of $L_X(x)$. Let $b \in L_Y(x)$. Then $a \in L_X(b)$ implies $a \in L_X(x)$. So $a \leq u$. Thus $u \in Y$ is an upper bound of $L_X(b)$. Since X is a regular subspace of Y it follows that $b \leq u$. So u is an upper bound for $L_Y(x)$. Since Y is a regular subspace of Z this implies $x \leq u$. So x is a lower bound for $U_Y(y)$. Since Y is a regular subspace of Z this implies $x \leq y$. Thus x is the least upper bound of $L_X(x)$. Hence X is a regular subspace of Z. $\square$

Using the same notation as above, we have the following corollary.

Corollary 5.6.9 *Let X, Y, Z be order unit Banach spaces with the same order unit. Suppose that X is a regular subspace of Y. If Z is the Dedekind completion of Y then it is also the Dedekind completion of X.*

Proof Any upper bounded subset of X is also an upper bounded subset of Y. Hence it has a least upper bound in Z.

Since X is a regular subspace of Z, for each $c \in Z$ the least upper bound, in Z, of

$$L_X(c) = \{a \in X : a \leq c\}$$

is c. $\square$

Definition 5.6.10 An ordered pair (W, j) is a Dedekind completion of an order unit vector space U if (1) W is the self-adjoint part of a commutative monotone complete C^*-algebra, (2) j is a unital linear bipositive map from U into W and (3) $j(U)$ is a regular subspace of W.

We have seen from Exercise 5.6.6 that when $V = A(K)$ is an order unit Banach space it posses a Dedekind completion, which we denote by $\tilde{V}$. Then this Dedekind completion is unique. (See [123].) More precisely, if $V^\#$ is an isomorphic copy of V and α is a bipositive, linear, unital bijection of V onto $V^\#$ then α has a unique extension $\tilde{\alpha}$, where $\tilde{\alpha}$ is a bipositive, linear, unital bijection of $\tilde{V}$ onto $\widetilde{V^\#}$. Further, $\tilde{\alpha}$ preserves all infima and suprema that exist in V. In particular, $\tilde{\alpha}$ is normal. In fact, if $\widetilde{V^\#}$ is a Dedekind completion of $V^\#$, then by Proposition 2.3.9, there exist a linear bipositive map $\tilde{\alpha}$ from $\tilde{V}$ into $\widetilde{V^\#}$ such that $\tilde{\alpha}|_V = \alpha$ and a linear bipositive map

$\widetilde{\alpha^{-1}}$ from $\widetilde{V^\#}$ into $\tilde{V}$ such that $\widetilde{\alpha^{-1}}\,|_{V^\#} = \alpha^{-1}$. Since V and $V^\#$ are regular subspaces, $\tilde{\alpha}\circ\widetilde{\alpha^{-1}} = \iota$ and $\widetilde{\alpha^{-1}}\circ\tilde{\alpha} = \iota$. So, $\tilde{\alpha}$ is a normal bipositive bijection of $\tilde{V}$ onto $\widetilde{V^\#}$. See, for an account of injectivity in Banach spaces, page 88 Corollary 2 and Chapter 3 Section 11 of [98]. (See also [54, 72, 94, 108, 149].)

The following theorem says that, given a C^*-algebra A, its regular σ-completion is unique.

Theorem 5.6.11 *Let A be a unital C^*-algebra and π a $*$-isomorphism of A onto $A^\#$. Let B be a regular σ-completion of A and $B^\#$ a regular σ-completion of $A^\#$. Then π has an extension which is an isomorphism of B onto $B^\#$.*

Proof Let α be the restriction of π to A_{sa}.

We have $A_{sa} \subset B_{sa} \subset \widetilde{B_{sa}}$. By Corollary 5.6.9, we may replace $\widetilde{B_{sa}}$ by $\widetilde{A_{sa}}$. Then B_{sa} is the σ-completion of A_{sa} inside $\widetilde{A_{sa}}$. Then $\tilde{\alpha} : \widetilde{A_{sa}} \mapsto \widetilde{A^\#_{sa}}$ is normal and so is its inverse. So it maps the σ-completion of A_{sa} in $\widetilde{A_{sa}}$ to the σ-completion of $A^\#_{sa}$ in $\widetilde{A^\#_{sa}}$. Thus it is a bipositive linear bijection of B_{sa} onto $B^\#_{sa}$. Now let $\tilde{\pi} : B \mapsto B^\#$ be defined by

$$\tilde{\pi}(x + iy) = \tilde{\alpha}(x) + i\tilde{\alpha}(y) \text{ when } x, y \text{ are in } B_{sa}.$$

Finally, by applying Corollary 5.3.8, $\tilde{\pi}$ is an isomorphism from B onto $B^\#$. $\square$

Remark 5.6.12 A straight forward modification of this approach gives the uniqueness of regular completions. We shall see in Sect. 5.8, that for every small C^*-algebra its regular σ-completion is monotone complete and hence is its regular completion. This will suffice for most of our applications. However Hamana showed [61] that for every C^*-algebra A, its regular σ-completion is embedded in a regular completion. We give details in a later chapter.

Definition 5.6.13 For each unital C^*-algebra A we denote its regular σ-completion by $\hat{A}$ and identify A with a subalgebra of $\hat{A}$.

In the following theorem and its corollaries we may take B to be $\hat{A}$ but, of course, they apply more generally.

Theorem 5.6.14 *Let the unital C^*-algebra A be a regular subalgebra of B. Let I be a closed proper (two sided) ideal of B. Then $I \cap A$ is a closed proper (two sided) ideal of A.*

Proof Clearly $I \cap A$ is a closed (two sided) ideal in A and $1 \notin I \cap A$.

Assume that $I \cap A = \{0\}$. Let $h : B \to B/I$ be the quotient homomorphism and let h_0 be the restriction of h to A. Then h_0 is a $*$-isomorphism of A into B/I. Hence it is an isometry and its restriction to A_{sa} is a bipositive order isomorphism into B_{sa}.

Let b be any self-adjoint element of I. Whenever $a \in A_{sa}$ and $a \le b$ we have

$$h_0(a) \le h(b) = 0.$$

So $a \leq 0$. But, by regularity, b is the least upper bound of the elements of A_{sa} dominated by b. So $b \leq 0$. Similarly $-b \leq 0$ and so $b = 0$. Thus $I = \{0\}$, contrary to hypothesis.   □

Corollary 5.6.15 *Let A be a simple unital C^*-algebra and a regular subalgebra of B. Then B is simple.*

Proof If B is not a simple C^*-algebra then it has a proper closed ideal I. But then, by Theorem 5.6.14, A is not simple.   □

Corollary 5.6.16 *Let π be a $*$-homomorphism of B into a C^*-algebra C. Let π_0 be the restriction of π to A. If π_0 is an isomorphism of A into C then π is an isomorphism of B into C.*

Proof The kernel of π_0 is the intersection of the kernel of π with A.   □

Corollary 5.6.17 *Let $\hat{A}$ have a non-trivial centre. Then there exist proper closed ideals I_1 and I_2 of A such that $I_1 \cap I_2 = \{0\}$.*

Proof Since the centre of $\hat{A}$ is monotone σ-complete and non-trivial it contains a non-trivial projection e. Let $I_1 = A \cap \hat{A}e$ and $I_2 = A \cap \hat{A}(1 - e)$.   □

Corollary 5.6.18 *If A is prime, then $\hat{A}$ has trivial centre.*

In a following section we shall consider the class of *Small C^*-algebras*. Let H be a separable, infinite dimensional Hilbert space. Then $L(H)$ and all its C^*-subalgebras are small. But there are many small C^*-algebras which are not subalgebras of $L(H)$. Each small C^*-algebra has a separable state space and hence has a faithful state.

We shall see in Sect. 5.8 that when A is small its regular σ-completion $\hat{A}$ is also small. So this monotone σ-complete algebra has a faithful state and hence is monotone complete. *It follows that the regular σ-completion of A is also its regular completion.*

In particular, when A is simple or prime $\hat{A}$ is a monotone complete factor.

When A is a separable C^*-algebra and Q the state space of $\hat{A}$ then ∂Q is separable (in fact, "hyperseparable") [177]. So when A is simple then $\hat{A}$ is a monotone complete factor for which ∂Q is separable. The only infinite dimensional von Neumann factor whose pure state space is separable is $L(H)$ [174]. But $L(H)$ is not simple. So when A is separable, simple and infinite dimensional then $\hat{A}$ is a wild factor. (It can be shown to be of Type III.) This gives a method for constructing examples of small wild factors. In particular if A is a UHF or Glimm algebra then $\hat{A}$ is a wild factor.

5.6.1   Open Problem

Let A_1 and A_2 be separable, unital, infinite dimensional C^*-algebras which are both simple. Suppose their pure state spaces have no isolated points. Then, how different can $\widehat{A}_1$ and $\widehat{A}_2$ be? Do they take the same value in the weight semigroup? Do they have the same spectroid? If so, is there any other invariant which distinguishes between them? Or are they in fact isomorphic?

5.7   Separable State Spaces and Embeddings in $L(H)$

Let $L(H)$ be the algebra of bounded operators on a separable (infinite dimensional) Hilbert space. Let $S(L(H))$ be its state space; the set of all states, equipped with the $\sigma(L(H)^*, L(H))$-topology. By Proposition 5.2.18 this state space has a countable dense subset.

Akemann [1] showed that any von Neumann algebra with a separable state space has a faithful normal representation on a separable Hilbert space. He posed the question: if a C^*-algebra has a separable state space, does it have a faithful representation on a separable Hilbert space ?

The answer is "no". For a counter example, let $\widehat{A}$ be the regular σ-completion of a UHF algebra; this has a separable state space. (So by Proposition 5.2.19 it is monotone complete.) It was shown in [178], as a consequence of a more general result, that $\widehat{A}$ is a wild factor which cannot be realised as a subalgebra of $L(H)$.

We observe that if a unital C^*-algebra, A, has a separable state space then so does $\widehat{A}$, and $\widehat{A}$ is monotone complete. See Lemma 1.1 in [179]. In particular when A has a separable state space then $\widehat{A}$ is the regular completion of A.

But having a separable state space is equivalent to being realisable as a linear subspace of $L(H)$. More precisely:

The unital algebra A has a separable state space if, and only if, for some separable Hilbert space H, there exists a linear order isomorphism ϕ from A_{sa} into $L(H)_{sa}$ with $\phi(1) = 1$.

We recall that, just like von Neumann algebras, monotone complete factors can be divided into Types I, II and III. All the Type I factors are von Neumann. If a Type II factor has a faithful state then it is von Neumann [175, 176]. So there are NO wild factors of Type II which have a separable state space. What about Type III? If a Type III factor can be realised as a subalgebra of $L(H)$ (H separable) then it is a von Neumann algebra [119]. (See also [130] and [39].) So a wild factor never has a faithful representation on a separable Hilbert space.

More generally, let A be monotone complete and (as a C^*-algebra) have a faithful representation on a separable Hilbert space. Then A is a von Neumann algebra if, and only if, its centre is von Neumann. See Corollary 3.6 in [179].

But we have seen above that wild Type III factors, with separable state spaces, do exist. In fact we shall see that they exist in huge abundance.

It is convenient to work with the class of small C^*-algebras [137, 144]. A unital C^*-algebra A is *small* if, and only if, it satisfies any one of the following:

(i) For each n, $M_n(A)$ has a separable state space.
(ii) There exists a unital complete isometry of A into $L(H)$ (H separable).

(iii) There exists a unital completely bipositive map of A into $L(H)$ (H separable).

For more information on small algebras, where these three conditions are shown to be equivalent, see below. But these sections can be omitted on a first reading. However it is useful to be aware of the following points.

Obviously $L(H)$, where H is separable, is small. Every closed $*$-subalgebra of a small C^*-algebra is, itself, small.

Let A be a wild factor with a separable state space. We have seen that it must be of Type III. This implies that it is isomorphic to $M_n(A)$ for each n. So it must be small by (i). It can never be a subalgebra of $L(H)$ (H separable). But by (iii) it can be realised as an operator system on a separable Hilbert space.

Proposition 5.7.1 *Let A be a commutative unital C^*-algebra with separable state space S. Then A is small.*

Proof Let $\{s_n : n = 1, 2 \ldots\}$ be a dense subset of S. Let $\phi : A \to \ell^\infty$ be defined by

$$\phi(f) = (f(s_n)) \, (n = 1, 2 \ldots).$$

Then ϕ is a *unital* bipositive linear map of A into ℓ^∞. Since A is commutative, it follows from Theorem 5.1.5 [38] that ϕ is completely positive. Since the restriction of ϕ to A_{sa} is an isometry, $\phi[A]$ is an operator system. By Lemma 5.1.4 [38], since A is commutative, ϕ^{-1} is completely positive. So, by (iii), A is small. $\square$

5.7.1 Open Problem

Let A be a unital C^*-algebra with a separable state space. Is A small? If A is commutative or a factor (monotone complete with trivial centre) the answer is "yes". But, in general, the answer is unknown. The problem can be reduced to the question: does $A \otimes M_2(\mathbb{C})$ have a separable state space when A has a separable state space. A key difficulty is that a closed subspace of a separable space may fail to be separable.

5.8 Small C^*-Algebras and Completely Positive Maps

Small C^*-algebras were defined above. In fact we gave three definitions! *We adopt (ii) as the official definition and show that this is equivalent to both (i) and (iii).*

This Sects. 5.8 and 5.9 can be omitted on a first reading of this book. But in Sect. 5.9 we show that the regular σ-completion of a small C^*-algebra is small. Hence it has a faithful state. So it is monotone complete. So the regular σ-completion of a small algebra is its monotone completion.

We introduced the notion of small C^*-algebras in [144] (see also [66]).

We shall give a characterisation of small C^*-algebras in Theorem 5.8.9 (see [137]).

Let ϕ be a linear map from a C^*-algebra A into another C^*-algebra B. Let us recall that the linear map $\phi_n : M_n(A) \mapsto M_n(B)$ is defined by

$$\phi_n(\mathbf{x}) = \phi_n((x_{ij})_{1 \leq i,j \leq n}) = (\phi(x_{ij}))_{1 \leq i,j \leq n} \ (\mathbf{x} = (x_{ij})_{1 \leq i,j \leq n} \in M_n(A)).$$

See Sect. 2.4. If ϕ_n is positive, then we call the map ϕ, n-positive. The map ϕ is called *completely positive* when ϕ is n-positive for all $n \in \mathbb{N}$. If the map ϕ_n is an isometry for all $n \in \mathbb{N}$, then we call ϕ a *complete isometry*. If A and B are unital and $\phi(1_A) = 1_B$, then ϕ_n is unital for every $n \in \mathbb{N}$. Furthermore, if ϕ is completely positive (respectively, a complete isometry) then we call ϕ a unital completely positive map (respectively a unital complete isometry).

Lemma 5.8.1 *Let A be a unital C^*-algebra and let H be a Hilbert space. Let $\tau : A \mapsto L(H)$ be a unital completely positive linear map.*

(1) *(Stinespring's theorem [155]) There is a $*$-representation $\{\pi, K\}$ of A, where K is a Hilbert space, and a bounded linear operator v from H into K such that $\tau(x) = v^*\pi(x)v$ for all $x \in A$. On noting that v is an isometry from H into K, if we identify H with vH, we may assume that $H \subset K$. Let P be the projection from K onto H. Then*

$$\tau(x) = P\pi(x)|_H \ \text{for all } x \in A. \tag{*}$$

Conversely, any map of the above form (*) is always a completely positive linear map from A into $L(H)$.

(2) *(The Schwarz-Kadison inequality [83]) We have $\tau(x^*)\tau(x) \leq \tau(x^*x)$ for all $x \in A$.*

(3) *(The Choi multiplicative domain theorem [24]) Let A and B be unital C^*-algebras and let $\tau : A \mapsto B$ be a* unital *completely positive linear map. Let us define multiplicative domains by*

$$A_\tau^l = \{a \in A : \tau(a^*a) = \tau(a^*)\tau(a)\} \text{ and } A_\tau^r = \{a \in A : \tau(aa^*) = \tau(a)\tau(a^*)\}.$$

Then we have $A_\tau^l = \{a \in A : \tau(xa) = \tau(x)\tau(a) \text{ for all } x \in A\}$ and $A_\tau^r = \{a \in A : \tau(ax) = \tau(a)\tau(x) \text{ for all } x \in A\}$, that is, A_τ^l and A_τ^r are closed subalgebras of A.

Proof

(1) See for example, Theorem 3.6, page 194 [161].

(2) By (1), for every $x \in A$, we have $\tau(x^*)\tau(x) = P\pi(x)^*P\pi(x)P \leq P\pi(x^*x)P = \tau(x^*x)$.

(3) We may assume that B acts on a Hilbert space H. By (1), there exists a representation $\{\pi, K\}$ of A such that $K \supset H$ and $\tau(a) = P_H\pi(a)|_H$ for all $a \in A$. Take $a \in A_\tau^l$. Then, we have

$$P_H\pi(a^*)\pi(a)|_H = P_H\pi(a^*a)|_H = \tau(a^*a) = \tau(a^*)\tau(a) = P_H\pi(a^*)P_H\pi(a)|_H ,$$

which implies that

$$P_H \pi(a^*)(1 - P_H)\pi(a)\,|_H = ((1 - P_H)\pi(a)P_H)^*((1 - P_H)\pi(a)P_H)\,|_H = 0.$$

Hence, for every $a \in A$, $\pi(a)\,|_H = P_H\pi(a)\,|_H$. So we have, for every $a \in A$,

$$\tau(xa) = P_H\pi(xa)\,|_H = P_H\pi(x)\pi(a)\,|_H = P_H\pi(x)P_H\pi(a)\,|_H = \tau(x)\tau(a).$$

By similar reasoning, the claim for $\mathcal{A}_\tau^r$ also holds true. This completes the proof. See [24]. □

Corollary 5.8.2 *Let A and B be unital C^*-algebras. Let $\tau : A \mapsto B$ be a unital completely positive linear map. Suppose that τ is a Jordan homomorphism, that is, $\tau(a^2) = \tau(a)^2$, for all $a \in A_{sa}$. Then, τ is a $*$-homomorphism.*

Proof Note first that τ is a $*$-map. By our assumption, for any $a \in A_{sa}$, $a \in A_\tau^l \cap A_\tau^r$. So, we have $A = A_\tau^l = A_\tau^r$. □

The next technical lemma is useful later. It is cumbersome to state but its proof is easy.

Lemma 5.8.3 *For a C^*-algebra C and each pair $m, n \in \mathbb{N}$, the map $\pi_C^{(m,n)} : M_m(M_n(C)) \mapsto M_{mn}(C)$ defined by deleting the inner parentheses, that is,*

$$\pi_C^{(m,n)} : (\mathbf{T}_{ij})_{1 \le i,j \le m} \longmapsto \begin{pmatrix} t_{11}^{(11)} & \cdots & t_{11}^{(1n)} & t_{1m}^{(11)} & \cdots & t_{1m}^{(1n)} \\ \vdots & & \vdots & \vdots & & \vdots \\ t_{11}^{(n1)} & \cdots & t_{11}^{(nn)} & t_{1m}^{(n1)} & \cdots & t_{1m}^{(nn)} \\ \vdots & & & \vdots & & \\ t_{m1}^{(11)} & \cdots & t_{m1}^{(1n)} & t_{mm}^{(11)} & \cdots & t_{mm}^{(1n)} \\ \vdots & & \vdots & \vdots & & \vdots \\ t_{m1}^{(n1)} & \cdots & t_{m1}^{(nn)} & t_{mm}^{(n1)} & \cdots & t_{mm}^{(nn)} \end{pmatrix}$$

for all $(\mathbf{T}_{ij}) \in M_m(M_n(C))$, where each $\mathbf{T}_{ij} = (t_{ij}^{(kl)})_{1 \le k,l \le n}$ with $t_{ij}^{(kl)} \in C$ ($1 \le i,j \le m; 1 \le k,l \le n$), is a $*$-isomorphism from $M_m(M_n(C))$ onto $M_{mn}(C)$.

Proposition 5.8.4 *Let A be a unital C^*-algebra with a system $\{e_{ij}\}_{1 \le i,j \le n}$ of $n \times n$ matrix units and let B be a unital C^*-algebra with a system of $n \times n$ matrix units $\{f_{ij}\}_{1 \le i,j \le n}$. Let $\phi : e_{11}Ae_{11} \longmapsto f_{11}Bf_{11}$ be a unital completely positive linear map. Then*

(1) *There exists a unique unital completely positive linear map $\Phi : A \mapsto B$ such that $\Phi|_{e_{11}Ae_{11}} = \phi$ and $\Phi(e_{ij}) = f_{ij}$ for all i and j with $1 \le i,j \le n$.*
(2) *If ϕ is a $*$-homomorphism (clearly ϕ is completely positive), then so is Φ.*

(3) *If A and B are monotone $(\sigma\text{-})$complete and the map ϕ is a $(\sigma\text{-})$normal completely positive map, then Φ is also a completely positive $(\sigma\text{-})$normal map.*

Proof Let us define the map $\Phi : A \mapsto B$ by

$$\Phi(x) = \sum_{i,j=1}^{n} f_{i1}\phi(e_{1i}xe_{j1})f_{1j} \text{ for every } x \in A.$$

Let $\pi_A : A \mapsto M_n(e_{11}Ae_{11})$ and $\pi_B : B \mapsto M_n(f_{11}Bf_{11})$ be canonical maps defined, as before, by $\pi_A(x) = (e_{1i}xe_{j1})_{1\leq i,j\leq n}$ for $x \in A$ and $\pi_B(y) = (f_{1i}yf_{j1})_{1\leq i,j\leq n}$ for $y \in B$. Then, as before, on noting that

$$\pi_B^{-1}((b_{ij})_{1\leq i,j\leq n}) = \sum_{i,j=1}^{n} f_{i1}b_{ij}f_{1j} \text{ for every } (b_{ij}) \in M_n(f_{11}Bf_{11}),$$

we have $\Phi = \pi_B^{-1} \circ \phi_n \circ \pi_A$. We shall show that ϕ_n is completely positive (and so Φ is also completely positive). To do this, first of all, note that $\pi_B^{(m,n)} \circ (\phi_n)_m = \phi_{mn} \circ \pi_A^{(m,n)}$ by Lemma 5.8.3. Since ϕ_{mn} is positive, $(\phi_n)_m$ is positive, that is, ϕ_n is m positive for all m. So, ϕ_n is completely positive, which implies that Φ is also completely positive. Since ϕ is unital, Φ is also unital. Calculation shows

$$\Phi(x) = \sum_{i,j=1}^{n} f_{i1}\phi(e_{1i}xe_{j1})f_{1j}$$

$$= f_{11}\phi(x)f_{11} = \phi(x) \text{ for every } x \in e_{11}Ae_{11}.$$

Moreover, on noting that $\phi(e_{11}) = f_{11}$ (because ϕ is unital), calculation shows that, for any pair k and l with $1 \leq k,l \leq n$,

$$\Phi(e_{kl}) = \sum_{i,j=1}^{n} f_{i1}\phi(e_{1i}e_{kl}e_{j1})f_{1j}$$

$$= f_{k1}\phi(e_{11})f_{1l} = f_{kl}.$$

Hence Φ satisfies our requirements. Next, we shall show that Φ is unique. Take any completely positive map Ψ satisfying $\Psi|_{e_{11}Ae_{11}} = \phi$ and $\Psi(e_{ij}) = f_{ij}$ for all i,j with $1 \leq i,j \leq n$. Since $\Psi(e_{ij})^*\Psi(e_{ij}) = f_{ji}f_{ij} = f_{jj} = \Psi(e_{jj}) = \Psi((e_{ij})^*e_{ij})$, e_{ij} is in the multiplicative domain of Ψ and so by Lemma 5.8.1(3), Choi's multiplicative domain theorem, we have $\Psi(xe_{ij}) = \Psi(x)\Psi(e_{ij})$ and $\Psi(e_{ij}x) = \Psi(e_{ij})\Psi(x)$ for all

$x \in A$ and i, j with $1 \le i, j \le n$. Hence we have

$$\Psi(x) = \sum_{i,j=1}^{n} \Psi(e_{i1}e_{1i}xe_{j1}e_{1j})$$

$$= \sum_{i,j=1}^{n} f_{i1}\Psi(e_{1i}xe_{j1})f_{1j}$$

$$= \sum_{i,j=1}^{n} f_{i1}\phi(e_{1i}xe_{j1})f_{1j}$$

$$= \Phi(x) \text{ for all } x \in A.$$

Suppose that ϕ is a $*$-homomorphism. Since ϕ is completely positive and linear, it is clear that Φ is also a $*$-linear map by the above argument. We show the map Φ is multiplicative. To do this, take x and y in A. Then we have

$$\phi(e_{1i}xye_{j1}) = \sum_{k=1}^{n} \phi(e_{1i}xe_{k1}e_{1k}ye_{j1})$$

$$= \sum_{k=1}^{n} \phi(e_{1i}xe_{k1})\phi(e_{1k}ye_{j1})$$

$$= \sum_{k=1}^{n} \phi(e_{1i}xe_{k1})f_{1k}f_{k1}\phi(e_{1k}ye_{j1}) \text{ for each pair } i, j \in \{1, \cdots, n\}.$$

Hence it follows that

$$\Phi(xy) = \sum_{i,j=1}^{n} \sum_{k=1}^{n} f_{i1}\phi(e_{1i}xe_{k1})f_{1k}f_{k1}\phi(e_{1k}ye_{j1})f_{1j} = \Phi(x)\Phi(y).$$

So Φ is a $*$-homomorphism.

(Please note that, as usual, we use i for $\sqrt{-1}$. We also use i, j, k to represent integer variables. This dual use of "i" does not normally cause any difficulties.)

Suppose that A and B are monotone σ-complete and the completely positive map $\phi : e_{11}Ae_{11} \mapsto f_{11}Bf_{11}$ is σ-normal. We shall show that Φ is σ-normal. To show this, take any norm bounded increasing sequence (a_m) of elements in A such that $a_m \uparrow a$ in A_{sa}.

For a C^*-algebra C with a system of matrix units $\{p_{ij}\}$, and for a pair i and j with $1 \le i, j \le n$, by Lemma 2.4.3(1), $p_{1i}xp_{j1} = \sum_{k=0}^{3} i^k(p_{j1} + i^k p_{i1})^* x(p_{j1} + i^k p_{i1})$.

Hence, by Corollary 2.1.23, we have $LIM_m e_{1i}a_m e_{j1} = e_{1i}ae_{j1}$ in A and so, by Propositions 2.2.25 or 2.2.26 we have the order limit is in $e_{11}Ae_{11}$. Since ϕ is σ-normal and positive, by Definition 2.2.20, $LIM_m\phi(e_{1i}a_m e_{j1}) = \phi(e_{1i}ae_{j1})$ in $f_{11}Bf_{11}$

for each pair i and j. On the other hand, since Φ is positive, $(\Phi(a_m))$ is norm bounded and increasing and there exists $b \in B_{sa}$ such that $\Phi(a_m) \uparrow b$ in B_{sa}. By the same reasoning as above, for each pair i and j, it follows that $LIM_m \phi(e_{1i}a_m e_{j1}) = f_{1i} b f_{j1}$ in $f_{11} B f_{11}$. Hence it follows that $f_{1i} b f_{j1} = f_{1i} \Phi(e_{1i} a e_{j1}) f_{j1}$ for every pair i and j and so we have $b = \Phi(a)$.

We used sequences but everything can easily be generalised to nets.   □

Definition 5.8.5 (See [38]) Let V be a complex vector space. Let $M_n(V)$ denotes the linear space of all $n \times n$ matrices $\mathbf{v} = (v_{ij})_{1 \leq i,j \leq n}$ with $v_{ij} \in V$. Take any $\phi \in L(V, W)$ where W is another complex vector space. We define a linear map from $M_n(V)$ into $M_n(W)$ by $\phi_n(\mathbf{v}) = (\phi(v_{ij}))_{1 \leq i,j \leq n}$ $(\mathbf{v} = (v_{ij}) \in M_n(V))$.

Let H be a Hilbert space and let V be a unital $*$-subspace which is closed in $L(H)$ with respect to the norm topology. We call V an *operator system on the Hilbert space* H.

We introduce the order "$\leq$" on the system V by the positive cone $V^+ = V \cap L(H)^+$.

Take any $n \in \mathbb{N}$. As before, we define the canonical map $\pi : M_n(L(H)) \longmapsto L(H) \otimes M_n(\mathbb{C})$ defined by $\pi((a_{ij})) = \sum_{i,j=1}^n a_{ij} \otimes e_{ij}((a_{ij}) \in M_n(L(H)))$. Then the map $\pi |_V : M_n(V) \longmapsto V \otimes M_n(\mathbb{C})$ induces a topological isomorphism.

We can introduce an order on $M_n(V)$ by $M_n(V) \cap M_n(L(H))^+$ and the space $M_n(V)$ is order isomorphic, as an ordered Banach space, with a unit $1_n = (\delta_{i,j} 1_H)$, to an operator system $V \otimes M_n(\mathbb{C})$ on $H \otimes \mathbb{C}^n$. For any $\mathbf{v} \in M_n(V)_{sa}$, we have $\mathbf{v} = (\mathbf{v} + \|\mathbf{v}\| \cdot 1_n) - \|\mathbf{v}\| \cdot 1_n$ and so

$$M(V)_{sa} = M_n(V)^+ - M_n(V)^+,$$

where $M_n(V)^+ = M_n(V) \cap M_n(L(H))^+$.

Let H_1 and H_2 be Hilbert spaces and let $V_i \subset L(H_i)$ be operator systems for $i = 1, 2$. Suppose $\phi : V_1 \longmapsto V_2$ is a linear map. We say that ϕ is n-positive if ϕ_n is positive, that is, $\phi_n(\mathbf{v}) \geq 0$ if $\mathbf{v} \in M_n(V)^+$. We call ϕ completely positive if ϕ is n-positive for every $n \in \mathbb{N}$. The map ϕ is an n-isometry if ϕ_n is an isometry. We call ϕ a complete isometry if for every $n \in \mathbb{N}$, ϕ is an n-isometry.

The following technical lemma is borrowed from [38] and will be used to show Lemma 5.8.7.

Lemma 5.8.6 *Let C be a unital C^*-algebra and take any $a \in C$ and put $\tilde{a} = \begin{pmatrix} 0 & a \\ a^* & 0 \end{pmatrix} \in M_2(C)_{sa}$. Then $\|a\| = \|\tilde{a}\|$. Also $\|a\| \leq 1$ if, and only if, $-1_2 \leq \tilde{a}$, where $1_2 = \begin{pmatrix} 1 & 0 \\ 0 & 1 \end{pmatrix}$ is the unit of $M_2(C)$, that is, if, and only if, $\begin{pmatrix} 1 & a \\ a^* & 1 \end{pmatrix} \geq 0$.*

Proof Clearly $\tilde{a}^* \tilde{a} = \begin{pmatrix} aa^* & 0 \\ 0 & a^*a \end{pmatrix}$. So we have $\|\tilde{a}\|^2 = \sup\{\|aa^*\|, \|a^*a\|\} = \|a\|^2$ and so $\|\tilde{a}\| = \|a\|$ follows.

Suppose that $\|a\| \leq 1$. Then $\|\tilde{a}\| \leq 1$ and so $\begin{pmatrix} 1 & a \\ a^* & 1 \end{pmatrix} = 1_2 + \tilde{a} \geq 0$.

Conversely if $1_2 + \tilde{a} \geq 0$. Then we have

$$1_2 - \tilde{a} = \begin{pmatrix} 1 & -a \\ -a^* & 1 \end{pmatrix} = \begin{pmatrix} 1 & 0 \\ 0 & -1 \end{pmatrix} \begin{pmatrix} 1 & a \\ a^* & 1 \end{pmatrix} \begin{pmatrix} 1 & 0 \\ 0 & -1 \end{pmatrix} \geq 0.$$

Hence it follows that $-1_2 \leq \tilde{a} \leq 1_2$, that is, $\|a\| = \|\tilde{a}\| \leq 1$.   □

Lemma 5.8.7 *If ϕ is a unital linear map from a unital C^*-algebra A into a unital C^*-algebra B, then ϕ is a complete isometry if, and only if, ϕ is completely bipositive.*

Proof Suppose that ϕ is a (complete) isometry. Take any $a \in A$ with $a \geq 0$. We shall show $\phi(a) \geq 0$. To do this, we may assume $\|a\| \leq 1$. Then, $\|a - 1\| \leq 1$ and so $\|\phi(a) - 1\| \leq 1$.

Take any $\sigma \in S_B$, the state space of B. Let $\sigma \circ \phi(a) = \alpha + i\beta$ with $\alpha, \beta \in \mathbb{R}$.

Let $a_m = a - \alpha 1 + i m \beta 1$ for each positive integer m. Then we have $\sigma \circ \phi(a_m) = i(1 + m)\beta$, which implies that

$$(1 + m)|\beta| \leq \|\phi(a_m)\| = \|a_m\| \leq \sqrt{\|a - \alpha 1\|^2 + m^2 \beta^2}.$$

Hence it follows that $(1 + 2m)\beta^2 \leq \|a - \alpha 1\|^2$ for all m and so $\beta = 0$ follows. So, $\sigma(\phi(a)) \in \mathbb{R}$ for all $\sigma \in S_B$, that is, $\phi(a) \in B_{sa}$. The fact that $\|\phi(a) - 1\| \leq 1$ now tells us that $-1 \leq 1 - \phi(a) \leq 1$. So $\phi(a) \geq 0$ follows.

Let $a \in A$ such that $\phi(a) \geq 0$. To show that $a \geq 0$, we may assume that $\|a\| \leq 1$. Then $a = a_1 + i a_2$ where a_1 and a_2 are self-adjoint. So $\phi(a) = \phi(a_1) + i\phi(a_2)$. Since $\phi(a)$, $\phi(a_1)$ and $\phi(a_2)$ are self-adjoint, it follows that $\phi(a_2) = 0$. So, $0 = \|\phi(a_2)\| = \|a_2\|$, that is, $a_2 = 0$, which implies $a = a_1 \in A_{sa}$. Since $0 \leq \phi(a) \leq 1$, $1 \geq \|1 - \phi(a)\| = \|\phi(1 - a)\| = \|1 - a\|$. Hence it follows that $\|1 - a\| \leq 1$ and $a \geq 0$ follows. So ϕ is a completely bipositive map.

Conversely suppose that ϕ is completely bipositive. We shall show that ϕ is a complete isometry. To do this, take any $a \in A$ and take any positive real number ε. Put $b = \frac{a}{\|\phi(a)\| + \varepsilon} \in A$. Then $\|\phi(b)\| = \frac{\|\phi(a)\|}{\|\phi(a)\| + \varepsilon} < 1$, which implies that

$$\phi_2\left(\begin{pmatrix} 1 & b \\ b^* & 1 \end{pmatrix}\right) = \phi_2\left(\begin{pmatrix} 1 & \frac{a}{\|\phi(a)\| + \varepsilon} \\ \frac{a^*}{\|\phi(a)\| + \varepsilon} & 1 \end{pmatrix}\right)$$

$$= \begin{pmatrix} 1 & \frac{\phi(a)}{\|\phi(a)\| + \varepsilon} \\ \frac{\phi(a^*)}{\|\phi(a)\| + \varepsilon} & 1 \end{pmatrix} \geq 0.$$

Since ϕ_2 is bipositive, we have $\begin{pmatrix} 1 & \frac{a}{\|\phi(a)\| + \varepsilon} \\ \frac{a^*}{\|\phi(a)\| + \varepsilon} & 1 \end{pmatrix} \geq 0$, that is, $\|a\| \leq \|\phi(a)\| + \varepsilon$

for all such ε, which implies that $\|a\| \leq \|\phi(a)\|$. A similar argument shows that

$\|\phi(a)\| \leq \|a\|$, by putting $b = \frac{a}{\|a\|}$ and using the positivity of ϕ_2. Hence ϕ is an isometry. By Lemma 5.8.3, $\pi_A^{(2,n)}$ is a $*$-isomorphism of $M_2(M_n(A))$ onto $M_{2n}(A)$ such that $\pi_B^{(2,n)} \circ (\phi_n)_2 = \phi_{2n} \circ \pi_A^{(2,n)}$ and so, $(\phi_n)_2$ is also bipositive for every $n \in \mathbb{N}$. By making use of this and Lemma 5.8.6, the same reasoning as above also tells us that ϕ_n is an isometry. Since n is arbitrarily chosen, it follows that ϕ is a complete isometry. $\qquad \square$

Let us recall that a C^*-algebra A is small if there exists a separable Hilbert space H and a unital complete isometry from A into $L(H)$.

Next, we shall give a useful alternative characterisation of small C^*-algebras. To do this we need the following lemma (See Lemma 2 and Theorem 1 in [137].):

Lemma 5.8.8 *Let B be a unital C^*-algebra. Suppose that there exist a separable Hilbert space H and a unital isometry $\Psi : B \mapsto L(H)$. Then S_B is $\sigma(B^*, B)$-separable.*

Proof Since $\Psi : B \mapsto L(H)$ is a unital isometry it is bipositive onto $\Psi[B]$. So the state space of B may be identified with the state space of $\Psi[B]$.

For each state ϕ of $L(H)$ let $R\phi$ be the restriction of ϕ to $\Psi[B]$. Then R is a continuous affine map of the state space of $L(H)$ into the state space of $\Psi[B]$.

Let η be a state of $\Psi[B]$. Then by the Hahn-Banach Theorem, η can be extended to a self-adjoint ϕ, where $\|\eta\| = \|\phi\| = \phi(1) = 1 = \eta(1)$. So ϕ (Lemma III.3.2 of [161] or Lemma 3.1.4 of [121]) is a state of $L(H)$ such that $R\phi = \eta$. So R is a continuous map from the state space of $L(H)$ onto the state space of $\Psi[B]$.

But, by Proposition 5.2.18, $L(H)$ has a separable state space. So S_B is the continuous image of a separable space. Hence S_B is separable. $\qquad \square$

Theorem 5.8.9 *Let A be a unital C^*-algebra. Then A is small if, and only if, for each $n \in \mathbb{N}$, the state space of $M_n(A)$ is separable.*

Proof If A is small, then for each n, there exists a unital isometry from $M_n(A)$ into $L(K)$ for some separable Hilbert space K. By Lemma 5.8.8, the state space of $M_n(A)$ is separable.

Conversely, suppose the state space of $M_n(A)$ is separable for each $n \in \mathbb{N}$. We shall show A is small. To do this, take any $n \in \mathbb{N}$. Note that $M_n(A) \cong A \otimes M_n(\mathbb{C})$ via the canonical map Π defined, in Sect. 2.4 (see the fourth paragraph after Definition 2.4.2), by

$$\Pi((t_{ij})) = \sum_{i,j=1}^{n} t_{ij} \otimes e_{ij}^{(n)} \text{ for } (t_{ij}) \in M_n(A).$$

Note that for each map $\Phi : A \mapsto B$ with a C^*-algebra B, we have $\Phi_n = \Pi^{-1} \circ (\Phi \otimes \iota_{M_n(\mathbb{C})}) \circ \Pi$ and so we may assume that $M_n(A) = A \otimes M_n(\mathbb{C})$. Since $S(M_n(A))$, the state space of $M_n(A)$, is separable, there is a dense subset $\{\varphi_{(n,p)} : p \in \mathbb{N}\}$ in $S(M_n(A))$. Let $\{\pi_{(n,p)}, H_{(n,p)}, \xi_{(n,p)}\}$ be the Gelfand-Naimark-Segal representation of

$M_n(A)$ defined by $\varphi_{(n,p)}$. Let us note that for all $a \in A$ and for all i and j,

$$(a \otimes 1_{\mathbb{C}^n})(1_A \otimes e_{ij}^{(n)}) = (1_A \otimes e_{ij}^{(n)})(a \otimes 1_{\mathbb{C}^n}).$$

So for all $(a_{ij}) \in M_n(A)$, $1 \leq i,j \leq n$, we can write, on noting $\varphi_{(n,p)}((a_{ij})) = (\pi_{(n,p)}((a_{ij}))\xi_{(n,p)}, \xi_{(n,p)})$,

$$\varphi_{(n,p)}((a_{ij})) = \sum_{1 \leq i,j \leq n} (\pi_{(n,p)}(a_{ij} \otimes e_{ij}^{(n)})\xi_{(n,p)}, \xi_{(n,p)})$$

$$= \sum_{1 \leq i,j \leq n} (\pi_{(n,p)}(a_{ij} \otimes 1_{\mathbb{C}^n})\pi_{(n,p)}(1_A \otimes e_{1j}^{(n)})\xi_{(n,p)}, \pi_{(n,p)}(1_A \otimes e_{1i}^{(n)})\xi_{(n,p)}).$$

Let $\sigma_{(n,p)}$ be the $*$-representation of A on $H_{(n,p)}$ defined by

$$\sigma_{(n,p)}(a) = \pi_{(n,p)}(a \otimes 1_{\mathbb{C}^n}) \text{ for each } a \in A.$$

Let $\{\pi, K\}$ be the direct sum of $\{(\sigma_{(n,p)}, H_{(n,p)}) : n, p \in \mathbb{N}\}$. Since $\{\varphi_{(1,p)} : p \in \mathbb{N}\}$ is $\sigma(A^*, A)$-dense in S_A, π is faithful. The closed subspace of K, generated by $\{\pi_{(n,p)}(1_A \otimes e_{ij}^{(n)})\xi_{(n,p)} : i,j = 1, \cdots, n, n \in \mathbb{N}, p \in \mathbb{N}\}$, shall be denoted by H. Then H is a separable Hilbert space. Let Q be the orthogonal projection on H and let us define the map Γ by $\Gamma(a) = Q\pi(a)Q|_H$ for each $a \in A$. Clearly Γ is a unital completely positive linear map by Lemma 5.8.1 (1).

Take any $n \in \mathbb{N}$ and suppose that $(a_{ij}) \in M_n(A)$ satisfies $\Gamma_n((a_{ij})) \geq 0$, that is, $(\Gamma(a_{ij})) \geq 0$. We shall show $(a_{ij}) \geq 0$. To do this, take any $\{x_1, \cdots, x_n\} \subset L(H)$ and $\xi \in H$. Then we have $\sum_{1 \leq i,j \leq n}(x_i^*\Gamma(a_{ij})x_j\xi, \xi) \geq 0$, that is, $\sum_{1 \leq i,j \leq n}(x_i^*Q\pi(a_{ij})Qx_j\xi, \xi) \geq 0$. Note that $\pi_{(n,p)}(1_A \otimes e_{ij}^{(n)})H_{(n,p)} \subset H_{(n,p)}$ and so $\pi_{(n,p)}(1_A \otimes e_{ij}^{(n)})H \subset H$ for all i and j with $1 \leq i,j \leq n$. Put $\xi = \xi_{(n,p)}$ and $x_j = \pi_{(n,p)}(1_A \otimes e_{1j}^{(n)})|_H$ $(j = 1, \cdots, n)$. Since $\pi|_{H_{(n,p)}} = \sigma_{(n,p)}$, it follows that $Q\pi_{(n,p)}(1_A \otimes e_{1j}^{(n)})\xi_{(n,p)} = \pi_{(n,p)}(1_A \otimes e_{1j}^{(n)})\xi_{(n,p)}$ $(j = 1, \cdots, n)$ and so we have

$$\sum_{1 \leq i,j \leq n} (\sigma_{(n,p)}(a_{ij})\pi_{(n,p)}(1_A \otimes e_{1j}^{(n)})\xi_{(n,p)}, \pi_{(n,p)}(1_A \otimes e_{1i}^{(n)})\xi_{(n,p)}) \geq 0,$$

which implies that $\varphi_{(n,p)}((a_{ij})) \geq 0$ for all $p \in \mathbb{N}$. Since $\{\varphi_{(n,p)} : p \in \mathbb{N}\}$ is dense in $S(M_n(A))$, we also have that $\varphi((a_{ij})) \geq 0$ for all $\varphi \in S(M_n(A))$. So, we have $(a_{ij}) \in M_n(A)_{sa}$ and $(a_{ij}) \geq 0$. Therefore, Γ_n is bipositive for every $n \in \mathbb{N}$. Hence by Lemma 5.8.7, Γ is a complete isometry.   □

The following technicalities will not be used in this chapter but since they are an easy, "real variable", analogue of Lemma 5.8.7 this is a convenient place for them.

Proposition 5.8.10 *Let A and B be partially ordered vector spaces with order units and each equipped with the order unit norm. Let $\phi : A \mapsto B$ be a linear map with $\phi(1) = 1$. Then ϕ is positive if, and only if, $||\phi|| \leq 1$.*

Proof First suppose $||\phi|| \leq 1$. Let $0 \leq x \leq 1$. Then $0 \leq 1-x \leq 1$. So $||1-\phi(x)|| \leq 1$. Then

$$-1 \leq 1 - \phi(x) \leq 1.$$

So $\phi(x) \geq 0$. From this it follows that ϕ is positive.

Conversely, suppose that ϕ is positive. Let $-1 \leq x \leq 1$. Then $-1 \leq \phi(x) \leq 1$. So $||\phi|| \leq 1$. $\qquad\square$

Corollary 5.8.11 *Let A and B be as above. Let $\phi : A \mapsto B$ be a positive linear map with $\phi(1) = 1$. Let ϕ be an isometry. Then ϕ^{-1} is positive.*

Proof Apply the proposition to ϕ^{-1}. $\qquad\square$

5.9 Small C^*-Algebras and Regular σ-Completions

In this section we shall see that the regular (σ-)completion of a small C^*-algebra is also small. Since each small C^*-algebra has a separable state space, by Proposition 5.2.19, it has a faithful state. So if a small C^*-algebra is monotone σ-complete then it must be monotone complete. *In particular the regular σ-completion coincides with the regular completion.*

First of all, we shall state and prove the following technical lemma and its corollary which are based on [61].

Lemma 5.9.1 *Let C be a unital C^*-algebra and let $\hat{C}$ be the regular σ-completion of C. Take any non-zero projection d in C. Then $d\hat{C}d$ is a regular σ-completion of dCd.*

Proof Clearly $d\hat{C}d$ is monotone σ-complete and is monotone σ-generated by dCd. To show that dCd is a regular subalgebra of $d\hat{C}d$, take any $y \in (d\hat{C}d)_{sa}$. Then by Lemma 2.1.36, one has a bounded subset L of $L(y) \subset C_{sa}$ such that $y = \sup L$ in C_{sa}. Then by Lemma 2.1.9, $y = dyd = \sup dLd$ in C_{sa}. Since $dLd \subset L(dyd) \cap (dCd)_{sa}$, it follows that $y = \sup\{b \in (dCd)_{sa} : b \leq y\}$ in $(d\hat{C}d)_{sa}$. Hence $d\hat{C}d$ is a regular σ-completion of dCd. $\qquad\square$

Corollary 5.9.2 *Let B be a unital C^*-algebra with the regular σ-completion $\hat{B}$. For each $n \in \mathbb{N}$, the algebra $M_n(\hat{B})$ is the regular σ-completion of $M_n(B)$.*

Proof Let us put $C = M_n(B)$ and let $\{e_{ij}\}_{1 \leq i,j \leq n}$ be the standard system of matrix units for C. Then clearly $\{e_{ij}\}_{1 \leq i,j \leq n}$ is also a system of matrix units in $\hat{C}$. On writing the $n \times n$ matrix whose $(1, 1)$ entry is $b \in \hat{B}$ and other entries are all 0 by $\begin{pmatrix} b & 0 \\ 0 & 0 \end{pmatrix}$ and

$\mathbf{e}_{11} = e \in C$, it follows that the reduced algebra $eM_n(\hat{B})e = \left\{ \begin{pmatrix} b & 0 \\ 0 & 0 \end{pmatrix} : b \in \hat{B} \right\}$ is a

regular σ-completion of the reduced algebra $eCe = \left\{ \begin{pmatrix} c & 0 \\ 0 & 0 \end{pmatrix} : c \in B \right\}$, that is also

$*$-isomorphic to $e\hat{C}e$ by Lemma 5.9.1. Hence by the uniqueness of the regular σ-completions, there exists a $*$-isomorphism $\pi : e\hat{C}e \mapsto eM_n(\hat{B})e$ such that $\pi(\mathbf{a}) = \mathbf{a}$

for all $\mathbf{a} = \begin{pmatrix} a & 0 \\ 0 & 0 \end{pmatrix}$ with $a \in B$. Hence by Proposition 5.8.4, there exists a unique

$*$-isomorphism Φ from $\hat{C}$ onto $M_n(\hat{B})$ such that $\Phi|_{e\hat{C}e} = \pi$ and $\Phi(\mathbf{e}_{ij}) = \mathbf{e}_{ij}$ for every pair i and j with $1 \leq i,j \leq n$. Since for each $x \in C = M_n(B)$, $\mathbf{e}_{1i}x\mathbf{e}_{j1} \in eCe$ with $e = \mathbf{e}_{11}$, it follows that

$$\Phi(x) = \sum_{i,j=1}^{n} \mathbf{e}_{i1} \pi(\mathbf{e}_{1i}x\mathbf{e}_{j1})\mathbf{e}_{1j} = \sum_{i,j=1}^{n} \mathbf{e}_{i1}\mathbf{e}_{1i}x\mathbf{e}_{j1}\mathbf{e}_{1j} = \sum_{i,j=1}^{n} \mathbf{e}_{ii}x\mathbf{e}_{jj} = x$$

for every $x \in M_n(B)$ and so $\Phi|_C = \iota_C$. This implies that $M_n(\hat{B})$ is the regular σ-completion of $M_n(B)$. $\square$

Definition 5.9.3 (See, for example, Definition XV.1.2 (Volume III) of [162] and [38]) A C^*-algebra B is *injective* if every completely positive map ϕ from any unital *operator* subsystem (norm closed self-adjoint subspace with the identity of D) C of any unital C^*-algebra D into B can be extended to a completely positive map $\hat{\phi}$ from D into B.

Lemma 5.9.4 *Let H be a Hilbert space and let $L(H)$ be the type I von Neumann factor of all bounded linear operators on H. Then $L(H)$ is injective.*

Proof This is well-known. See, the proof of Theorem XV.1.1 (Volume III) of [162].
                                                                                                $\square$

Theorem 5.9.5 *Let A be a unital C^*-algebra. Suppose that A is small. Then the regular σ-completion $\hat{A}$ is also small and so $\hat{A}$ has a faithful state, which implies that $\hat{A}$ is monotone complete and $(\hat{A}, \iota)$ is the regular completion of A.*

Proof Let $\phi : A \mapsto L(H)$ be any unital complete isometry, where H is a separable Hilbert space. By Lemma 5.9.4, there exists a unital completely positive contractive linear map $\hat{\phi}$ from $\hat{A}$ into $L(H)$ such that $\hat{\phi}|_A = \phi$. Take any $n \in \mathbb{N}$ and consider the map $\hat{\phi}_n : M_n(\hat{A}) \mapsto M_n(L(H))$.

We shall show that $\hat{\phi}_n$ is bipositive. To do this, take any $\mathbf{a} \in M_n(\hat{A})_{sa}$ with $\hat{\phi}_n(\mathbf{a}) \geq 0$. Note that $\hat{\phi}_n|_{M_n(A)} = \phi_n$ and so ϕ_n is an isometry and so it is bipositive (see Lemma 5.8.7). For every $\mathbf{b} \in M_n(A)_{sa}$ with $\mathbf{a} \leq \mathbf{b}$, we have $\phi_n(\mathbf{b}) \geq \hat{\phi}_n(\mathbf{a}) \geq 0$ and so $\phi_n(\mathbf{b}) \geq 0$. Since ϕ_n is bipositive, we have $\mathbf{b} \geq 0$, that is, for every $\mathbf{b}$ with $\mathbf{b} \geq \mathbf{a}$ and $\mathbf{b} \in M_n(A)_{sa}$. Note that $\mathbf{a} = \inf\{\mathbf{b} \in M_n(A)_{sa} : \mathbf{b} \geq \mathbf{a}\}$ in $M_n(\hat{A})_{sa}$ by Corollary 5.9.2 and we have that $\mathbf{a} \geq 0$. So, $\hat{\phi}_n$ is bipositive for every $n \in \mathbb{N}$. Again by Lemma 5.8.7, it follows that $\hat{\phi}$ is a unital complete isometry. So, $\hat{A}$ is also small.

This implies that the regular σ-completion $\hat{A}$ has a faithful state. Hence, as remarked in Lemma 3.1.1, $\hat{A}$ is monotone complete and so $(\hat{A}, \iota)$ is a regular completion of A. This completes the proof.   □

The next Corollary is a direct consequence of Theorem 5.9.5.

Corollary 5.9.6 *Let A be a unital C^*-algebra which has a faithful representation on a separable Hilbert space, in particular, when A is norm separable. Then the regular σ-completion $\hat{A}$ is a small C^*-algebra and so $(\hat{A}, \iota)$ is a regular completion of A.*

We shall use ℓ^2-summable sequences in A (see Proposition 2.1.38) to state and prove an infinite version, of Proposition 5.8.4, which is based on [63].

Proposition 5.9.7 *Let A and B be monotone σ-complete unital C^*-algebras. Let $\{e_{ij} : i,j \in \mathbb{N}\}$ (respectively $\{f_{ij} : i,j \in \mathbb{N}\}$) be a system of matrix units for A (respectively for B) such that $(\sum_{i=1}^{n} e_{ii}) \uparrow 1_A$ in A_{sa} (respectively, $(\sum_{i=1}^{n} f_{ii}) \uparrow 1_B$ in B_{sa}). Suppose that there exists a unital completely positive linear map ϕ from $e_{11}Ae_{11}$ into $f_{11}Bf_{11}$. Then there exists a unique unital completely positive linear map $\Phi : A \mapsto B$ such that $\Phi|_{e_{11}Ae_{11}} = \phi$ and $\Phi(e_{ij}) = f_{ij}$ for all $i,j \in \mathbb{N}$. If ϕ is a σ-normal $*$-homomorphism, then so is Φ. In particular, if ϕ is a $*$-isomorphism from $e_{11}Ae_{11}$ onto $f_{11}Bf_{11}$, then Φ is a $*$-isomorphism from A onto B.*

Proof We shall prove Proposition 5.9.7 for a σ-normal $*$-homomorphism ϕ. By working through Exercise 5.9.8, the reader will be able to prove the general result. Let us put $e_n = \sum_{i=1}^{n} e_{ii}$ and $f_n = \sum_{i=1}^{n} f_{ii}$ for each n. Then e_nAe_n and f_nBf_n satisfy the conditions of Proposition 5.8.4 and so one finds a unique completely positive linear map $\Phi(n)$ from e_nAe_n into f_nBf_n such that $\Phi(n)|_{e_{11}Ae_{11}} = \phi$ and $\Phi(n)(e_{ij}) = f_{ij}$ for all i,j with $1 \le i,j \le n$. If ϕ is a (σ-normal) $*$-homomorphism, then $\Phi(n)$ is also a σ-normal $*$-homomorphism from e_nAe_n into f_nBf_n which is defined by

$$\Phi(n)(e_n x e_n) = \sum_{i,j=1}^{n} f_{i1}\phi(e_{1i}xe_{j1})f_{1j} \tag{#}$$

for all $x \in A$.

We shall show $\sum_{i=1}^{\infty} f_{i1}\left\{\sum_{j=1}^{\infty} \phi(e_{1i}xe_{j1})f_{1j}\right\}$ does exist in B for each $x \in A$. To do this, take $x \in A$ and first of all, for each i, we shall check that $\{\phi(e_{1i}xe_{j1})^*\}_{j=1}^{\infty}$ is ℓ^2-summable in B. Since ϕ is a $*$-homomorphism, this is a direct consequence of the following calculation

$$\sum_{j=1}^{n} \phi(e_{1i}xe_{j1})\phi(e_{1i}xe_{j1})^* = \phi\left(\sum_{j=1}^{n} e_{1i}xe_{jj}x^*e_{i1}\right) \le \phi(e_{1i}xx^*e_{i1}) \text{ for all } n.$$

On noting that $\{f_{1j}\}_{j=1}^{\infty}$ is also ℓ^2-summable in B, Lemma 2.1.38 tells us that $LIM_{n\to\infty}\sum_{j=1}^{n}\phi(e_{1i}xe_{j1})f_{1j}$, say $b(i)$, does exist in B. By Corollary 2.1.23,

$$b(i)f_n = \sum_{j=1}^{n}\phi(e_{1i}xe_{j1})f_{1j}$$

and

$$f_{i1}b(i) = LIM_{n\to\infty}\sum_{j=1}^{n}f_{i1}\phi(e_{1i}xe_{j1})f_{1j}$$

for each i and n. Next we shall show that $\{b(i)\}$ is also ℓ^2-summable in B. In fact, since ϕ is a $*$-homomorphism,

$$f_n\left(\sum_{i=1}^{k}b(i)^*b(i)\right)f_n = \sum_{i=1}^{k}\left(\sum_{s,t=1}^{n}f_{s1}\phi(e_{1s}x^*e_{i1})\phi(e_{1i}xe_{t1})f_{1t}\right)$$

$$= \sum_{s,t=1}^{n}f_{s1}\phi\left(e_{1s}x^*\left(\sum_{i=1}^{k}e_{ii}\right)xe_{t1}\right)f_{1t}$$

$$= \Phi(n)(e_nx^*e_kxe_n) \leq \Phi(n)(e_nx^*xe_n) \leq \|x\|^2 f_n$$

for all n, which implies, by Lemma 2.1.24, that $\sum_{i=1}^{k}b(i)^*b(i) \leq \|x\|^2 1_B$ for all k and so $\{b(i)\}$ is ℓ^2-summable. By Lemma 2.1.38, $LIM_{k\to\infty}\sum_{i=1}^{k}f_{i1}b(i)$ does exist in B. For each $x \in A$, let

$$\Phi(x) = \sum_{i=1}^{\infty}f_{i1}\left\{\sum_{j=1}^{\infty}\phi(e_{1i}xe_{j1})f_{1j}\right\} \in B.$$

Clearly for each n in $\mathbb{N}$ and for all $x \in A$, $\Phi(x)$ satisfies

$$f_n\Phi(x)f_n = \sum_{i=1}^{n}f_{i1}\left(\sum_{j=1}^{n}\phi(e_{1i}xe_{j1})f_{1j}\right) = f_n\Phi(n)(e_nxe_n)f_n. \qquad (\#\#)$$

Since $\Phi(n)$ is unital and linear by Lemma 2.1.24 and Proposition 2.1.22 and positive by Lemma 2.1.24, Φ is also unital, linear and positive. Moreover, as was noted in Proposition 5.8.4, $\Phi(n)$ is also a σ-normal map from e_nAe_n into f_nBf_n. To show that Φ is σ-normal, take any increasing norm bounded sequence (a_n) such that $a_n \uparrow a$ in A_{sa} for some $a \in A_{sa}$. Recall that, by Propositions 2.1.10 and 2.2.25, for any fixed $m \in \mathbb{N}$, $e_ma_ne_m \uparrow e_mae_m$ in $(e_mAe_m)_{sa}$. Since Φ is positive and unital, $(\Phi(a_n))$ is also norm bounded and increasing. Since B is monotone σ-complete, $\Phi(a_n) \uparrow b$ in

B_{sa} for some $b \in B_{sa}$. This implies, again by Propositions 2.1.10 and 2.2.25, that $f_m\Phi(a_n)f_m \uparrow f_m b f_m$ in $(f_m B f_m)_{sa}$ for each $m \in \mathbb{N}$. Since (##) for each $x \in A$, we have $f_k\Phi(x)f_k = \Phi(k)(e_k x e_k)$ for all k, $f_m\Phi(a_n)f_m = \Phi(m)(e_m a_n e_m) \uparrow \Phi(m)(e_m a e_m) = f_m\Phi(a)f_m$ in $(f_m B f_m)_{sa}$ for all m and so it follows that $f_m b f_m = f_m\Phi(a)f_m$ for each m. Hence we have, by Lemma 2.1.24, $b = \Phi(a)$ and so Φ is σ-normal.

We shall show that $\Phi(xy) = \Phi(x)\Phi(y)$ for all $x, y \in A$. Indeed, take any m, n in $\mathbb{N}$ with $n < m$. We have then, by (##),

$$f_m\Phi(xe_ny)f_m = \Phi(m)(e_m x e_m)f_n\Phi(m)(e_m y e_m) = f_m\Phi(x)f_n\Phi(y)f_m,$$

which implies that $f_m(\Phi(xe_ny) - \Phi(x)f_n\Phi(y))f_m = 0$ for all m.

Put $a = \Phi(xe_ny) - \Phi(x)f_n\Phi(y)$. Then, for all m, $f_m a f_m = 0$. Taking adjoints gives $f_m a^* f_m = 0$. Hence it follows, by Lemma 2.1.24 $a = 0$, that is, $\Phi(xe_ny) = \Phi(x)f_n\Phi(y)$ for all n. Since Φ is σ-normal, we have $\Phi(xy) = \Phi(x)\Phi(y)$. So, Φ is a $*$-homomorphism from A into B. The uniqueness can be shown as in Proposition 5.8.4.

For a general completely positive linear map ϕ, see Exercise 5.9.8. □

Exercise 5.9.8 Suppose that ϕ is a completely positive unital linear map from $e_{11}Ae_{11}$ into $f_{11}Bf_{11}$.

(1) Prove that $\{\phi(e_{1i}xe_{n1})^*\}_{n=1}^\infty$ is ℓ^2-summable in B. Use Lemma 5.8.1 (2).
(2) Let $b(i) = \sum_{n=1}^\infty \phi(e_{1i}xe_{n1})f_{1n} \in B$. Then $\{b(i)\}$ is ℓ^2-summable in B. Use the fact that $f_{1i}f_{j1} = \delta_{i,j}f_{11}$ and $\phi(x)\phi(y) = \phi(x)f_{11}\phi(y)$.
(3) By making use of (1) and (2), define the map Φ from A into B in a way which makes sense and show that the map Φ is completely positive. Here we may assume that $M_m(A) = A \otimes M_m(\mathbb{C})$ and $M_m(B) = B \otimes M_m(\mathbb{C})$. Show that $\Phi \otimes \iota_m$ is positive by making use of (##) for $\phi \otimes \iota_m$.

Theorem 5.9.9 *Let A be a unital monotone σ-complete C^*-algebra with an $\aleph_0 \times \aleph_0$-system of matrix units $\{e_{ij} : i, j \in \mathbb{N}\}$ such that $\sum_{i=1}^n e_{ii} \uparrow 1$ in A_{sa}. Suppose that $e_{11}Ae_{11}$ has a faithful representation on a separable Hilbert space. Then A is small.*

Proof We may assume that $e_{11}Ae_{11}$ acts on a separable Hilbert space H. Let B be the C^*-subalgebra of $L(H \otimes \ell^2(\mathbb{N}))$ generated by $e_{11}Ae_{11} \otimes L(\ell^2(\mathbb{N}))$. Since B acts faithfully on a separable Hilbert space $H \otimes \ell^2(\mathbb{N})$, it is small. We shall show that the regular σ-completion $\hat{B}$ of B is $*$-isomorphic to A, making use of ideas of Hamana[63].

Let $\{g_{ij} : i, j \in \mathbb{N}\}$ be the standard system of matrix units for $L(\ell^2(\mathbb{N}))$ and let $f_{ij} = 1_H \otimes g_{ij}$ for each pair i and j. Then, clearly, $f_{ij} \in B$ for each i and j and $\sum_{i=1}^n f_{ii} \uparrow 1_B$ in B_{sa}. Indeed, take any $t \in B_{sa}$ such that $\sum_{i=1}^n f_{ii} \leq t$ for all n. Since $\sum_{i=1}^n f_{ii}$ converges strongly to 1, it follows that $1 \leq t$, which implies that $\sum_{i=1}^n f_{ii} \uparrow 1$ in B_{sa}. Let C be the regular σ-completion of B. Then, by a general property of the regular σ-completion, $\sum_{i=1}^n f_{ii} \uparrow 1$ in C_{sa} and $\{f_{ij}\}$ is a system of matrix units for C. Moreover, $f_{11}Bf_{11} = e_{11}Ae_{11} \otimes f_{11}$ tells us that $f_{11}Bf_{11}$ is monotone σ-complete. Since, by Lemma 5.9.1, $f_{11}Cf_{11}$ is $*$-isomorphic to the regular σ-completion of $f_{11}Bf_{11}$ and $f_{11}Bf_{11}$ is monotone σ-complete, it follows that $f_{11}Cf_{11}$ is $*$-isomorphic to $e_{11}Ae_{11}$.

Hence by Proposition 5.9.7 C is $*$-isomorphic to A. Since B is small, C is also small, which means that A is small. This completes the proof.   □

We shall give here a non-commutative example. Let $H = \ell^2$ be the separable infinite dimensional Hilbert space with canonical basis and let $\{e_{mn}\}$ be the standard system of matrix units for $L(H)$.

Given any monotone complete C^*-algebra D and any von Neumann algebra E, Hamana [63] gives a construction of a corresponding monotone complete tensor product algebra $A = (D\overline{\otimes}E, \bullet)$. See also [65]. Here the product is not straightforward. But when D is commutative and $E = L(H)$, for H separable, this tensor product has an easy description as a quotient of a "conventional" tensor product [140]. In fact this approach works even when D is not commutative.

Let D be commutative and $A = (D\overline{\otimes}L(H), \bullet)$, where D is a monotone complete (commutative) C^*-algebra acting on a separable Hilbert space, with the system of matrix units $\{E_{mn}\}$ defined by $E_{mn} = 1_A \otimes e_{mn}$ for each pair m and n. Then $E_{11}AE_{11} \cong D$ and $\{E_{mn}\}$ satisfies the condition in Theorem 5.9.9. Hence A is small. It will follow from this that all the factors we construct in Chap. 7 are small, although these algebras never act on a separable Hilbert space. We shall discuss details of these algebras later.

Chapter 6
Generic Dynamics

We gave an introduction to generic dynamics at the beginning of this book. Here we shall give a brief account which emphasises orbit equivalence. This has fruitful applications to monotone complete C^*-algebras. Many open problems and interesting possibilities for further development exist.

The important survey by Weiss [165], which is mainly concerned with generic dynamics over complete metric spaces, includes topics not touched on here. See also contributions by Goldets, Kulagin and Sinel'shchikov e.g. [53] and [52]. On the other hand we include some of our recent results which are applicable to non-metrisable, separable compact Hausdorff spaces [145]. As we saw in Chap. 4, there is an abundance of such spaces. Generic dynamics can be developed in different ways. Here, our approach to generic dynamics is intended to be elementary. It is aimed at the applications to monotone complete C^*-algebras.

6.1 Basics

Since no prior knowledge of dynamical systems is required, we start with routine preliminaries and some illustrative examples.

Let G be a countable group. Let X be a non-empty set. Then an *action* of G on X is a group homomorphism ε from G into the group of all bijections of X onto X. We normally write ε_g instead of $\varepsilon(g)$. Then we shall call the triple (X, G, ε) a *G-system*. We also call ε the *action* of G on X. We call the system *non-degenerate* if ε is injective; when ε is injective then $\varepsilon[G]$ is isomorphic to G. In this chapter, we shall normally confine our attention to non-degenerate systems.

When $x_0 \in X$ the set $\{\varepsilon_g(x_0) : g \in G\}$ is the *orbit* of x_0 under the action of G; we shall frequently denote it by $G[x_0]$.

A subset Z of X is *G-invariant* if $\varepsilon_g[Z] = Z$ for each $g \in G$. Clearly each orbit is G-invariant.

© Springer-Verlag London 2015

K. Saitô, J.D.M. Wright, *Monotone Complete C*-algebras and Generic Dynamics*,
Springer Monographs in Mathematics, DOI 10.1007/978-1-4471-6775-4_6

When there is no risk of ambiguity, we shall often simplify our notation by writing "g" instead of "ε_g".

Given $F \subset X$, the *saturation* of F by G is the smallest G-invariant subset of X which contains F; this is $\bigcup_{g \in G} \varepsilon_g[F]$. We shall denote it by $G[F]$. Given $Y \subset X$, the largest subset of Y which is G-invariant is $\bigcap_{g \in G} g[Y] = X \backslash G[X \backslash Y]$.

Exercise 6.1.1 Let $G[x_0]$ be an orbit such that $\varepsilon_g(x_0) \neq x_0$ for each $g \in G$, except the identity element of G. Show that for $g \neq \iota$, ε_g leaves no point of the orbit fixed. Deduce that the action ε is non-degenerate.

An orbit $G[x_0]$ is said to be *free* if $\varepsilon_g(x_0) \neq x_0$ for each $g \in G$, except the identity ι.

If every point in X has a free orbit, then G is said to *act freely* on X.

As in the rest of this book, *all topological spaces are required to be Hausdorff*, but we sometimes emphasise this by stating that a space is Hausdorff.

When X is a (Hausdorff) topological space and ε is a homomorphism of G into $Homeo(S)$ (the group of all homeomorphisms of X onto itself) we shall call (X, G, ε) a *dynamical system*.

When we say "G *acts as a group of homeomorphisms of* X", we mean there is a non-degenerate dynamical system (X, G, ε) where $\varepsilon[G]$ is a group of homeomorphisms of X.

Example 6.1.2 Let X be the circle $\{z \in \mathbb{C} : |z| = 1\}$ and let $\rho : X \mapsto X$ be the operation of rotation through an angle θ. Clearly ρ is a homeomorphism. First suppose θ is a rational multiple of π. Then after a finite number of rotations we get back to our starting point; that is ρ^n is the identity map for some n. Now suppose that θ is an *irrational* multiple of π. Let x_0 be any point on the circle. Then $\{\rho^n(x_0) : n \in \mathbb{Z}\}$ is a dense subset of the circle. (This is "Kronecker's theorem in one dimension". See [71, Chapter 23].) So $n \mapsto \rho^n$ is an action of $\mathbb{Z}$ with a dense orbit. On putting $\varepsilon(n) = \rho^n$ we see that $(X, \mathbb{Z}, \varepsilon)$ is a dynamical system. This may be thought of as an example of "topological dynamics".

Example 6.1.3 To do "classical ergodic theory" observe that (the circumference of) the circle X comes equipped with a Lebesgue measure, μ, which is rotation invariant. In particular, ρ maps μ-null sets to μ-null sets. So ρ induces an automorphism of $L^\infty(X, \mu)$ by $h_\rho([f]) = [f \circ \rho]$. (Here $[f]$ is the equivalence class of bounded μ-measurable functions which differ from f only on a null set.) Let S_L be the spectrum of the commutative algebra $L^\infty(X, \mu)$. Since this algebra is monotone complete, S_L is extremally disconnected and $C(S_L) = L^\infty(X, \mu)$. Because of the duality between compact Hausdorff spaces and unital commutative C^*-algebras, h_ρ corresponds to a homeomorphism of S_L; this homeomorphism we denote by $\tilde{\rho}$. Then $n \mapsto \tilde{\rho}^n$ is an action of $\mathbb{Z}$ on the compact Hausdorff, extremally disconnected space S_L. But this action *cannot* have a dense orbit, because S_L is not separable; see Corollary 4.2.19. We shall come back to this point later.

Let us recall that a *Polish space* is a topological space which is homeomorphic to a complete separable metric space. For example $(-\pi, \pi)$ is not a complete metric space when equipped with the usual metric of the real numbers. But $(-\pi, \pi)$ is homeomorphic to $\mathbb{R}$ (consider $\lambda \mapsto tan\frac{\lambda}{2}$) which is a complete separable metric space. So $(-\pi, \pi)$ is a Polish space. A Polish space is said to be *perfect* if it has no isolated points.

For a topological space X, a subset Y, is a G_δ-*set* if Y is the intersection of countably many open subsets of X. Let A be a subspace of a Polish space B. Then A is a G_δ-subspace of B, if, and only if, A is a Polish space in the relative topology induced by B. See [167, Theorem 24.12] or [161, Theorem A.1, page 375]. Furthermore, each Polish space is a G_δ subspace of its Stone-Czech compactification. See the discussion in Sect. 6.5.

We discussed Baire spaces in Chap. 4. By the Baire Category Theorem, every Polish space and every compact (Hausdorff) space is a Baire space. A dense, G_δ-subset of a Baire space, is a Baire space. From this it follows that every G_δ-subset of a compact space is a Baire space.

We recall that the abelian group $\bigoplus \mathbb{Z}_2$ is called the Dyadic Group.

Example 6.1.4 Let $\mathbb{Z}_2$ be the two element field (integers mod 2), equipped with the discrete topology. Then the product $\prod_{n=1}^{\infty} \mathbb{Z}_2$ is homeomorphic to the Cantor set. It also has a natural (additive) group structure which turns it into a compact abelian group. Let $\bigoplus \mathbb{Z}_2$ be the Dyadic Group embedded as a subgroup of $\prod_{n=1}^{\infty} \mathbb{Z}_2$. So each element of $\prod_{n=1}^{\infty} \mathbb{Z}_2$ is an infinite sequence of zeroes and ones; whereas the elements of $\bigoplus \mathbb{Z}_2$ are those sequences which end in an infinite string of zeroes.

For $y \in \bigoplus \mathbb{Z}_2$ and $x \in \prod_{n=1}^{\infty} \mathbb{Z}_2$ let $\varepsilon_y(x) = y + x$. Then $(\prod_{n=1}^{\infty} \mathbb{Z}_2, \bigoplus \mathbb{Z}_2, \varepsilon)$ is a dynamical system with a dense orbit. This is because the orbit $\{\varepsilon_y(0) : y \in \bigoplus \mathbb{Z}_2\}$ coincides with $\bigoplus \mathbb{Z}_2$, which is a dense subset of the Cantor space. Clearly the action of the Dyadic Group is free.

There is a strong uniqueness theorem for countable group actions on perfect Polish spaces. We shall make this precise later. But, roughly, it says the following. Consider a non-degenerate dynamical system (X, G, ε), where X is a perfect Polish space and $\varepsilon[G]$ is an infinite countable group of homeomorphisms with a dense orbit. Then, modulo meagre sets, this dynamical system is equivalent to the one in Example 6.1.4. A consequence of this result is that Examples 6.1.2 and 6.1.4 are, modulo meagre sets, equivalent. In particular, however complicated a group G may be, (X, G, ε) is equivalent to a canonical system arising from an action of $\mathbb{Z}$.

Now suppose S is not Polish but is a compact (Hausdorff) extremally disconnected space which is separable and has no isolated points. Then there is no longer a uniqueness theorem. But we can still show the following. Suppose we are given a non-degenerate dynamical system (S, G, ε), where $\varepsilon[G]$ is an infinite countable group of homeomorphisms with at least one free dense orbit. Then this dynamical system is equivalent (modulo meagre sets) to one arising from an action of the Dyadic Group on S. Unlike the situation for perfect Polish spaces, it is *NOT* obvious that this dynamical system is equivalent to a system coming from an action of $\mathbb{Z}$. However it is true. So we give a proof. But there remains an unanswered question here (see Sect. 6.7).

In contrast to the situation for Polish spaces, there are $2^{\mathbb{R}}$ "essentially different" dynamical systems $(S, \mathbb{Z}, \varepsilon)$ where the spaces S are compact, separable and extremally disconnected and $\varepsilon[\mathbb{Z}]$ has a dense orbit.

Given a dynamical system there is an associated relation of *orbit equivalence*.

Definition 6.1.5 Let (Y, G, ε) be a non-degenerate dynamical system. Let x and y be in Y. Then x is *orbit equivalent* to y if, for some $g \in G$, $\varepsilon_g(x) = y$. It is clear that this is an equivalence relation and that the corresponding equivalence classes are just the orbits of $\varepsilon[G]$. When it is clear which action of G is being considered, we write

$$x \sim_G y$$

when x is orbit equivalent to y. We identify the relation $\sim_G$ with its graph in $Y \times Y$. This is $\{(x, \varepsilon_g(x)) : g \in G\}$.

For a large collection of separable topological spaces, Y, there is a natural method for constructing a monotone complete C^*-algebra associated with (Y, G, ε); we shall discuss this in the next chapter. This construction depends only on the orbit equivalence relation. Since different groups can give rise to the same orbit equivalence relation, it is useful to know when this happens.

These methods give rise to $2^{\mathbb{R}}$ small factors each taking different values in the classification semigroup. But, when Y is a perfect Polish space and ε is an ergodic action of G we get a canonical monotone complete factor which does not depend on the choice of Y or the choice of G.

6.2   Extending Continuous Functions

We gather together some useful topological results. For extremally disconnected spaces, the most important of these is Theorem 6.2.7.

Throughout this section, K is a Hausdorff space and D is a dense subset of K, equipped with the relative topology. It is easy to see that K has no isolated points if, and only if D has no isolated points.

Let us recall that a Hausdorff topological space T is *extremally disconnected* if the closure of each open subset is still an open set.

We shall see that when K is compact and extremally disconnected then, whenever Z is a compact Hausdorff space and $f : D \mapsto Z$ is continuous, there exists a unique extension of f to a continuous function $F : K \mapsto Z$. In other words, K is the Stone-Czech compactification of D.

For any Hausdorff space K, the closed subsets of D, in the relative topology, are all of the form $F \cap D$ where F is a closed subset of K. For any $A \subset K$, we denote the closure of A (in the topology of K) by $cl(A)$. For $A \subset D$, we note that the closure of this set in the relative topology of D is $cl(A) \cap D$. We denote this by $cl_D(A)$. We normally use $intA$ for the interior of A but, for clarity, sometimes use $int_K A$. When $A \subset D$, its interior with respect to the relative topology is denoted by $int_D A$.

The following lemmas are standard point-set topology.

Lemma 6.2.1 *Let K be a Hausdorff space and D a dense subset of K.*

(i) *For any open subset U of K we have $cl(U) = cl(U \cap D)$.*

(ii) *Let U, V be open subsets of K. Then $V \subset cl(U)$ if, and only if,*

$$V \cap D \subset cl(U \cap D) \cap D$$

$$= cl_D(U \cap D).$$

(iii) *Let U be an open subset of K. Then $D \cap int(clU) = int_D(cl_D(U \cap D))$.*

(iv) *If U is a regular open subset of K then $U \cap D$ is a regular open subset of D in the relative topology of D. Conversely, if E is a regular open subset of D in the relative topology, then $E = V \cap D$ where V is a regular open subset of K.*

(v) *Let C be a clopen subset of D. Then there exists W, an open subset of K, with $D \subset W$, and E a clopen subset of W, in the relative topology of W, such that $E \cap D = C$.*

(vi) *Let (C_n) be a sequence of pairwise disjoint, clopen subsets of D. Then there exists Y, a dense G_δ-subset of K with $D \subset Y$. Furthermore there exists (F_n), a sequence of pairwise disjoint, clopen subsets of Y, in the relative topology of Y, such that $C_n = F_n \cap D$ for each n.*

Proof

(i) This follows from Lemma 4.3.1.

(ii) If $V \subset cl(U)$ then $V \cap D \subset cl(U) \cap D$. So $V \cap D \subset cl(U \cap D)$ by (i). Hence $V \cap D \subset cl(U \cap D) \cap D$.

Conversely, suppose that $V \cap D \subset cl(U \cap D) \cap D$. Then

$$V \cap D \subset cl(U) \cap D \subset cl(U).$$

Since $cl(V) = cl(V \cap D)$ it follows that $cl(V) \subset cl(U)$. Hence $V \subset cl(U)$.

(iii) $int_K(clU) = \cup\{V : V \subset clU$ and V is open in $K\}$.

So $D \cap int_K(clU) = \cup\{V \cap D : V \subset clU$ and V is open in $K\}$.

By applying (ii) it follows that

$$D \cap int_K(clU) = \cup\{V \cap D : V \cap D \subset Cl(U \cap D) \cap D \text{ and } V \text{ is open in } K\}.$$

But the right hand side of this equation is just

$$\cup\{W : W \subset cl_D(U \cap D) \text{ and } W \text{ is open in the relative topology of } D\}.$$

(iv) First suppose that U is a regular open set. Then $U = int_K(clU)$.

By (iii), $D \cap U = int_D(cl_D(U \cap D))$. But this just says that $D \cap U$ is a regular open set in the relative topology of D.

Conversely, let us now suppose that E is a regular open set in the relative topology of D. Then, from the definition of the relative topology, $E = D \cap U$ where U is an open subset of K.

Then $D \cap U = int_D(cl_D(U \cap D))$. So applying (iii), $D \cap U = D \cap int_K(clU)$. Let $V = int_K(clU)$. Then $D \cap U = D \cap V$.

By applying (ii), we find that $V \subset clU$ and, also, $U \subset clV$. Hence $clV = clU$.

But $V = int_K(clU)$. So $V = int_K(clV)$. Thus V is a regular open subset of K.

Then V is a regular open subset of K and $D \cap V = D \cap U = E$.

(v) Since C and $D \backslash C$ are clopen subsets of D, there exist regular open subsets of K, U and V, with $C = U \cap D$ and $D \backslash C = V \cap D$. Then $U \cap V \cap D = \varnothing$. Since D is dense in K it follows that the open set $U \cap V$ is empty. Let $W = U \cup V$. Put $E = U$.

(vi) For each n, let W_n and E_n correspond to C_n as W and E correspond to C in (v). Let $Y = \bigcap_{n \geq 1} W_n$ and let $F_n = Y \cap E_n$. Then Y is a G_δ-subset of K with $D \subset Y$. Since each E_n is a clopen subset of W_n, F_n is a clopen subset of Y, in the relative topology of Y.

For $m \neq n$, $F_m \cap F_n \cap D = C_m \cap C_n = \varnothing$. Since D is dense in Y, $F_m \cap F_n$ is empty. □

For any topological space Y we recall that $RegY$ denotes the Boolean algebra of regular open subsets of Y.

Let $H : \mathcal{P}(K) \mapsto \mathcal{P}(D)$ be defined by $H(A) = A \cap D$. Here $\mathcal{P}(X)$ is the collection of all subsets of a set X.

Lemma 6.2.2 *The function H, when restricted to $RegK$, becomes a Boolean isomorphism of $RegK$ onto $RegD$.*

Proof By Lemma 6.2.1, part (iv), H maps $RegK$ onto $RegD$.

Suppose that H is a homomorphism. If $H(U) = \varnothing$ then $U \cap D = \varnothing$. Since D is dense in K and U is open, this implies that $U = \varnothing$. So H is a Boolean isomorphism.

It remains to show that H is, indeed, a homomorphism.

Let U, V be regular open sets. Then, trivially, $H(U \cap V) = H(U) \cap H(V)$.

The least upper bound of U, V in $RegK$ is $int_K(cl(U \cup V))$. It now follows from Lemma 6.2.1, part (iii), that H maps the least upper bound of U, V in $RegK$ to the least upper bound of $H(U), H(V)$ in $RegD$. Since $H(K) = D$ it now follows that H is a Boolean homomorphism; so H is a Boolean isomorphism. □

Lemma 6.2.3 *A Hausdorff topological space T is extremally disconnected if, and only if, each regular open set is closed, and hence clopen.*

Proof First suppose the space to be extremally disconnected and let U be an open subset. Then clU is open.

So $int_K(clU) = clU$. Hence, when U is a regular open set, $U = clU$. Thus each regular open set is clopen.

Conversely, suppose that each regular open set is closed (and so clopen). Let U be an open set. Let $W = int_K clU$. Then W is an open set with $U \subset W \subset clU$. On taking closure, we see that $clU = clW$. It follows that $W = int_K(clW)$, that is, W is a regular open set. So W is clopen. Hence $W = clW = clU$. Thus T is extremally disconnected. □

Corollary 6.2.4 *Let D be a dense subset of a compact Hausdorff extremally disconnected space S. Let D be equipped with the relative topology. Then D is an extremally disconnected space.*

Proof Let V be a regular open subset of D. Then, by Lemma 6.2.1, part (iv), there exists U, a regular open subset of S, such that $V = U \cap D$. By Lemma 6.2.3 U is a clopen subset of S. Hence V is a clopen subset of D in the relative topology. By using Lemma 6.2.3, we see that D is an extremally disconnected space. □

Lemma 6.2.5 *Let D be an extremally disconnected topological space. Also let D be a completely regular (Hausdorff) space. Then βD, its Stone-Czech compactification, is extremally disconnected.*

Proof Let W be a regular open subset of βD. Then $W \cap D$ is a regular open subset of D, by Lemma 6.2.2.

So $W \cap D$ is a clopen subset of D, in the relative topology. It follows that the characteristic function of this set is a continuous map from D into the two point compact space $\{0, 1\}$. So, by the fundamental property of the Stone-Czech compactification, it has a unique extension to a continuous function from βD to $\{0, 1\}$. Thus there exists a clopen set E in βD such that $E \cap D = W \cap D$. By Lemma 6.2.1 (i), this gives

$$clW = cl(W \cap D) = cl(E \cap D) = clE = E.$$

So $int(clW) = E$.

Since W is a regular open set this gives $W = E$. Hence, by Lemma 6.2.3, βX is extremally disconnected. □

Lemma 6.2.6 *Let D be a dense subspace of a compact Hausdorff extremally disconnected space Z. When A is a clopen subset of D in the relative topology, then clA is a clopen subset of Z. Let A and B be disjoint clopen subsets of D, in the relative topology. Then clA and clB are disjoint clopen subsets of Z.*

Proof By Corollary 6.2.4, D is extremally disconnected. So, by Lemma 6.2.3 and Lemma 6.2.2, there is a unique clopen subset of Z, say, U, such that $A = U \cap D$. On applying Lemma 6.2.1 we find that $clU = cl(U \cap D) = clA$. Since U is clopen, $U = clA$.

Let $V = clB$. Then U, V are clopen sets such that $H(U) = A$ and $H(V) = B$. So $H(U \cap V) = A \cap B = \emptyset$. But H is a Boolean isomorphism, so $U \cap V = \emptyset$. In other words, clA and clB are disjoint clopen sets. □

The following theorem was given by Gillman and Jerison [51] as a by-product of other results. The argument given here may be slightly easier and more direct.

Theorem 6.2.7 *Let D be a dense subspace of a compact Hausdorff extremally disconnected space S. Then S is the Stone-Czech compactification of D. More precisely, there exists a unique homeomorphism from βD onto S which restricts to the identity homeomorphism on D.*

Proof Since D is a subspace of the compact Hausdorff space, S, D is completely regular and hence has a well defined Stone-Czech compactification. By the fundamental property of βD, there exists a unique continuous surjection α from βD onto S, which restricts to the identity on D.

Let a and b be distinct points in βD. Then there exist disjoint clopen sets U and V such that $a \in U$ and $b \in V$. Let $A = U \cap D$ and $B = V \cap D$ then $U = cl_{\beta D}A$ and $V = cl_{\beta D}B$. So $\alpha[U] \subset cl_S A$ and $\alpha[V] \subset cl_S B$. By Lemma 6.2.6, $cl_S A$ and $cl_S B$ are disjoint. Hence $\alpha(a)$ and $\alpha(b)$ are distinct points of S. Thus α is injective. It now follows from compactness, that α is a homeomorphism.   □

Corollary 6.2.8 *Let D be a dense subspace of a compact Hausdorff extremally disconnected space S. Let φ be a homeomorphism of D onto itself. Then there exists a unique homeomorphism Φ of S onto itself which is an extension of φ.*

Proof Since S is the Stone-Czech compactification of D, there is a continuous function $\Phi : S \mapsto S$ which extends ϕ. Since D is dense in S, the continuous function Φ is uniquely determined. By compactness, the range of Φ is a closed subspace of S. So the range of Φ contains the closure of D, that is, Φ maps S onto S. Similarly we can extend ϕ^{-1} to a continuous Ψ mapping S onto S. But $\Psi\Phi(d) = d$ for each d in D. So, by continuity, $\Psi\Phi$ is the identity map on S. Similarly, $\Phi\Psi$ is the identity map on S. So Φ is a homeomorphism.   □

When working with Polish spaces we shall need a different extension theorem, applicable to complete metric spaces.

Theorem 6.2.9 (Lavrentiev's Theorem) *Let X_1 and X_2 be complete metric spaces. Let A_1 be a dense subset of X_1 and A_2 a dense subset of X_2. Let σ be a homeomorphism of A_1 onto A_2. Then there exist Z_1 and Z_2 where, for $j = 1, 2$, Z_j is a dense G_δ-subset of X_j and $A_j \subset Z_j$. Further, there exists an homeomorphism $\sigma^\#$ from Z_1 onto Z_2 which extends σ.*

For a proof see [97, Chapter 2, page 429] or [167, Theorem 24.9]

6.3   Ergodic Discrete Group Actions on Topological Spaces

In this section, Y is a Hausdorff topological space which has no isolated points. In most of our applications of this section, Y will also be completely regular; for example, a compact Hausdorff space with no isolated points, or a dense subset of such a space. As before, G is a countable infinite group.

Lemma 6.3.1 *Let G be a group of homeomorphisms of Y.*

(i) *If there exists $x_0 \in Y$ such that the orbit $G[x_0]$ is dense in Y then every G-invariant open subset of Y is either empty or dense.*

(ii) *If every non-empty open G-invariant subset of Y is dense then, for each x in Y, the orbit $G[x]$ is either dense or nowhere dense.*

Proof

(i) Let U be a G-invariant open set which is not empty. Since $G[x_0]$ is dense, for some $g \in G$, we have $g(x_0) \in U$. But U is G-invariant. So $x_0 \in U$. Hence $G[x_0] \subset U$. So U is dense in Y.

(ii) Suppose y is an element of Y such that $G[y]$ is not dense in Y. Then $Y \backslash clG[y]$ is a non-empty G-invariant open set. So it is dense in Y. So $clG[y]$ has empty interior.   □

Definition 6.3.2 When G is a group acting as homeomorphisms of Y, its action is said to be *ergodic* if each G-invariant open subset of Y is either empty or dense in Y.

Proposition 6.3.3 *Let D be a dense subspace of Y. Let $\varepsilon : G \mapsto Homeo(Y)$ be an injective homomorphism (so a non-degenerate action of G). Let D be G-invariant. Let $\varepsilon^D(g)$ be the restriction of $\varepsilon(g)$ to D. Then ε^D is an injective homomorphism of G into Homeo(D), so a non-degenerate action of G on D. The action ε is ergodic on Y if, and only if, ε^D acts ergodically on D.*

Proof By Lemma 6.2.1 (i), for any open $U \subset Y$, $clU = cl(U \cap D)$.

Suppose that ε^D acts ergodically on D. Let U be any non-empty, G-invariant open subset of Y. Then $U \cap D$ is a G-invariant, open subset of D. So

$$D = cl_D(U \cap D) = D \cap clU.$$

It now follows that U is dense in Y. So ε acts ergodically on Y.

Conversely, suppose that ε acts ergodically on Y. A non-empty G-invariant subset of D which is open in the relative topology of D, is of the form $U \cap D$ where U is a non-empty open set. Suppose $U \cap D$ is G-invariant. Then it is straight forward to verify that $U \cap D = G[U] \cap D$. By hypotheses, each non-empty G-invariant open subset of Y is dense in Y. So

$$Y = clG[U] = cl(G[U] \cap D) = cl(U \cap D).$$

So ε^D acts ergodically on D.   □

Lemma 6.3.4 *Let Y be an extremally disconnected space. Let G be a group acting as homeomorphisms of Y. Then the action of G is ergodic, if, and only if, the only G-invariant clopen subsets are Y and $\varnothing$.*

Proof Let U be a G-invariant open set. Then clU and $Y \backslash clU$ are G-invariant clopen sets. Then U is neither empty nor dense, if, and only if, clU and $Y \backslash clU$ are non-trivial clopen sets.   □

In Example 6.1.3 it can be shown that the action $n \mapsto \bar{\rho}^n$ is an ergodic action of $\mathbb{Z}$ on S_L. Since S_L is not separable, no orbit can be dense so, by Lemma 6.2.1 (ii), every orbit is nowhere dense.

For the first time in this section, we now require Y to be a Baire space.

Proposition 6.3.5 *Let Y be a Baire space and G a countable group of homeomorphisms of Y. Then the following properties are equivalent.*

(i) *The action of G on Y is ergodic.*
(ii) *When E is a G-invariant Borel subset of Y then either E is meagre or its compliment is meagre.*

Proof First suppose that (ii) holds. Let U be a G-invariant open subset of Y. By (ii) either U is meagre or $Y \backslash U$ is meagre. Since Y is a Baire space, meagre open subsets are empty. So either U is empty or the closed set $Y \backslash U$ has empty interior. So U is empty or it is dense in Y. Thus the action is ergodic.

Now suppose that (i) holds. Let E be a G-invariant Borel subset of Y. Each Borel subset of Y has the Baire Property, see Chap. 4, Sect. 6.1. So there exists an open set U and a meagre set M, such that

$$E = U \triangle M = (U \backslash M) \cup (M \backslash U).$$

(The symmetric difference operation, $\triangle$, was introduced in the section on Boolean algebras in Chap. 4.)

If U is empty then E is meagre, and we are done. So now suppose U is not empty. We have

$$U = U \triangle M \triangle M = E \triangle M \subset E \cup M.$$

Since E is G-invariant, for each $g \in G$, we have

$$g[U] \subset E \cup g[M].$$

So

$$G[U] \subset E \cup G[M].$$

We know that $G[U]$ is a non-empty, G-invariant open set and so, by (i), it is dense in Y. So $Y \backslash G[U]$ is a closed nowhere dense set. Thus $G[M] \cup (Y \backslash G[U])$ is a meagre set, say, M_1. Also

$$Y = G[U] \cup (Y \backslash G[U]) \subset E \cup M_1.$$

Hence the compliment of E in Y is meagre.   □

6.4 Induced Actions

Let K be a compact Hausdorff space. Then, see Lemma 4.2.20 in Chap. 4, K is separable if, and only if $C(K)$ is isomorphic to a closed (unital) $*$-subalgebra of ℓ^∞.

Let X be a completely regular (Hausdorff) Baire space.

For the definition of the regular σ-completion of an arbitrary C^*-algebra see Definition 2.1.37. In Sect. 4.2, we discussed Baire measurability and regular σ-completions of commutative algebras. We saw that the regular σ-completion of $C_b(X)$ can be identified with the monotone σ-complete C^*-algebra $B^\infty(X)/M_0(X)$, where $B^\infty(X)$ is the algebra of bounded Baire measurable functions on X and $M_0(X)$ is the ideal of all f in $B^\infty(X)$ for which $\{x : f(x) \neq 0\}$ is meagre. Let S_1 be the spectrum (Gelfand-Naimark structure space) of $B^\infty(X)/M_0(X)$ i.e. this algebra can be identified with $C(S_1)$.

Let $j : C_b(X) \mapsto B^\infty(X)/M_0(X)$ be the natural embedding. This is an injective (isometric) $*$-homomorphism.

Suppose that X is separable. Then, by Corollary 4.2.24, $B^\infty(X)/M_0(X) = B(X)/M(X)$. So $C(S_1)$ is monotone complete and thus S_1 is extremally disconnected. It follows from Proposition 4.2.22 that S_1 is separable and $C(S_1)$ supports a faithful state. Putting things another way, we can identify S_1 with the spectrum of $B(X)/M(X)$.

Let us recall that a Hausdorff space T is said to be *topologically complete* if it is completely regular and is a G_δ-subset of βT, its Stone-Czech compactification. In particular all Polish spaces have this property. When T is topologically complete, $\beta T \setminus T$ is a meagre, F_σ-subspace of βT. So it is trivial to observe that $B(T)/M(T) = B(\beta T)/M(\beta T)$ whenever T is topologically complete; in particular when T is Polish. More generally, if K is compact and T is a dense G_δ-subset of K then $B(T)/M(T) = B(K)/M(K)$. (See the exercises at the end of this section.)

Now, without any separability restrictions on X, it follows from Corollary 4.2.11, that $B(X)/M(X)$ is monotone complete, with $C_b(X) \simeq C(\beta X)$ embedded as a regular subalgebra. In the following, K is required to be compact (Hausdorff). But the preceding remarks should make it clear that by replacing K by βX, much of the following work can be generalised to non-compact spaces.

Let S be the spectrum of $B(K)/M(K)$. We can identify $B(K)/M(K)$ with $C(S)$, where S is a compact (Hausdorff) extremally disconnected space. Then $j : C(K) \mapsto C(S)$ is an (isometric) injective $*$-homomorphism. By the usual duality between compact Hausdorff spaces and commutative (unital) C^*-algebras, there is a continuous surjection ρ from S onto K such that $j(f) = f \circ \rho$ for each f in $C(K)$.

Since $C(K)$ is embedded in $C(S)$ as a regular subalgebra, for each self-adjoint b in $C(S)$, the set

$$\{j(a) : a \in C(K)_{sa} \text{ and } j(a) \leq b\}$$

has b as its least upper bound in $C(S)_{sa}$. (We recall that $C(S)_{sa} = C_{\mathbb{R}}(S)$ that is, the self-adjoint part of $C(S)$ can be identified with the algebra of real valued continuous functions on S.)

Lemma 6.4.1 *Let Z be a subset of S. Then $\rho[Z]$ is dense in K if, and only if, Z is dense in S.*

Proof First let $\rho[Z]$ be dense in K. Suppose Z is not dense in S. Then there exists a non-empty clopen set E which is disjoint from clZ.

Let $j(a) \leq \chi_E$. Then $j(a)(s) \leq 0$ for $s \in Z$. So $a(\rho(s)) \leq 0$ for $\rho(s) \in \rho[Z]$. Hence $a \leq 0$. But this implies $\chi_E \leq 0$ which is a contradiction. So Z must be dense in S.

Conversely, let Z be dense in S. Suppose $\rho[Z]$ is not dense in K. So there exists a non-empty open set $U \subset K$, where U is disjoint from $\rho[Z]$. Since ρ is surjective, $\rho^{-1}[U]$ is a non-empty open subset of S. Since $\rho^{-1}[U]$ is disjoint from Z this contradicts Z being dense in S. □

Let K be a compact (Hausdorff) space. Let $Homeo(K)$ be the group of all homeomorphisms from K onto K. Let $AutC(K)$ be the group of all $*$-automorphisms of $C(K)$. For $\phi \in Homeo(K)$ let $h_\phi(f) = f \circ \phi$ for each $f \in C(K)$. Then $\phi \mapsto h_\phi$ is a bijection from the group $Homeo(K)$ onto $AutC(K)$ which switches the order of multiplication. In other words it is a group anti-isomorphism.

Let θ be a homeomorphism of K onto K. As above, let h_θ be the corresponding $*$-automorphism of $C(K)$. Also $f \mapsto f \circ \theta$ induces an automorphism $\widehat{h_\theta}$ of $B(K)/M(K)$. Since $B(K)/M(K)$ can be identified with $C(S)$, there exists $\hat{\theta}$ in $Homeo(S)$ corresponding to $\widehat{h_\theta}$. Clearly, $\widehat{h_\theta}$ restricts to the automorphism, h_θ, of $C(K)$.

Lemma 6.4.2 *The automorphism $\widehat{h_\theta}$ is the unique automorphism of $C(S)$ which is an extension of h_θ. Hence $\hat{\theta}$ is uniquely determined by θ. Furthermore, the map $\theta \mapsto \hat{\theta}$ is an injective group homomorphism from $Homeo(K)$ into $Homeo(S)$.*

Proof Let H be an automorphism of $B(K)/M(K) = C(S)$, which is an extension of h_θ. Let b be a self-adjoint element of $B(K)/M(K)$. Then, for $a \in C_{\mathbb{R}}(K)$, $j(a) \leq b$ if, and only if, $Hj(a) \leq Hb$ i.e. $j(h_\theta a) \leq Hb$.

So Hb is the supremum of $\{j(h_\theta(a)) : a \in C_{\mathbb{R}}(K), j(a) \leq b\}$. Hence $H = \widehat{h_\theta}$. That is, $\widehat{h_\theta}$ is the unique extension of h_θ to an automorphism of $C(S)$.

Let h_1 and h_2 be in $AutC(K)$. Then for $a \in C(K)$, we have

$$\widehat{h_1 h_2}(j(a)) = j(h_1 h_2(a)) = \widehat{h_1} j(h_2(a)) = \widehat{h_1}\widehat{h_2}(j(a)).$$

By uniqueness, it now follows that $\widehat{h_1 h_2} = \widehat{h_1}\widehat{h_2}$. Hence $h \mapsto \hat{h}$ is an injective group homomorphism of $AutC(K)$ into $AutC(S)$. So the map $\theta \mapsto \hat{\theta}$ is the composition of a group anti-isomorphism with an injective group homomorphism composed with a group anti-isomorphism. So it is an injective group homomorphism. □

Corollary 6.4.3 *For each $s \in S$, $\theta(\rho s) = \rho(\hat{\theta}s)$.*

Proof For $a \in C(K)$, $s \in S$,

$$a \circ \theta(\rho s) = h_\theta(a)(\rho s) = \widehat{h_\theta}(j(a))(s) = j(a)(\hat\theta s) = a(\rho(\hat\theta s)).$$

Hence $\theta(\rho s) = \rho(\hat\theta s)$.   □

Throughout this chapter, unless we specify otherwise, G is a countably infinite group. Let $\varepsilon : G \mapsto Homeo(K)$ be a homomorphism into the group of homeomorphisms of the compact space K. That is, ε is an action of G on K. For each $g \in G$, let $\hat\varepsilon_g$ be the homeomorphism of S onto S induced by ε_g. Then $\hat\varepsilon$ is the action of G on S induced by ε.

We shall normally assume that K has no isolated points. By Corollary 4.2.14, this is equivalent to S having no isolated points.

Proposition 6.4.4 *Let x_0 be a point in K. Let $s_0 \in S$ such that $\rho s_0 = x_0$. Then the orbit $\{\varepsilon_g(x_0) : g \in G\}$ is dense in K if, and only if, the orbit $\{\hat\varepsilon_g(s_0) : g \in G\}$is dense in S.*

Proof By Corollary 6.4.3, $\varepsilon_g(x_0) = \rho(\hat\varepsilon_g(s_0))$. The proposition now follows from Lemma 6.4.1.   □

Proposition 6.4.5 *Let G, K, S and ε be as above. Let $\hat\varepsilon$ be the action of G on S induced by the action ε on K. Let $s_0 \in S$ and let $x_0 = \rho s_0$. Let the orbit $\{\varepsilon_g(x_0) : g \in G\}$ be free then $\{\hat\varepsilon_g(s_0) : g \in G\}$ is a free orbit in S.*

Proof By Corollary 6.4.3, $\varepsilon_g(\rho s_0) = \rho(\hat\varepsilon_g s_0)$. That is, $\varepsilon_g(x_0) = \rho(\hat\varepsilon_g s_0)$.

Let $h \in G$ such that $\hat\varepsilon_h s_0 = s_0$. Then $\rho(\hat\varepsilon_h s_0) = \rho(s_0)$. So $\varepsilon_h(x_0) = x_0$. Since the orbit $\{\varepsilon_g(x_0) : g \in G\}$ is free, h is the neutral element of G. So $\{\hat\varepsilon_g(s_0) : g \in G\}$ is a free orbit in S.   □

Corollary 6.4.6 *Let the action of ε on K have a dense orbit. Then the extremally disconnected space S is separable and*

$$B^\infty(K)/M_0(K) = B(K)/M(K).$$

Proof The countable set $\{\hat\varepsilon_g(s_0) : g \in G\}$ is dense in S. So S is separable. Since K also has a dense orbit, the rest follows from Corollary 4.2.24.   □

The following lemma does not require X to be compact; being a Baire space is enough.

Lemma 6.4.7 *Let X be a Baire space and G a countable group acting as homeomorphisms of X. Let $x_0 \in X$ be such that the orbit $G[x_0]$ is both dense and free. Then there exists a G-invariant Y, which is a dense G_δ subset of X, such that, for $g \neq 1$, ε_g has no fixed point in Y. Also $x_0 \in Y$.*

Proof Fix $g \neq \iota$, let $K_g = \{x \in X : g(x) = x\}$. Then K_g, the fix-point set of g, is closed. Let U be the interior of K_g. Then the orbit $G[x_0]$ is disjoint from K_g. So its closure is disjoint from U. But since the orbit is dense, this means that K_g has empty interior.

Let $Z = \bigcup \{K_g : g \in G, g \neq \iota\}$. Then Z is the union of countably many, closed nowhere dense sets. A calculation shows that

$$\varepsilon_h[K_g] = K_{hgh^{-1}}$$

and from this it follows that Z is G-invariant. Put $Y = X \backslash Z$. Then Y has all the required properties.   □

Remark 6.4.8 Let D be a dense subset of a compact Hausdorff space K. Let α be a homeomorphism of D onto D. Then, in general, α need not extend to a homeomorphism of K. But, from the fundamental properties of the Stone-Czech compactification, α does extend to a unique homeomorphism of βD, say θ_α. Let S_b be the Gelfand-Naimark structure space of $B(\beta D)/M(\beta D)$. Then, from the results of this section, θ_α induces a homeomorphism $\widehat{\theta}_\alpha$ of S_b. Let S be the structure space of $B(K)/M(K)$.

By Lemma 6.2.2, the Boolean algebra $RegK$ is isomorphic to $RegD$. Again by Lemma 6.2.2, $Reg\beta D$ is isomorphic to $RegD$. So, by Proposition 4.2.13, the Boolean algebra of projections in $B(\beta D)/M(\beta D)$ is isomorphic to the Boolean algebra of projections in $B(K)/M(K)$. It follows that S_b is homeomorphic to S.

Hence each homeomorphism of D induces a canonical homeomorphism of S. So each action of G, as homeomorphisms of D, induces, canonically, an action of G as homeomorphisms of S.

Exercise 6.4.9

(a) Let X be a Baire space and Y a dense G_δ-subset of X. For each $f \in B(X)$ let Rf be the restriction of f to Y. Show that R maps $B(X)$ onto $B(Y)$ and Rf is in $M(Y)$ precisely when $f \in M(X)$. Deduce that R induces an isomorphism of $B(X)/M(X)$ onto $B(Y)/M(Y)$ (so we may identify these algebras). As we have seen earlier $B(X)/M(X)$ is monotone complete and so isomorphic to $C(S)$, where S is compact and extremally disconnected. As remarked above S can be identified with the Stone space of the (complete) Boolean algebra of regular open subsets of X (or, equivalently, of Y).

(b) Let θ be a homeomorphism of Y onto itself. Show that the automorphism of $B(Y)$ defined by $f \mapsto f \circ \theta$ induces an automorphism H_θ of $C(S)$. Now suppose that θ can be extended to a homeomorphism θ_X of X onto itself. Show that θ_X induces the same automorphism of $C(S)$ as is induced by θ.

(c) Let Y_1 and Y_2 be dense G_δ-subsets of X. Let θ be a homeomorphism of Y_2 onto Y_1. Show that the map from $B(Y_1)$ to $B(Y_2)$, defined by $f \mapsto f \circ \theta$, induces an isomorphism of $B(Y_1)/M(Y_1)$ onto $B(Y_2)/M(Y_2)$. By (a) it follows that this can be regarded as an isomorphism of $B(X)/M(X)$ onto $B(X)/M(X)$; hence it induces an automorphism of $C(S)$.

(d) Let S be compact Hausdorff and extremally disconnected. Let α and β be homeomorphisms of S. Let h_α and h_β be the corresponding automorphisms of $C(S)$.

Show that there is a clopen set $K \subset S$ such that $\alpha|_K = \beta|_K$ if, and only if, there is a projection χ_K such that $\chi_K h_\alpha = \chi_K h_\beta$. That is for each $f \in C(S)$, we have

$$\chi_K h_\alpha(f) = \chi_K h_\beta(f).$$

6.5 Ergodicity on Special Spaces

In generic dynamics we are especially interested in dense G_δ-subspaces of separable compact Hausdorff spaces. In particular each Polish space is of this form.

Lemma 6.5.1 *Let X be a metrisable space. The following are equivalent:*

(i) *The space X is homeomorphic to a complete metric space.*
(ii) *The space X is a G_δ-subspace of βX.*
(iii) *Whenever X is densely embedded in a (Hausdorff) Baire space, it is a G_δ-subspace.*

Proof See [167, Theorem 24.13]. □

It is clear that if a topological space has a countable base then it is separable. For metrisable spaces the converse is well-known but, in general, separability is a weaker property than having a countable base. We shall make use of an intermediate property – *hyperseparability*.

Definition 6.5.2 Let X be a (Hausdorff) topological space. Then X is *hyperseparable* if there exists a sequence of open sets (V_n) such that, whenever W is a non-empty open subset of X, then, for some n, $\varnothing \neq V_n \subset W$.

By taking a point from each non-empty V_n we see that a hyperseparable space is separable. It is easy to check that a dense subspace of a hyperseparable space is, itself, hyperseparable. Obviously the existence of a countable base implies hyperseparability.

Lemma 6.5.3 *Let X be a Polish space. Then the Boolean algebra RegX has a countable order dense set. Let S_X be the spectrum of $B(X)/M(X)$. Then S_X is hyperseparable. Furthermore, X has no isolated points if, and only if, S_X has no isolated points.*

Proof Let (U_n) be a countable base for X. Put $V_n = int(cl U_n)$. So V_n is a regular open set.

Let W be a non-empty regular open subset of X. Then, for some n, $\varnothing \neq U_n \subset W$. So

$$\varnothing \neq V_n = int(clU_n) \subset int(clW) = W.$$

So $RegX$ has a countable subset, (V_n) which is order dense.

By Proposition 4.2.13 $RegX$ is isomorphic to the Boolean algebra of clopen subsets of S_X. So the Boolean algebra of clopen subsets of S_X has a countable order dense set, (K_n). Let $W \subset S_X$ be a non-empty open set. Then W contains a non-empty clopen set. So, for some n, $\varnothing \neq K_n \subset W$. So S_X is hyperseparable.

Now S_X has an isolated point if, and only if, the Boolean algebra $RegX$ has an atom. Clearly $RegX$ has an atom if, and only if, X has a minimal regular open set V.

Let V be a minimal regular open set. Suppose V is not a one point set. So it contains distinct points a and b. Since X is metrisable we can find open sets A and B such that $a \in A \subset V$ and $b \in B \subset V$ and also clA is disjoint from clB. But then $int(clA)$ is a non-empty regular open set, strictly smaller than V. It follows that X can only have a minimal regular open set if X has an isolated point. □

Corollary 6.5.4 *Let X be any perfect Polish space. Then S_X is homeomorphic to $S_\mathbb{R}$.*

Proof By Corollary 4.2.14 S_X has no isolated points; equivalently $RegS_X$ has no atoms.

But, see [153, page 155], there is exactly one Boolean algebra which is complete, non-atomic and with a countable order dense subset. So S_X is homeomorphic to $S_\mathbb{R}$.□

An alternative argument can be based on the following. By Lemma 6.5.3, $RegS_X$ has a countable order dense set. This countable set generates a countable Boolean subalgebra B. Let F be the Stone space of B.

Each countable Boolean algebra is a quotient of the free Boolean algebra on countably many generators. So, by duality, F is a closed subspace of $2^\mathbb{N}$.

Any atom of B would be an atom of $RegS_X$. So B has no atoms. Thus F is a compact, zero-dimensional metric space with no isolated points. So by [167, Corollary 30.4] F is homeomorphic to $2^\mathbb{N}$. From this and order-density we get $C(S_X) \cong B(2^\mathbb{N})/M(2^\mathbb{N})$. This gives $S_X \cong S_{2^\mathbb{N}}$.

The following result applies to perfect Polish spaces, to $S_\mathbb{R}$ and to dense G_δ-subsets of $S_\mathbb{R}$.

Proposition 6.5.5 *Let X be a hyperseparable (Hausdorff) Baire space with no isolated points. Let G act as homeomorphisms of X. Then the following are equivalent:*

(i) *Every G-invariant open subset of X is empty or dense.*

(ii) *Every G-invariant Borel subset of X is either meagre or is the compliment of a meagre set.*

(iii) *There exists a dense, G-invariant, G_δ subset Y of X such that $G[y]$ is a dense orbit for each $y \in Y$.*

(iv) *There exists an $x_0 \in X$ such that $\{\varepsilon_g(x_0) : g \in G\}$ is dense in X.*

Proof By Proposition 6.3.5, (i) and (ii) are equivalent. Clearly (iii) implies (iv). By Lemma 6.3.1, (iv) implies (i).

It now suffices to show (i) implies (iii). Let (V_n) be a sequence of non-empty open sets such that each non-empty open set, W, contains some V_n. Then

$$\bigcup_{g \in G} g^{-1}[V_n]$$

is a non-empty, G-invariant open set. So, by (i), it is dense in X. Since X is a Baire space, the G_δ-set

$$Y = \bigcap_{n=1}^{\infty} \bigcup_{g \in G} g^{-1}[V_n]$$

is a dense subset of X.

Let y be any element of Y and let O be any non-empty open subset of X. Then, for some n, $V_n \subset O$. Then $g(y) \in V_n$ for some $g \in G$. Hence $G[y]$ is a dense orbit. □

6.5.1 Zero-Dimensional Spaces

Let us recall that a (Hausdorff) topological space is said to be *zero-dimensional* if it has a base of clopen subsets. A topological space is *totally disconnected* if its connected components consist of single points. Every zero-dimensional space is totally disconnected. We proved in Lemma 2.3.1, that the converse holds for compact Hausdorff spaces. But totally disconnected metric spaces which are not zero-dimensional do exist [167, Examples 29.8]. But in the literature "totally disconnected" is sometimes used where "zero-dimensional" is intended, see [157]. For our purposes these delicate distinctions do not matter. Roughly speaking, in generic dynamics it is enough to work with zero-dimensional spaces.

Proposition 6.5.6 *Let X be a perfect Polish space with G acting as homeomorphisms of X. Then there exists a dense G_δ-subset $Z \subset X$ such that Z is G-invariant, zero-dimensional and a perfect Polish space. If the action of G on X is ergodic then so, also, is its action on Z.*

Proof Let (U_n) be a base of open sets for the topology of X. Then $clU_n \backslash U_n$ is a closed nowhere dense set for each n. Take their union M_1. Now let M_2 be the G-saturation of M_1. Put $Z = X \backslash M_2$. Then Z is a dense G_δ-subset of X. So Z is a perfect Polish space. Also $U_n \cap Z = clU_n \cap Z$. So Z has a base of clopen sets.

Suppose G acts ergodically on X. By Proposition 6.5.5 (iii), there exists $Y \subset X$, where Y is a G-invariant, G_δ-subset of X and, for each y in Y, the orbit $G[y]$ is dense in X. By the Baire Category Theorem, $Y \cap Z$ is dense in X. So Z contains a dense orbit. So, by Proposition 6.5.5, G acts ergodically on Z. □

Exercise 6.5.7 Let X be a (Hausdorff) zero-dimensional space. Show that every subspace is zero-dimensional in the relative topology.

We shall make use of a weaker property than free actions.

Definition 6.5.8 Let X be a Hausdorff space with G acting as homeomorphisms of X. The action is said to be *pseudo free* if, for each $g \in G$, the fixed point set

$$K_g = \{x \in X : g(x) = x\}$$

is clopen. In particular, if $G[x_0]$ is an orbit such that the action of G, restricted to this orbit, is pseudo free then we say that the orbit is pseudo free.

Lemma 6.5.9 *Let X be a Hausdorff space with G acting as homeomorphisms of X. Let the action of G be pseudo free. Let $G[x_0]$ be a free, dense orbit in X. Then the action of G on X is free.*

Proof Let $g \in G$ such that $g(c) = c$ for some $c \in X$. Since the action is pseudo free, the fixed point set $K = \{x \in X : g(x) = x\}$ is clopen.

Since K is not empty and is open, for some $h \in G$, $h(x_0)$ is in K. So

$$g(h(x_0)) = h(x_0).$$

Hence

$$h^{-1}gh(x_0) = x_0.$$

But the orbit $G[x_0]$ is free. So $h^{-1}gh = \iota$. Thus $g = h\iota h^{-1} = \iota$. Hence the action of G on X is free. $\square$

If (X, G, ε) is a dynamical system where the action is free then this action is also pseudo free, because each fixed point set is empty or the whole of X.

Proposition 6.5.10 *Let X be a (Hausdorff) Baire space with G acting as homeomorphisms of X. Then there exists a G-invariant, dense G_δ subset Y such that, for each $g \in G$,*

$$F_g = \{x \in Y : g(x) = x\}$$

is a clopen subset of Y. When X is zero-dimensional, so, also, is Y.

Proof For each $g \in G$, let $K_g = \{x \in X : g(x) = x\}$. Then K_g is closed. Then $K_g \setminus \mathrm{int} K_g$ is a closed, nowhere dense set. Then

$$\bigcup K_g \setminus \mathrm{int} K_g$$

is a meagre F_σ subset of Y. Let M be the saturation of this set by G. Then this is also a meagre F_σ. Let $Y = X \backslash M$. Then, since X is a Baire space, Y is a dense G_δ-subset of Y.

For each $g \in G$,

$$F_g = Y \cap K_g = Y \cap int K_g.$$

So F_g is both open and closed in the relative topology of Y. $\qquad\qquad\qquad\square$

We shall need the following lemma shortly:

Lemma 6.5.11 *Let X be a Hausdorff space with no isolated points. Let Y be a dense subset of X. Let O be a non-empty open subset of X. Then $O \cap Y$ is an infinite set. In particular, every dense subset of X is infinite.*

Proof Suppose $O \cap Y$ is a non-empty finite set, say, $\{p_1, p_2, \cdots, p_n\}$.

Then $O \backslash \{p_1, p_2, \cdots, p_n\}$ is an open subset of X which is disjoint from Y. But Y is dense in X. Hence $O = \{p_1, p_2, \cdots, p_n\}$. So $\{p_1\}$ is an open subset of X. But X has no isolated points. So this is a contradiction. So $O \cap Y$ is infinite. On putting $O = X$ it follows that Y is infinite. $\qquad\qquad\qquad\square$

6.6 Orbit Equivalence: Zero-Dimensional Spaces

Let us recall, from Definition 6.1.5, that when a group acts on a space X, two points of X are *orbit equivalent* if they are in the same orbit.

In this section we shall focus on group actions on zero-dimensional spaces and prove a key technical result, Proposition 6.6.16. In Sect. 6.7, we shall apply it to generic dynamics over extremally disconnected spaces. In Sect. 6.8, we shall use it to show that when an infinite countable group of homeomorphisms acts ergodically on a perfect Polish space then the orbit equivalence relation is unique (modulo meagre sets).

For any perfect Polish space T, let S_T be the spectrum of $B^\infty(T)/M(T)$. Then, for every T, S_T is homeomorphic to $S_\mathbb{R}$. When an infinite countable group of homeomorphisms acts ergodically on S_T, then the orbit equivalence relation is unique (modulo meagre sets).

Throughout this section X is a separable, zero-dimensional (Hausdorff) space with no isolated points. Also G is a countably infinite group acting on X as homeomorphisms, that is, there exists a non-degenerate action $\varepsilon : G \mapsto Homeo(X)$. Furthermore, let D be a dense G-invariant subset of X.

We shall be particularly interested in dynamical systems with a dense orbit. In other words, when there exists $c \in X$, such that the orbit $G[c] = \{\varepsilon_g(c) : g \in G\}$ is dense in X. Since X has no isolated points, it follows from Lemma 6.5.11 that a dense orbit must be infinite.

To simplify our notation, we often write "g" for ε_g. The restriction of ε_g to a G-invariant subspace of X will also be denoted by "g".

Looking at things another way, we are, in effect, identifying G with the isomorphic group of homeomorphisms $\varepsilon[G]$.

Definition 6.6.1 Let Z be a G-invariant dense subset of X and let h be a bijection of Z onto itself. Then h is said to be *strongly G-decomposable* over Z if there exists a sequence of pairwise disjoint clopen subsets of Z, (A_j) where $Z = \bigcup A_j$, and a sequence (g_j) in G such that

$$h(x) = g_j(x) \text{ for } x \in A_j.$$

When this occurs, h is a continuous map from Z into Z. Since each g_j is a homeomorphism of Z, each $g_j[A_j]$ is clopen. Since h is a bijection, $(h[A_j])$ is a sequence of pairwise disjoint clopen subsets of Z whose union is Z. Hence h^{-1} is strongly G-decomposable over Z. So h and h^{-1} are both continuous. Thus h is a homeomorphism of Z onto Z.

We sometimes use a slightly weaker condition. Let h be a homeomorphism of X onto itself. Then h is *G-decomposable* (over X) if there exists a sequence of pairwise disjoint clopen subsets of X, (K_j) where $\bigcup K_j$ is dense in X, and there exists a sequence (g_j) in G such that

$$h(x) = g_j(x) \text{ for } x \in K_j.$$

When this holds, $\bigcup K_j$ is an open dense subset of X, hence its compliment is a closed nowhere dense set.

Let Γ be a countable group acting on X as homeomorphisms of X onto itself. If every $\gamma \in \Gamma$ is strongly G-decomposable over X and every $g \in G$ is strongly Γ-decomposable over X, then we say that the action of G on X and the action of Γ on X are *strongly equivalent*.

Exercise 6.6.2 Let G_1, G_2 and G_3 be countable groups acting as homeomorphisms of X. Let the action of G_1 on X be strongly equivalent to the action of G_2 on X. Let the action of G_2 on X be strongly equivalent to the action of G_3 on X. Show that the action of G_1 on X is strongly equivalent to the action of G_3 on X.

For a given dynamical system (X, G, ε), the corresponding orbit equivalence relation, $\sim_G$, is identified with its graph, $Gr(\varepsilon[G])$ that is:

$$\{(x, \varepsilon_g x) : x \in X, g \in G\} = \{(x, gx) : x \in X, g \in G\} = Gr(\varepsilon[G]).$$

When it is clear which action, ε, of G is intended, we shall use $Gr[G]$.

Lemma 6.6.3 *Let G and Γ be countably infinite groups acting on X as homeomorphisms of X onto itself. Suppose that the actions of G and Γ are strongly equivalent. Then $Gr(G) = Gr(\Gamma)$. In other words, the orbit equivalence relations with respect to G and Γ coincide.*

Proof Each element of $Gr(G)$ is of the form (x, gx). Since g is strongly Γ-decomposable over X, $gx = \gamma x$ for some $\gamma \in \Gamma$. So $Gr(G) \subseteq Gr(\Gamma)$. Similarly $Gr(\Gamma) \subseteq Gr(G)$. $\square$

The above lemma tells us that when G and Γ have strongly equivalent actions on X then orbit equivalence with respect to G is the same relation as orbit equivalence with respect to Γ. When the actions of G and Γ are free and X is a (Hausdorff) Baire space, we have a converse, modulo meagre sets.

Lemma 6.6.4 *Let X be a (Hausdorff) Baire space. Let Γ and G be countable groups of homeomorphisms of X, both of which act freely on X. Suppose that the orbit equivalence relations coincide, that is $Gr(G) = Gr(\Gamma)$. Then there exists Z, a dense G_δ subset of X which is invariant under the actions of both G and Γ, such that the actions of the two groups are strongly equivalent over Z.*

Proof Let $\Gamma \vee G$ be the countable group of homeomorphisms of X which is generated by G and Γ.

Take any $g \in G$ and fix it. Let $F^*(\gamma)$ be the closed set $\{x \in X : g(x) = \gamma(x)\}$. Let $F^\circ(\gamma)$ be the interior of $F^*(\gamma)$ so that $F^*(\gamma) \backslash F^\circ(\gamma)$ is a closed nowhere dense set.

Let $M_1 = \bigcup_{\gamma \in \Gamma} F^*(\gamma) \backslash F^\circ(\gamma)$. Then M_1 is a meagre F_σ-set. Let M_2 be the $\Gamma \vee G$-saturation of M_1 and let $Y_g = X \backslash M_2$. Because X is a Baire space Y_g is a dense G_δ-subset of X, which is invariant under the actions of Γ and G.

For each $\gamma \in \Gamma$, let $F(\gamma) = F^*(\gamma) \cap Y_g = F^\circ(\gamma) \cap Y_g$. So $F(\gamma)$ is a clopen subset of Y_g. Let $y \in Y_g$, then $g(y) = \gamma(y)$ for some γ. So $y \in F(\gamma)$ for some $\gamma \in \Gamma$. So Y_g is the union of the clopen sets $F(\gamma)$.

Suppose, for γ and σ in Γ, there exists $x \in F(\gamma) \cap F(\sigma)$. Then we have $\gamma(x) = g(x) = \sigma(x)$. But since the action of Γ is free on X, it follows that $\gamma = \sigma$. So $(F(\gamma))_{\gamma \in \Gamma}$ is a clopen partition of Y_g. So, g is strongly Γ-decomposable over Y_g. Now let $Y_1 = \bigcap_{g \in G} Y_g$. Then Y_1 is a G_δ-subset of X, which is dense because X is a Baire space. Each element of G is strongly Γ-decomposable over Y_1. Similarly, we can construct a $\Gamma \vee G$-invariant dense G_δ-subset Y_2 such that each element of Γ is strongly G-decomposable over Y_2. Put $Z = Y_1 \cap Y_2$. Then G and Γ are strongly equivalent over Z and Z is a dense G_δ, invariant subset of X. $\square$

Definition 6.6.5 Let X and Y be Hausdorff spaces. Let G and Γ be countable groups acting as homeomorphisms on, respectively, X and Y. We say the action of G on X and the action of Γ on Y are *(conjugately) strongly equivalent* if there exists a surjective homeomorphism $\pi : X \mapsto Y$ such that the group of homeomorphisms $\{\pi g \pi^{-1} : g \in G\} = \pi G \pi^{-1}$ is strongly equivalent to Γ.

Definition 6.6.6 Let T_1 and T_2 be Baire spaces. Let G_1 and G_2 be countable groups acting as homeomorphisms of, respectively, T_1 and T_2. We say the actions of G_1 and G_2 are *generically equivalent* if the following hold: There exists a G_j-invariant, dense G_δ-subset $X_j \subset T_j$ for $j = 1, 2$. There exists a homeomorphism π from X_1 onto X_2 such that the action of G_1 on X_1 and the action of G_2 on X_2 are *(conjugately) strongly equivalent*.

Exercise 6.6.7 Let G_j be a countable group acting on a Baire space T_j for $j = 1, 2, 3$. Suppose the action of G_1 on T_1 is generically equivalent to the action of G_2 on T_2 and suppose the action of G_2 on T_2 is generically equivalent to the action of G_3 on T_3. Show that the action of G_1 on T_1 is generically equivalent to the action of G_3 on T_3.

In the special situation where X is extremally disconnected we have the following useful result.

Lemma 6.6.8 *Suppose that X is compact and extremally disconnected. Let D be a dense subset of X. Let α be a homeomorphism of D onto D. Let α be strongly G-decomposable over D. Then there exists $\hat{\alpha}$, a unique extension of α to a homeomorphism of X onto itself. This $\hat{\alpha}$ is G-decomposable over X.*

Proof The existence and uniqueness of $\hat{\alpha}$, follows from Corollary 6.2.8, since X is the Stone-Czech compactification of D.

Let (A_j) be a sequence of pairwise disjoint clopen subsets of D. Then, by Lemma 6.2.6, (clA_j) is a sequence of pairwise disjoint clopen subsets of X. Let (g_j) be a sequence in G such that $\alpha(x) = g_j(x)$ for $x \in A_j$. Then, by continuity, $\hat{\alpha}(x) = g_j(x)$ for $x \in clA_j$. Also the open set $\bigcup clA_j$ is dense in X (because it contains D). $\qquad\qquad\square$

Following our usual notation for each homeomorphism θ of X onto itself, we define a homomorphism of the algebra of bounded continuous ($\mathbb{C}$−valued) functions on X, by $h_\theta(f) = f \circ \theta$.

Corollary 6.6.9 *Let X be compact and extremally disconnected. Let α be a homeomorphism of X onto itself. Then (i) there exists a sequence $(g(j))$ in G and a sequence of orthogonal projections (p_j) such that $p_j h_{g(j)} = p_j h_\alpha$ for each j, and $\bigvee_{j \geq 1} p_j = 1$ if, and only if, (ii) α is G-decomposable over X.*

Proof By Exercise 6.4.9 (d), $p_j h_{g(j)} = p_j h_\alpha$ is equivalent to $p_j = \chi_{K_j}$ and $g(j)|_{K_j} = \alpha|_{K_j}$ for a clopen K_j. The projections (p_j) are orthogonal if, and only if, the clopen sets (K_j) are pairwise disjoint. Also $\bigvee_{j \geq 1} p_j = 1$ if, and only if, the open set $\bigcup_{j \geq 1} K_j$ is dense in X. So (i) implies α is strongly G−decomposable over a dense open subset of X. This, in turn, implies (ii). The converse, (ii) implies (i), is equally straightforward provided we use Lemma 6.2.6. $\qquad\qquad\square$

Lemma 6.6.10 *Let X be a separable, zero-dimensional space with no isolated points. Let G and Γ be countable groups acting as homeomorphisms of X. Suppose also that the action of G is pseudo free. Let each $\gamma \in \Gamma$ be strongly G-decomposable over X. Furthermore, let there exist a countable dense subset $\Delta \subset X$ which is G-invariant, Γ-invariant and*

$$\Delta \times \Delta \subset Gr(G) \text{ and } \Delta \times \Delta \subset Gr(\Gamma).$$

Then there exists a dense G_δ-subset Y of X such that $\Delta \subset Y \subset X$, where Y is G- and Γ-invariant and the action of G on Y and the action of Γ on Y are strongly equivalent.

Proof Let $(d_1, d_2, \ldots)$ be an enumeration, without repetitions, of Δ. Fix $g \in G$. We shall decompose g using elements of Γ. Let $x_1 = d_1$. Since $(x_1, g(x_1)) \in \Delta \times \Delta \subset Gr(\Gamma)$, there exists a $\gamma_1 \in \Gamma$ such that $g(x_1) = \gamma_1(x_1)$. Since γ_1 is strongly G-decomposable over X, there exists a clopen neighbourhood K of x_1 and a $g_1 \in G$ such that $\gamma_1(x) = g_1(x)$ for all $x \in K$. Since $g(x_1) = g_1(x_1)$, we have $x_1 = g^{-1}g_1(x_1)$. Since the action of G on X is pseudo free, $\{y \in X \ : \ g^{-1}g_1(y) = y\}$ is clopen and contains x_1. Therefore, there exists a clopen neighbourhood K_1 of x_1 with $K_1 \subset K$ and $g(x) = g_1(x)$ for all $x \in K_1$. So, $g(x) = \gamma_1(x)$ for all $x \in K_1$.

If $K_1 = X$ then $g = \gamma_1$ and we stop. Otherwise $K_1^c \neq \varnothing$. Then, because Δ is dense in X, we have $K_1^c \cap \Delta \neq \varnothing$. Let p be the smallest whole number such that $d_p \in K_1^c \cap \Delta$.

Let $x_2 = d_p$. By repeating the above argument, we find a clopen neighbourhood K_2 of x_2, and a $\gamma_2 \in \Gamma$ such that $x_2 \in K_2 \subset K_1^c$ and $g(x) = \gamma_2(x)$ for all $x \in K_2$. By repeating these arguments, we find a sequence (finite or infinite) (K_n) of disjoint clopen subsets of X and a sequence (γ_n) in Γ such that $\Delta \subset \bigcup_{n \geq 1} K_n$ and, for each $x \in K_n$, $g(x) = \gamma_n(x)$.

Let $O_g = \bigcup_{n \geq 1} K_n$. Then clearly O_g is a dense open subset of X such that $O_g \supset \Delta$. Let $Y_0 = \bigcap_{g \in G} O_g$. Then $\Delta \subset Y_0$ and so Y_0 is a dense G_δ subset of X. Let $G \vee \Gamma$ be the (countable) group generated by G and Γ. (Here we are using "G" for the isomorphic group of homeomorphisms $\varepsilon[G]$; similarly for Γ). Let $Y = \bigcap\{\upsilon(Y_0) : \upsilon \in G \vee \Gamma\}$. Then clearly $\Delta \subset Y$. Hence Y is a dense G_δ subset of X containing Δ, which is G-invariant and Γ-invariant.

For each $g \in G$, $g|_Y$ is strongly Γ-decomposable over Y and for each $\gamma \in \Gamma$, $\gamma|_Y$ is strongly G-decomposable over Y. Hence the actions of G and Γ on Y are strongly equivalent. $\square$

The next two lemmas are easy technicalities. Unless we specify otherwise, we do not assume anything about X, other than being a zero-dimensional (Hausdorff) space with no isolated points. The countably infinite group G acts as homeomorphisms of X. We shall soon require the existence of a dense orbit for the action of G. So X must be separable.

Lemma 6.6.11 *Let Y and W be G-invariant subsets of X with $\varnothing \neq W \subset Y$. Let α be a homeomorphism of Y onto itself, which is strongly G-decomposable over Y. Let α_W be the restriction of α to W. Then α_W is a homeomorphism of W onto itself, which is strongly G-decomposable over W.*

Proof The homeomorphism α is strongly G-decomposable over Y. So $Y = \bigcup_{n \geq 1} A_n$ where (A_n) is a sequence of pairwise disjoint clopen subsets of Y; also, there exists a sequence (g_n) in G such that $\alpha(y) = g_n(y)$ for each $y \in A_n$.

Let $w \in W$. Then $w \in A_n \cap W$ for some n. Also $\alpha(w) = g_n(w)$. Since W is G-invariant, $\alpha(w)$ is in W. So α_W is an injection from W into W.

Now take $\upsilon \in W$. Let $x = \alpha^{-1}(\upsilon)$. Then $x \in A_m$ for some m. Also $\alpha(x) = g_m(x)$.

Thus $x = g_m^{-1}(v)$. Since W is G-invariant, x is in W. It follows that α_W is a bijection from W onto W.

Since $W = \bigcup_{n \geq 1} W \cap A_n$ it is now clear that α_W is strongly G-decomposable over W. So α_W is a homeomorphism of W onto W. $\square$

Lemma 6.6.12 *Let Y and Z be G-invariant subsets of X with $Y \cap Z \neq \emptyset$. Let α be a homeomorphism of Y onto itself, with α strongly G-decomposable over Y. Let β be a homeomorphism of Z onto itself, with β strongly G-decomposable over Z. Then $\beta\alpha$ restricted to $Y \cap Z$ is a homeomorphism of $Y \cap Z$ onto itself which is strongly G-decomposable over $Y \cap Z$.*

Proof By Lemma 6.6.11, α and β map $Y \cap Z$ onto itself. Also by the lemma, the restriction of α to $Y \cap Z$ is strongly G-decomposable; similarly for β. So it suffices to prove the result when $Y = Z$.

Let $\{A_i : i \in \mathbb{N}\}$ be a partition of Y into clopen sets and (g_i^α) a sequence in G which gives the G-decomposition of α. Similarly, let $\{B_j : j \in \mathbb{N}\}$ be a partition of Y into clopen sets and (g_j^β) a sequence in G which gives the G-decomposition of β. Then $\{A_i \cap \alpha^{-1}[B_j] : i \in \mathbb{N}, j \in \mathbb{N}\}$ is a partition of Y into clopen sets.

Let $s \in A_i \cap \alpha^{-1}[B_j]$. Then $\beta\alpha(s) = g_j^\beta(\alpha(s)) = g_j^\beta g_i^\alpha(s)$. $\square$

The point of the following lemma is that whenever a homeomorphism of D is G-decomposable over D then it can be extended to a homeomorphism of a G_δ set $Y \subset X$.

Lemma 6.6.13 *Let D be a G-invariant, dense subset of X. Let α be a homeomorphism of D onto itself, which is strongly G-decomposable over D. Then there exists a G-invariant, G_δ-set Y, with $D \subset Y \subset X$, and a homeomorphism $\alpha^\#$ of Y onto itself, such that $\alpha^\#$ is an extension of α and $\alpha^\#$ is strongly G-decomposable over Y.*

Proof There exists a sequence of pairwise disjoint clopen subsets of D, (C_n) and a sequence in G, (g_n) such that $D = \bigcup_{n \geq 1} C_n$ and

$$\alpha(z) = g_n(z) \text{ for } z \in C_n.$$

By using Lemma 6.2.1 (vi) we can find a G_δ-set Y_1, $D \subset Y_1 \subset X$, and a sequence of pairwise disjoint clopen subsets of Y_1, (A_n) such that $C_n = D \cap A_n$. Let $Y_2 = \bigcup_{n \geq 1} A_n \subset Y_1$. Then Y_2 is a relatively open subset of Y_1. So Y_2 is a G_δ subset of X which contains D. and which is the union of the disjoint clopen (relative to Y_1) subsets A_n.

Since α is a homeomorphism of D onto D, $(\alpha[C_n])$ is a sequence of pairwise disjoint clopen subsets of D whose union is D.

Again, by using Lemma 6.2.1 (vi) we can find a G_δ-set Y_3, $D \subset Y_3 \subset X$, and a sequence of pairwise disjoint clopen subsets of Y_3, (B_n) such that $D \cap B_n = \alpha[C_n] = g_n[C_n]$. We may assume, without loss of generality, that $Y_3 = \bigcup_{n \geq 1} B_n$.

Now let $Y = \bigcap_{g \in G} g[Y_2 \cap Y_3]$. Then Y is a G-invariant, G_δ-subset of X, with $D \subset Y \subset Y_2 \cap Y_3$.

Let $A_n^\# = Y \cap A_n$ for each n. Then $(A_n^\#)$ is a sequence of pairwise disjoint clopen subsets of Y with $C_n = D \cap A_n^\#$ and $Y = \bigcup_{n \geq 1} A_n^\#$.

Let $B_n^\# = Y \cap B_n$. Then $(B_n^\#)$ is a sequence of pairwise disjoint clopen subsets of Y with $g_n(C_n) = \alpha(C_n) = D \cap B_n^\#$ and $Y = \bigcup_{n \geq 1} B_n^\#$.

We now define β on Y by $\beta(y) = g_n(y)$ for $y \in A_n^\#$. Since the clopen sets $(A_n^\#)$ are pairwise disjoint, β is well-defined. Because Y is G-invariant, β maps Y into Y. Clearly β is continuous. Similarly we now define γ on Y by $\gamma(y) = g_n^{-1}(y)$ for $y \in B_n^\#$. Then γ is also a continuous map into Y. Then $\beta\gamma(z) = z$ for all z in D. Since D is dense in Y, it follows, by continuity, that $\beta\gamma$ is the identity map on Y. Similarly for $\gamma\beta$. So β is a homeomorphism of Y onto Y. On putting $\alpha^\# = \beta$ we have the required extension. $\qquad\qquad\square$

Corollary 6.6.14 *Let D be a G-invariant, dense subset of X. Let Γ be a countable group. Let η be a (non-degenerate) action of Γ as homeomorphisms of D. Suppose that η_γ is strongly G-decomposable over D for each $\gamma \in \Gamma$. Then there exists a G-invariant, G_δ-set Y, with $D \subset Y \subset X$ and an action η^Y of Γ as homeomorphisms of Y, with the following properties. For each $\gamma \in \Gamma$, η_γ^Y is an extension of η_γ and η_γ^Y is strongly G-decomposable over Y.*

Proof For each $\gamma \in \Gamma$, apply Lemma 6.6.13 to find a G-invariant, G_δ-set Y_γ with $D \subset Y_\gamma$ and a homeomorphism $\eta_\gamma^\#$ of Y_γ onto itself, such that $\eta_\gamma^\#$ is strongly G-decomposable over Y_γ and is an extension of η_γ. Let $Y = \bigcap_{\gamma \in \Gamma} Y_\gamma$. Clearly Y is a G-invariant, G_δ set.

By Lemma 6.6.11, each $\eta_\gamma^\#$ restricts to a homeomorphism of Y onto Y and is strongly G-decomposable over Y. Let η_γ^Y be the restriction of $\eta_\gamma^\#$ to Y.

If $\eta_\gamma^\#$ is the identity map on Y then η_γ is the identity map on D. Since η is a non-degenerate action $\gamma = \iota$. So the action η^Y is non-degenerate. $\qquad\square$

Lemma 6.6.15 *Let X be a zero-dimensional space with no isolated points. Let G be a countably infinite group which acts on X as homeomorphisms, with a dense orbit D. Let A and B be disjoint clopen subsets of D. Let $a \in A$ and $b \in B$. Then there exists a homeomorphism h from D onto D with the following properties. First h is strongly G-decomposable. Secondly h interchanges A and B and leaves each point of $D \setminus (A \cup B)$ fixed. Thirdly, $h(a) = b$. Fourthly $h = h^{-1}$.*

Proof Since D is a dense subset of X, D is zero-dimensional in the relative topology and has no isolated points.

Since a and b are in the same orbit of G, there exists g_1 in G such that $g_1(a) = b$. Then $A \cap g_1^{-1}[B]$ is a clopen neighbourhood of a which is mapped by g_1 into B. By Lemma 6.5.11, each non-empty, open subset of D (in the relative topology) is an infinite set. So we can find a strictly smaller clopen neighbourhood of a, say A_1. By dropping to a clopen sub-neighbourhood if necessary, we can also demand that $g_1[A_1]$ is a proper clopen subset of B. Let $B_1 = g_1[A_1]$.

The open sets A and B are infinite sets. Since they are subsets of D, they are both countably infinite. Enumerate them both. Let a_2 be the first term of the enumeration of A which is not in A_1 and let b_2 be the first term of the enumeration of B which is not in B_1. Then there exists g_2 in G such that $g_2(a_2) = b_2$. Now let A_2 be a clopen neighbourhood of a_2, such that A_2 is a proper subset of $A \setminus A_1$ and $g_2[A_2]$ is a proper subset of $B \setminus B_1$. Proceeding inductively, we obtain a sequence, (A_n) of disjoint clopen subsets of A; a sequence (B_n) of disjoint clopen subsets of B and a sequence (g_n) from G such that g_n maps A_n onto B_n. The first n points in the enumeration of B are in $\bigcup_{j=1}^{n} B_j$. So $B = \bigcup_{n \geq 1}^{\infty} B_n$. Similarly $A = \bigcup_{n \geq 1} A_n$. Let us define a function h on D by:

$$h(s) = \begin{cases} g_n(s) & \text{if } s \in A_n \\ g_n^{-1}(s) & \text{if } s \in B_n \\ s & \text{if } s \in D \setminus (A \cup B). \end{cases}$$

From its definition, $h = h^{-1}$. So h is a bijection of D onto D. By its construction, h is strongly G-decomposable over D and so is continuous from D onto D. Since $h = h^{-1}$ and h maps D into D, it follows that h is a homeomorphism of D onto D. It is now clear that h has all the other required properties. $\square$

The following proposition seems complicated to state. But the basic idea is straightforward. We are manufacturing a copy of the Dyadic Group, $\bigoplus \mathbb{Z}_2$ from the action of G and, at the same time splitting D into Dyadic pieces. In the following $\mathbb{Z}_2^n$ is $\{0, 1\}^n$, equipped with the natural (additive) group structure. So it can be identified with the direct sum of n copies of $\mathbb{Z}_2$.

Proposition 6.6.16 *Let X be a zero-dimensional Hausdorff space with no isolated points. Let G be a countably infinite discrete group and let $\epsilon : G \mapsto Homeo(X)$ be a non-degenerate action of G on X. Let $D = (s_0, s_1, \cdots)$ be a dense orbit. Let (D_n) be a decreasing sequence of clopen neighbourhoods of s_0 such that $s_n \notin D_n$ for each $n \in \mathbb{N}$. Then the following statements hold.*

(a) *There is a sequence $(h_k)(k = 1, 2 \ldots)$ of homeomorphisms of D onto D where $h_k = h_k^{-1}$. For $1 \leq k \leq n$, the h_k are mutually commutative. Each h_k is strongly G-decomposable over D.*

(b) *For each positive integer n, there exists a finite family of pairwise disjoint, clopen subsets of D,*

$$\{K^n(\alpha_1, \alpha_2, \ldots, \alpha_n) : (\alpha_1, \alpha_2, \ldots, \alpha_n) \in \mathbb{Z}_2^n\}$$

whose union is D.

(c) *Let $K^0(\varnothing) = D$. For $0 \leq p \leq n - 1$, $K^p(\alpha_1, \alpha_2, \ldots, \alpha_p) = K^{p+1}(\alpha_1, \alpha_2, \ldots, \alpha_p, 0) \cup K^{p+1}(\alpha_1, \alpha_2, \ldots, \alpha_p, 1)$.*

(d) *For $1 \leq p \leq n$, $K^p(0, 0, \ldots, 0) \subset D_p$ and $s_0 \in K^p(0, 0, \ldots, 0)$.*

(e) *Let $(\alpha_1, \alpha_2, \ldots, \alpha_p) \in \mathbb{Z}_2^p$ where $1 \leq p \leq n$. Then the homeomorphism $h_1^{\alpha_1} h_2^{\alpha_2} \ldots h_p^{\alpha_p}$ interchanges $K^p(\beta_1, \beta_2, \ldots, \beta_p)$ with $K^p(\alpha_1 + \beta_1, \alpha_2 + \beta_2, \ldots, \alpha_p + \beta_p)$.*

(**f**) *For each n,*

$$\{s_0, s_1, \ldots, s_n\} \subset \{h_1^{\alpha_1} h_2^{\alpha_2} \ldots h_n^{\alpha_n}(s_0) : (\alpha_1, \alpha_2, \ldots, \alpha_n) \in \mathbb{Z}_2^n\}.$$

(**g**) *For each $s \in D$, if $h_1^{\alpha_1} h_2^{\alpha_2} \ldots h_n^{\alpha_n}(s) = s$ then $\alpha_1 = \alpha_2 = \ldots = \alpha_n = 0$.*

Proof We give an inductive argument. First, let $A = D_1$ and let $B = D \backslash D_1$. By applying Lemma 6.6.15, there exists a homeomorphism h_1 of D onto itself, where h_1 interchanges D_1 and $D \setminus D_1$, and maps s_0 to s_1. (So (f) holds for $n = 1$.) Also $h_1 = h_1^{-1}$ and h_1 is strongly G-decomposable over D.

For any $s \in D$, $h_1(s)$ and s are elements of disjoint clopen sets. Hence (g) holds for $n = 1$.

Now let $K^1(0) = D_1$ and $K^1(1) = D \backslash D_1$.

Let us now suppose that we have constructed the homeomorphisms $h_1, h_2, \ldots, h_n$ and the clopen sets

$$\{K^p(\alpha_1, \alpha_2, \ldots, \alpha_p) : (\alpha_1, \alpha_2, \ldots, \alpha_p) \in \mathbb{Z}_2^p\} \text{ for } p = 1, 2, \ldots, n.$$

We now need to make the $(n + 1)$th step of the inductive construction.

For some $(\alpha_1, \alpha_2, \ldots, \alpha_n) \in \{0, 1\}^n$, $s_{n+1} \in K^n(\alpha_1, \alpha_2, \ldots, \alpha_n)$. Let $c = h_1^{\alpha_1} h_2^{\alpha_2} \ldots h_n^{\alpha_n}(s_{n+1})$. Then $c \in K^n(0, 0, \ldots, 0)$.

If $c \neq s_0$ let $b = c$. If $c = s_0$ then let b be any other element of $K^n(0, 0, \ldots, 0)$. Now let A be a clopen subset of $K^n(0, 0, \ldots, 0) \cap D_{n+1}$ such that $s_0 \in A$ and $b \notin A$. Let $B = K^n(0, 0, \ldots, 0) \setminus A$. We apply Lemma 6.6.15 to find a homeomorphism h of D onto itself, which interchanges A and B, leaves every point outside $A \cup B$ fixed, maps s_0 to b, and $h = h^{-1}$. Also, h is strongly G-decomposable over D.

Let $K^{n+1}(0, 0, \ldots, 0) = A$ and $K^{n+1}(0, 0, \ldots, 1) = B$. By construction, (d) holds for $p = n + 1$.

Let $K^{n+1}(\alpha_1, \alpha_2, \ldots, \alpha_n, 0) = h_1^{\alpha_1} h_2^{\alpha_2} \ldots h_n^{\alpha_n}[A]$ and $K^{n+1}(\alpha_1, \alpha_2, \ldots, \alpha_n, 1) = h_1^{\alpha_1} h_2^{\alpha_2} \ldots h_n^{\alpha_n}[B]$. Then (b) holds for $n + 1$ and (c) holds for $p = n$.

We now define h_{n+1} as follows. For $s \in K^n(\alpha_1, \alpha_2, \ldots, \alpha_n)$,

$$h_{n+1}(s) = h_1^{\alpha_1} h_2^{\alpha_2} \ldots h_n^{\alpha_n} h h_1^{\alpha_1} h_2^{\alpha_2} \ldots h_n^{\alpha_n}(s).$$

Claim 1 h_{n+1} commutes with h_j for $1 \leq j \leq n$.

To simplify our notation we shall take $j = 1$, but the calculation works in general, since each of $\{h_r : r = 1, 2, \ldots, n\}$ commutes with the others.

Let $s \in D$. Then $s \in K^n(\alpha_1, \alpha_2, \ldots, \alpha_n)$ for some $(\alpha_1, \alpha_2, \ldots, \alpha_n) \in \mathbb{Z}_2^n$. So $h_1(s) \in K^n(\alpha_1 + 1, \alpha_2, \ldots, \alpha_n)$. Then

$$h_{n+1}(h_1 s) = h_1^{\alpha_1+1} h_2^{\alpha_2} \ldots h_n^{\alpha_n} h h_1^{\alpha_1+1} h_2^{\alpha_2} \ldots h_n^{\alpha_n}(h_1 s).$$

So

$$h_{n+1}h_1(s) = h_1 h_1^{\alpha_1} h_2^{\alpha_2} \ldots h_n^{\alpha_n} h h_1^{\alpha_1} h_2^{\alpha_2} \ldots h_n^{\alpha_n} h_1 h_1(s)$$

$$= h_1 h_{n+1}(s).$$

From this we see that h_{n+1} commutes with h_1. Similarly h_{n+1} commutes with h_j for $2 \leq j \leq n$.

Claim 2 h_{n+1} is strongly G-decomposable over D.

By Lemma 6.6.12, $h_1^{\alpha_1} h_2^{\alpha_2} \ldots h_n^{\alpha_n} h h_1^{\alpha_1} h_2^{\alpha_2} \ldots h_n^{\alpha_n}$ is strongly G-decomposable over D. So, on restricting to the clopen set $K^n(\alpha_1, \alpha_2, \ldots, \alpha_n)$ this gives that h_{n+1} is *strongly G-decomposable over each* $K^n(\alpha_1, \alpha_2, \ldots, \alpha_n)$. Hence h_{n+1} is strongly G-decomposable over D.

So, by Claim 1 and Claim 2, (a) holds for $n + 1$. It is straightforward to show that (b),(c), (d) and (e) hold for $n + 1$.

Now consider (f). Either $c = s_0$ in which case, $s_0 = h_1^{\alpha_1} h_2^{\alpha_2} \ldots h_n^{\alpha_n}(s_{n+1})$ which gives $s_{n+1} = h_1^{\alpha_1} h_2^{\alpha_2} \ldots h_n^{\alpha_n}(s_0)$, or $c \neq s_0$, in which case

$$h_{n+1}(h_1^{\alpha_1} h_2^{\alpha_2} \ldots h_n^{\alpha_n}(s_{n+1})) = h(h_1^{\alpha_1} h_2^{\alpha_2} \ldots h_n^{\alpha_n}(s_{n+1})) = s_0.$$

This gives $s_{n+1} = h_{n+1} h_1^{\alpha_1} h_2^{\alpha_2} \ldots h_n^{\alpha_n}(s_0)$. Because the homeomorphisms commute, this gives (f) for $n + 1$.

Finally consider (g). Let $s \in D$ with $h_1^{\alpha_1} h_2^{\alpha_2} \ldots h_{n+1}^{\alpha_{n+1}}(s) = s$. If $\alpha_{n+1} = 0$ then (g) implies $\alpha_1 = \alpha_2 = \ldots = \alpha_n = 0$.

So now suppose $\alpha_{n+1} = 1$. Let $h_1^{\beta_1} h_2^{\beta_2} \ldots h_n^{\beta_n}(s) \in K^n(0, 0, \ldots, 0)$. Then, since the h_r all commute, we can suppose without loss of generality that $s \in K^n(0, 0, \ldots, 0)$. Then $h_{n+1}(s) = h_1^{\alpha_1} h_2^{\alpha_2} \ldots h_n^{\alpha_n}(s)$. But h_{n+1} maps $K^n(0, 0, \ldots, 0)$ to itself and $h_1^{\alpha_1} h_2^{\alpha_2} \ldots h_n^{\alpha_n}$ maps $K^n(0, 0, \ldots, 0)$ to $K^n(\alpha_1, \alpha_2, \ldots, \alpha_n)$. So $h_{n+1}(s) \in K^n(0, 0, \ldots, 0) \cap K^n(\alpha_1, \alpha_2, \ldots, \alpha_n)$. But this intersection is only non-empty if $\alpha_1 = \alpha_2 = \ldots = \alpha_n = 0$. So $h_{n+1}(s) = s$. But h_{n+1} acting on $K^n(0, 0, \ldots, 0)$, interchanges $K^{n+1}(0, 0, \ldots, 0)$ with $K^{n+1}(0, 0, \ldots, 1)$.

So $s \in K^{n+1}(0, 0, \ldots, 1) \cap K^{n+1}(0, 0, \ldots, 0)$, which is impossible. $\qquad\square$

Corollary 6.6.17 *Let* $(\alpha_1, \alpha_2, \ldots, \alpha_p) \in \mathbb{Z}_2^p$. *Let* T *be the set:*

$$T = \{h_1^{\lambda_1} \ldots h_n^{\lambda_n}(s_0) \in D : n \geq p \;\&\; \lambda_j = \alpha_j \text{ for } 1 \leq j \leq p\}.$$

Then $T = K^p(\alpha_1, \alpha_2, \ldots, \alpha_p)$ *which is a clopen subset of* D.

Proof Let $(\lambda_1, .., \lambda_n) \in \mathbb{Z}_2^n$. By Proposition 6.6.16 (d) $s_0 \in K^n(0, \ldots, 0)$. So, by Proposition 6.6.16 (e), $h_1^{\lambda_1} \ldots h_n^{\lambda_n}(s_0) \in K^n(\lambda_1, .., \lambda_n)$. By Proposition 6.6.16 (c), $K^n(\lambda_1, .., \lambda_n) \subset K^p(\lambda_1, .., \lambda_p)$ for $n \geq p$. So $T \subset K^p(\alpha_1, \alpha_2, \ldots, \alpha_p)$.

Conversely, take any $s \in K^p(\alpha_1, \alpha_2, \cdots, \alpha_p)$ and fix $n \geq p$. Then by Proposition 6.6.16 (f), $s = h_1^{\beta_1} \cdots h_n^{\beta_n}(s_0)$ for some $\beta_1, \ldots \beta_n$.

Since $h_1^{\beta_1} \ldots h_n^{\beta_n}(s_0) \in K^p(\alpha_1, .., \alpha_p)$ we have, as above,

$$h_1^{\beta_1} \ldots h_n^{\beta_n}(s_0) \in K^p(\beta_1, .., \beta_p).$$

So $K^p(\alpha_1, .., \alpha_p) \cap K^p(\beta_1, .., \beta_p)$ is not empty. So, by Proposition 6.6.16 (b),

$$(\alpha_1, .., \alpha_p) = (\beta_1, .., \beta_p).$$

Thus $K^p(\alpha_1, .., \alpha_p) \subset T$.   □

The countable group $\bigoplus \mathbb{Z}_2$ is a sub-group of the compact group $\prod_{n=1}^{\infty} \mathbb{Z}_2$. We emphasise that we give $\bigoplus \mathbb{Z}_2$ the relative topology it inherits from the compact space $\prod_{n=1}^{\infty} \mathbb{Z}_2$. Hitherto we have usually considered countable groups as being equipped with the discrete topology.

Let $\sigma : \bigoplus \mathbb{Z}_2 \mapsto D$ be defined by $\sigma(\lambda_1, .., \lambda_n, 0, 0, 0 \ldots) = h_1^{\lambda_1} \ldots h_n^{\lambda_n}(s_0)$.

Corollary 6.6.18 *The map σ is a bijection from $\bigoplus \mathbb{Z}_2$ onto D. Also σ is an open map.*

Proof By Proposition 6.6.16 (f) σ maps $\bigoplus \mathbb{Z}_2$ onto D. By Proposition 6.6.16 (g) σ is injective.

The product topology of $\prod_{n=1}^{\infty} \mathbb{Z}_2$ induces the topology of $\bigoplus \mathbb{Z}_2$. It now follows, by using Corollary 6.6.17, that the sets $\sigma^{-1}[K^p(\alpha_1, \alpha_2, \ldots, \alpha_p)]$ form a sub-base for the topology of $\bigoplus \mathbb{Z}_2$. Since each $K^p(\alpha_1, \alpha_2, \ldots, \alpha_p)$ is a clopen subset of D, σ is an open map.   □

Let us recall once more that $\bigoplus \mathbb{Z}_2$ is the direct sum of an infinite sequence of copies of $\mathbb{Z}_2$. (So each element of the group is an infinite sequence of zeroes and ones, with 1 occurring only finitely many times.) The natural action of $\bigoplus \mathbb{Z}_2$ on $\prod_{n=1}^{\infty} \mathbb{Z}_2$ gives rise to the dynamical system, $(\prod_{n=1}^{\infty} \mathbb{Z}_2, \bigoplus \mathbb{Z}_2, \varepsilon)$, described in Example 6.1.4.

We observe that in the following there is no requirement that X be a Baire space.

Theorem 6.6.19 *Let X be a zero-dimensional Hausdorff space with no isolated points. Let G be a countably infinite discrete group and let $\epsilon : G \mapsto Homeo(X)$ be an action of G on X which is non-degenerate and pseudo free. Let s_0 be a point in X such that the orbit $\{\varepsilon_g(s_0) : g \in G\} = D$, is dense. Let (D_n) be a strictly decreasing sequence of clopen neighbourhoods of s_0, in the relative topology of D such that*

$$\{s_0\} = \bigcap_{n \geq 1} D_n.$$

Then there exist Z, a G_δ subset of X with $D \subset Z$, and an action $\phi : \bigoplus \mathbb{Z}_2 \mapsto Homeo(Z)$ such that the following properties hold.

(1) *The orbit $\{\phi_\delta(s_0) : \delta \in \bigoplus \mathbb{Z}_2\}$ is free and coincides with the set D.*
(2) *The groups of homeomorphisms $\varepsilon[G]$ and $\phi[\bigoplus \mathbb{Z}_2]$ are strongly equivalent over Z, which is invariant under the action of both these groups.*

(3) *The orbit equivalence relations corresponding, respectively, to $\varepsilon[G]$ and $\phi[\bigoplus \mathbb{Z}_2]$ coincide on Z.*

(4) *The action ϕ is an isomorphism of $\bigoplus \mathbb{Z}_2$ into Homeo(Z).*

(5) *For each n let e_n be the element of $\bigoplus \mathbb{Z}_2$ which has 1 in the nth place and zero elsewhere. Let h_n be the restriction to D of ϕ_{e_n}. Then (h_n) satisfies all the conditions of Proposition 6.6.16. In particular,*

$$s_0 \in K^p(0) \subset D_p \text{ for } p = 1, 2 \dots .$$

Proof We use Proposition 6.6.16 to find a sequence (h_r) of homeomorphisms of D onto itself with the properties listed in that proposition.

For each $\alpha \in \bigoplus \mathbb{Z}_2$, there exist a natural number n and $(\alpha_1, \alpha_2, \dots, \alpha_n) \in \mathbb{Z}_2{}^n$ such that $\alpha = (\alpha_1, \alpha_2, \dots \alpha_n, 0, 0, \dots)$. We define $\psi(\alpha) = h_1^{\alpha_1} h_2^{\alpha_2} \dots h_n^{\alpha_n}$. Then this is a homomorphism of $\bigoplus \mathbb{Z}_2$ into *Homeo*(D). By Proposition 6.6.16 (g) ψ is injective.

By Corollary 6.6.14 we can find a G-invariant, G_δ-set V, with $D \subset V \subset X$ and an action ψ^V of $\bigoplus \mathbb{Z}_2$ as homeomorphisms of V with the following properties. For each $\gamma \in \bigoplus \mathbb{Z}_2$, ψ_γ^V is an extension of ψ_γ and ψ_γ^V is strongly G–decomposable over V. Also the orbit $\{\psi_\gamma^V(s_0) : \gamma \in \bigoplus \mathbb{Z}_2\} = \{\psi_\gamma(s_0) : \gamma \in \bigoplus \mathbb{Z}_2\}$. By Proposition 6.6.16 (g) this orbit is free. By Proposition 6.6.16 (f) the orbit coincides with D.

It is clear, from Definition 6.5.8, that the action of G, when restricted to V, is still pseudo free. So, by putting $X = V$ in Lemma 6.6.10, we find that there exists a G_δ-subset Z of V, with $D \subset Z \subset V$, such that Z is invariant under the actions of G and $\bigoplus \mathbb{Z}_2$. Furthermore, these actions are strongly equivalent on Z. Let ϕ_γ be the restriction of ψ_γ^V to Z. So ϕ is an action of $\bigoplus \mathbb{Z}_2$ as homeomorphisms of Z which satisfies properties (1) and (2). It follows from Lemma 6.6.3 and (2) that (3) is also satisfied.

The restriction of ϕ_γ to D is ψ_γ. So, when ϕ_γ is the identity on Z, then ψ_γ is the identity on D. Since ψ is injective this implies $\gamma = 0$. So (4) holds. Clearly, from our construction, (5) holds.   □

6.7 Orbit Equivalence: Extremally Disconnected Spaces

The main theorem looks very similar to the previous one but makes use of the fact that compact (Hausdorff) spaces are Baire spaces and that a compact, extremally disconnected space is the Stone-Czech compactification of any dense subspace. In this section S is a compact (Hausdorff) extremally disconnected space with no isolated points; in particular S is zero-dimensional. As before, G is a countably infinite group and $\varepsilon : G \mapsto Homeo(S)$ is a non-degenerate action.

Lemma 6.7.1 *Let D be a G-invariant, dense subset of S. Let the action of G be pseudo free over D. Then there exists a G_δ-set Y, with $D \subset Y \subset S$, such that the action is pseudo free over Y.*

Proof Since S is extremally disconnected, it now follows from Corollary 6.2.4, that so also is the dense subset D. Let $g \in G$ and let $K_g = \{s \in S : g(s) = s\}$. Since g is continuous, K_g is closed. By hypothesis, $K_g \cap D$ is a clopen subset of D. So, by Lemma 6.2.6, $cl(K_g \cap D)$ is a clopen subset of S. Let $F_g = K_g \setminus cl(K_g \cap D)$. So F_g is a closed set which is disjoint from D and so has empty interior. Let $F = \bigcup_{g \in G} F_g$. Then F is a meagre F_σ-set which is disjoint from D. Since D is G-invariant, the saturation $G[F]$ is also disjoint from D. Let Y be the G_δ-set $S \setminus G[F]$. We have $Y \cap K_g = Y \cap cl(K_g \cap D)$ which is clopen in Y for each g.   □

Theorem 6.7.2 *Let S be a compact Hausdorff extremally disconnected space with no isolated points. Let G be a countably infinite group. Let $\varepsilon : G \mapsto Homeo(S)$ be a non-degenerate action of G as homeomorphisms of S. Let s_0 be a point in S such that the orbit $\{\varepsilon_g(s_0) : g \in G\} = D$, is dense and pseudo free. Then there exist an action $\eta : \bigoplus \mathbb{Z}_2 \mapsto Homeo(S)$ and Z, a dense G_δ-subset of S with $D \subset Z$, such that the following properties hold.*

(1) *The orbit $\{\eta_\delta(s_0) : \delta \in \bigoplus \mathbb{Z}_2\}$ is free and coincides with the set D.*
(2) *The groups of homeomorphisms $\varepsilon[G]$ and $\eta[\bigoplus \mathbb{Z}_2]$ are strongly equivalent over Z, which is invariant under the action of both these groups.*
(3) *The orbit equivalence relations corresponding, respectively, to $\varepsilon[G]$ and $\eta[\bigoplus \mathbb{Z}_2]$ coincide on Z.*
(4) *The action η is an isomorphism of $\bigoplus \mathbb{Z}_2$ into $Homeo(S)$.*
(5) *For each n let e_n be the element of $\bigoplus \mathbb{Z}_2$ which has 1 in the nth place and zero elsewhere. Let h_n be the restriction to D of ϕ_{e_n}. Then (h_n) satisfies all the conditions of Proposition 6.6.16.*

Proof By Lemma 6.7.1, there exists a G-invariant Y, such that the action of G is pseudo free on Y. Also Y is a dense G_δ-subset of S, and $D \subset Y$.

Since Y is zero-dimensional and has no isolated points we may apply Theorem 6.6.19 with X replaced by Y. We find a G-invariant, G_δ-set Z, with $D \subset Z \subset Y$; also an action $\phi : \bigoplus \mathbb{Z}_2 \mapsto Homeo(Z)$ with the five properties listed in Theorem 6.6.19.

By Corollary 6.2.8, each h in $Homeo(Z)$ has a unique extension to $\hat{h}$ in $Homeo(S)$. We now define $\eta : \bigoplus \mathbb{Z}_2 \mapsto Homeo(S)$ by $\eta_\delta = \hat{\phi}_\delta$.   □

Suppose the action of G on S has a dense, pseudo free orbit D. The following theorem says there exists a dense, G-invariant, G_δ-subset Y, $D \subset Y$ with the following property. Orbit equivalence on Y with respect to the action of G, $\sim_G$, is the same as orbit equivalence with respect to an action of $\mathbb{Z}$. In other words G orbit equivalence is generated by a single homeomorphism of S.

Theorem 6.7.3 *Let S be a compact Hausdorff extremally disconnected space with no isolated points. Let G be a countably infinite group. Let $\varepsilon : G \mapsto Homeo(S)$ be a non-degenerate action of G as homeomorphisms of S. Let s_0 be a point in S such that the orbit $\{\varepsilon_g(s_0) : g \in G\} = D$, is dense and pseudo free. Then there exists* *a homeomorphism θ from S onto itself such that the action $\eta : \mathbb{Z} \mapsto Homeo(S)$ defined by*

$$\eta(n) = \theta^n$$

is generically equivalent to the action of G. There exists a dense G_δ-set $S_0 \subset S$, where S_0 is G-invariant and $\theta[S_0] = S_0$, such that, on S_0, orbit equivalence with respect to the G-action coincides with orbit equivalence with respect to the $\mathbb{Z}$-action.

Proof It follows from the preceding theorem that we may replace G by $\bigoplus \mathbb{Z}_2$. There exists a sequence of homeomorphisms of D, $(h_r)(r = 1, 2 \ldots)$, which satisfies all the conditions of Proposition 6.6.16. Each h_r has a unique extension to a homeomorphism of S, which we shall again denote by h_r. The action of $\bigoplus \mathbb{Z}_2$ on S is:

$$\eta(\alpha_1, \ldots \alpha_n, 0, 0, \ldots) = h_1^{\alpha_1} \ldots h_n^{\alpha_n}.$$

Then $D = \{h_1^{\alpha_1} \ldots h_n^{\alpha_n}(s_0) : (\alpha_1, \ldots \alpha_n, 0, 0, \ldots) \in \bigoplus \mathbb{Z}_2\}$.

Let $E_1 = K^1(0)$ and $F_1 = K^1(1)$. Let $E_{j+1} = K^{j+1}(\mathbf{1}, 0)$ where $\mathbf{1} \in \mathbb{Z}_2^j$ and let $F_{j+1} = K^{j+1}(\mathbf{0}, 1)$ where $\mathbf{0} \in \mathbb{Z}_2^j$.

We observe that if $K^n(\alpha_1, \alpha_2, \ldots, \alpha_n)$ has non-empty intersection with $K^{n+p}(\beta_1, \beta_2, \ldots, \beta_{n+p})$ then $\alpha_1 = \beta_1, \ldots, \alpha_n = \beta_n$. From this it follows that $(E_n)(n = 1, 2 \ldots)$ is a sequence of pairwise disjoint clopen subsets of D. Similarly, $(F_n)(n = 1, 2 \ldots)$ is a sequence of pairwise disjoint clopen subsets of D. Let $E = \bigcup_{n=1} E_n$ and $\bigcup_{n=1} F_n = F$. We claim that $E = D$ and $D \backslash \{s_0\} = F$.

By Corollary 6.6.17 and Corollary 6.6.18, $\sigma^{-1}[K^n(\alpha_1, \ldots, \alpha_n)]$ is the set of all sequences in $\bigoplus \mathbb{Z}_2$ whose first n terms are $(\alpha_1, \ldots, \alpha_n)$. Sequences beginning with zero correspond to $K^1(0) = E_1$. Any other sequence in $\bigoplus \mathbb{Z}_2$ begins with p successive ones followed by a zero. So they are in $\sigma^{-1}[E_{p+1}]$. So $E = D$.

Now consider $\sigma^{-1}[F_1]$. It contains any sequence beginning with one. But s_0 corresponds to $(0, 0, \ldots.)$ and is not in any of the $\sigma^{-1}[F_k]$. All remaining sequences begin with a string of p zeroes with one in the $(p + 1)$th place. Such a sequence is in $\sigma^{-1}[F_{p+1}]$. So $F = D \backslash \{s_0\}$.

For $s \in E_n$ let $\theta(s) = h_1 h_2 \ldots h_n(s)$. Then θ is a continuous map of E_n onto F_n. From this it is straightforward to see that θ is a continuous bijection of D onto $D \backslash \{s_0\}$. Similarly, θ^{-1} is a continuous bijection from $D \backslash \{s_0\}$ onto D.

Since S has no isolated points, $D \backslash \{s_0\}$ is dense in S. But, by Theorem 6.2.7, S can be identified with the Stone-Czech compactification of any dense subset of itself. Applying this to θ^{-1} and θ we find continuous extensions which are homeomorphisms of S onto S and which are inverses of each other. We abuse notation and denote the extension of θ to the whole of S by θ. Then $j \mapsto \theta^j$ is the $\mathbb{Z}$-action considered here; let Δ be the subgroup of $Homeo(S)$ generated by θ.

On applying Lemma 6.2.6, we see that $(clE_n)(n = 1, 2 \ldots)$ is a sequence of pairwise disjoint clopen subsets of S. So its union is a dense open subset of S which we shall denote by O_1. By continuity, for $s \in clE_n$ we have $\theta(s) = h_1^1 h_2^1 \ldots h_n^1(s)$.

Similarly, $(clF_n)(n = 1, 2 \ldots)$ is a sequence of pairwise disjoint clopen subsets of S whose union, O_2, is also dense in S.

Let Γ be the countable subgroup of $Homeo(S)$ generated by θ and the homeomorphisms (h_r). Let S_0 be the intersection $\bigcap \{\gamma[O_1 \cap O_2] : \gamma \in \Gamma\}$. Then S_0 is a dense G_δ-subset of S which is invariant under the action of Γ. From the definition of θ, it is clear that θ is strongly G-decomposable over S_0. (Recall that we have identified G with $\bigoplus \mathbb{Z}_2$.) Similarly, θ^{-1} is also strongly G-decomposable over S_0. Hence each element of Δ is strongly G-decomposable over S_0.

Let $H(\Delta)$ be the group of all homeomorphisms h, of S onto S, such that h is strongly Δ-decomposable with respect to a finite partition of S into clopen sets. We shall show that h_1 is in $H(\Delta)$.

For $s \in E_1 = K^1(0)$ we have $\theta(s) = h_1(s)$ and, for $s \in F_1 = K^1(1)$, $\theta^{-1}(s) = h_1(s)$. We observe that clE_1 and clF_1 are disjoint clopen sets whose union is S. But h_1 restricted to clE_1 coincides with θ and h_1 restricted to clF_1 coincides with θ^{-1}. So $h_1 \in H(\Delta)$.

We now suppose that $h_1, h_2, \ldots, h_n$ are in $H(\Delta)$. We wish to show $h_{n+1} \in H(\Delta)$. Let $s \in K^{n+1}(\beta, 0)$ where $\beta \in \mathbb{Z}_2^n$. By Proposition 6.6.16 (e) $h_1^{\beta_1+1} \ldots h_n^{\beta_n+1}(s) \in K^{n+1}(1, 0) = E_{n+1}$. So, from the definition of θ,

$$\theta(h_1^{\beta_1+1} \ldots h_n^{\beta_n+1}(s)) = h_1 \ldots h_n h_{n+1} h_1^{\beta_1+1} \ldots h_n^{\beta_n+1}(s).$$

Making use of commutativity of the h_r, we get

$$h_{n+1}(s) = h_1^{\beta_1} \ldots h_n^{\beta_n} \theta h_1^{\beta_1+1} \ldots h_n^{\beta_n+1}(s).$$

Then, by using continuity, this holds for each $s \in clK^{n+1}(\beta, 0)$. By a similar argument, for $s \in clK^{n+1}(\beta, 1)$ we get $h_{n+1}(s) = h_1^{\beta_1} \ldots h_n^{\beta_n} \theta^{-1} h_1^{\beta_1+1} \ldots h_n^{\beta_n+1}(s)$.

Since $\{clK^{n+1}(\alpha) : \alpha \in \mathbb{Z}_2^{n+1}\}$ is a finite collection of disjoint clopen sets whose union is S, it follows that $h_{n+1} \in H(\Delta)$.

So, by induction, $G \subset H(\Delta)$.

It now follows that, on S_0, the G-action and the $\mathbb{Z}$-action generated by θ are strongly equivalent. So the corresponding orbit equivalence relations coincide on S_0. $\qquad\qquad\square$

Remark 6.7.4 In the above, D is not a subset of S_0. It is not obvious that the $\mathbb{Z}$-action on S_0, implemented by θ, has any dense orbit in S_0. Is it possible to modify the construction of θ so that it becomes a bijection of D onto itself?

Exercise 6.7.5 Show that D is disjoint from S_0.

Hint: Show that $O_1 \cap O_2 \cap D = D \backslash \{s_0\}$. Deduce $s_0 \notin O_1 \cap O_2$. Deduce $h_1^{\alpha_1} \ldots h_n^{\alpha_n}(s_0) \notin S_0$ for any $(\alpha_1, \ldots \alpha_n) \in \mathbb{Z}_2^n$.

6.8 Orbit Equivalence: Perfect Polish Spaces

Let X be a perfect Polish space. So X is homeomorphic to a complete separable metric space and X has no isolated points. Let G be a countably infinite group acting ergodically on X. Then, given the results of Sects. 6.6 and 6.7, it is not surprising that, modulo meagre sets, the orbit equivalence relation $\sim_G$ arises from an action of $\bigoplus \mathbb{Z}_2$. But much more is true. Modulo meagre sets, the orbit equivalence relation, $\sim_G$, is *unique*.

Theorem 6.8.1 *Let X be a perfect Polish space, G a countably infinite group and $\beta : G \mapsto Homeo(X)$ a non-degenerate action which is ergodic. Then the action β is generically equivalent to the natural action of the Dyadic Group, $\bigoplus \mathbb{Z}_2$ on $\prod \mathbb{Z}_2$.*

Proof By Proposition 6.5.5 there exists a dense, G-invariant $X_1 \subset X$ such that $G[x]$ is a dense orbit for each $x \in X_1$ and X_1 is a G_δ-subset of X.

By Proposition 6.5.6 there exists a dense, G-invariant $X_2 \subset X$ such that X_2 is zero-dimensional and X_2 is a G_δ-subset of X.

By Proposition 6.5.10 there exists a dense, G-invariant $X_3 \subset X$, where X_3 is a G_δ-subset of X, such that, for each $g \in G$,

$$F_g = \{x \in X : g(x) = x\}$$

is a clopen subset of X_3.

Since X is a Baire space, the G_δ set $T = X_1 \cap X_2 \cap X_3$ is dense in X. Clearly it is also G-invariant. Since it is a dense G_δ-subset of the Baire space X, it follows from Lemma 6.5.1 that T is a Polish space. Since T is a dense subspace of X, which has no isolated points, T does not have any isolated points, that is T is a perfect Polish space. Since T is a subspace of the zero-dimensional space X_2, it too is zero-dimensional.

Let ϵ_g be the restriction to T of β_g for each $g \in G$. Since T is a dense subspace of X and β is a non-degenerate action, it follows that ϵ is also non-degenerate. For each $x \in X$ the orbit $G[x]$ is dense. The action of G on T is pseudo free because $T \subset X_3$. Fix s_0 in T and let $D = G[s_0]$.

Since T is a Polish space we can give it a complete metric ρ. We can now find a decreasing sequence of clopen neighbourhoods of s_0, (S_n) with each S_n contained in a ball (centred on s_0) of radius $1/2^n$. Let $D_n = S_n \cap D$ for each n.

The dynamical system (T, G, ε) satisfies the hypotheses of Theorem 6.6.19. So there exists Z, where Z is a G-invariant, G_δ-subset of T, and $D \subset Z$, and there exists an action $\phi : \bigoplus \mathbb{Z}_2 \mapsto Homeo(Z)$ with the properties listed in Theorem 6.6.19.

Let $\sigma : \bigoplus \mathbb{Z}_2 \mapsto D$ be defined by $\sigma(\lambda_1, .., \lambda_n, 0, 0, 0 \ldots) = h_1^{\lambda_1} \ldots h_n^{\lambda_n}(s_0)$. Then, see Corollary 6.6.18, σ is a bijection and an open map. We claim that σ is continuous.

Let $(\mathbf{x}_n)$ be a sequence in $\bigoplus \mathbb{Z}_2$ converging to $\mathbf{x} = (\alpha_1, \ldots, \alpha_p, 0, 0, ..)$.

Fix N. Then for all large enough n, the first N terms of $\mathbf{x}_n$ and $\mathbf{x}$ coincide. It follows from Corollary 6.6.17 that $\sigma(\mathbf{x}_n) \in K^N(\alpha_1, \ldots, \alpha_p, 0, \ldots, 0)$ for large n.

Thus $h_1^{\alpha_1} \ldots h_p^{\alpha_p}(\sigma(\mathbf{x}_n)) \in K^N(0) \subset D_N$ for all large n. Hence $\rho(h_1^{\alpha_1} \ldots h_p^{\alpha_p}(\sigma(\mathbf{x}_n)), s_0) < 1/2^N$ for all large n. Thus $h_1^{\alpha_1} \ldots h_p^{\alpha_p}(\sigma(\mathbf{x}_n)) \to s_0$. By applying the homeomorphism $h_1^{\alpha_1} \ldots h_p^{\alpha_p}$ to this convergent sequence we find

$$(\sigma(\mathbf{x}_n)) \to h_1^{\alpha_1} \ldots h_p^{\alpha_p}(s_0).$$

It follows that σ is a homeomorphism of $\bigoplus \mathbb{Z}_2$ onto D.

By Lavrentiev's Theorem, see Theorem 6.2.9, the homeomorphism σ can be extended to a homeomorphism $\sigma^\#$ of V onto W, where V, W are dense G_δ-subsets of $\prod \mathbb{Z}_2$ and Z, respectively. Since we could replace W by $\bigcap_{g \in G} g[W]$ we may assume that W is G-invariant. Since the G-action on Z and the $\bigoplus \mathbb{Z}_2$ action on Z are strongly equivalent on Z, they are strongly equivalent on W. It now follows from the properties of σ that the β action of G on X is generically equivalent to the canonical action of $\bigoplus \mathbb{Z}_2$ on $\prod \mathbb{Z}_2$. $\qquad\square$

In Sect. 6.7 we showed that when a countably infinite group acts on a separable, compact, extremally disconnected space with no isolated points, the corresponding orbit equivalence relation is hyperfinite (the union of an increasing sequence of finite equivalence relations) modulo meagre sets. But for perfect Polish spaces, we have a stronger result – uniqueness.

6.9 Automorphisms and the Dixmier Algebra

We have made frequent references to the duality between compact Hausdorff spaces and commutative unital C^*-algebras. Let A be a commutative C^*-algebra with spectrum E; so A is isomorphic to $C(E)$. We have seen that, for each $\theta \in Homeo E$, we can define a $*$-automorphism by $h_\theta(f) = f \circ \theta$. Moreover every $*$-automorphism of $C(E)$ arises in this way. An action of G on A is a group homomorphism of G into $Aut A$, the group of all $*$-automorphisms.

When ε is an action of G as homeomorphisms of E, then we can define an action ε' as homomorphisms of A by $\varepsilon'_g(f) = f \circ \varepsilon_{g^{-1}} = f \circ \varepsilon_g^{-1}$. (We use g^{-1} instead of g, because we want ε' to be a group homomorphism, not an anti-homomorphism). So $\varepsilon'_g = h_{g^{-1}} = h_g^{-1}$. It is straightforward to show that each action of G (as automorphisms of A) arises in this way from an action of G on E.

We can set up a correspondence between properties of an action ε on E, and of the dual action ε' on A. For example, when E is extremally disconnected, the action ε is ergodic precisely when ε' has the following property: if p is a projection and $\varepsilon'_g(p) = p$ for each $g \in G$, then $p = 0$ or $p = 1$. This follows from Lemma 6.3.4.

We shall see later that the action ε is free on some dense G_δ-subset of (an extremally disconnected) E precisely when ε' is "strictly outer". That is, whenever p is a non-zero projection and ε'_g leaves each point of pA fixed, then $g = \iota$. When A is a von Neumann algebra, a strictly outer action on A is called a free action [162, Definition XI.2.24].

We have $\varepsilon'_g = \varepsilon'_\iota$ if, and only if $f(\varepsilon_g(x)) = f(x)$ for each $f \in C(E)$ and each $x \in E$, that is, $\varepsilon_g(x) = \varepsilon_\iota(x)$ for every x. So the action ε is non-degenerate precisely when its dual ε' is non-degenerate.

Now suppose that A is monotone complete or, equivalently E is extremally disconnected. Let G and Γ be countably infinite groups with, respectively, non-degenerate actions ε and η on E. Let $\gamma \in \Gamma$. We say that η'_γ is G-decomposable over E with respect to the action ε' when there exists a sequence of orthogonal projections (p_n) and a sequence $(g(n))$ in G, such that $\bigvee_{n \geq 1} p_n = 1$ and $p_n \varepsilon'_{g(n)} = p_n \eta'_\gamma$ for each n. By Exercise 6.4.9 (d) this holds precisely when η_γ is G-decomposable over E.

Definition 6.9.1 Let $A, E, G, \Gamma, \varepsilon$ and η be as above. The systems (A, G, ε') and (A, Γ, η') are said to be *equivalent* if, for each $\gamma \in \Gamma$, η'_γ is G-decomposable and for each $g \in G$, ε'_g is Γ-decomposable. It follows from the remarks above that this holds precisely when the actions ε and η are equivalent over S.

Let X be a Polish space and $RegX$ the complete Boolean algebra of regular open subsets of X. Then, by Lemma 6.5.3, $RegX$ has a countable order dense subset and it can be identified with the Boolean algebra of projections in $B(X)/M(X)$. Let S_X be the Boolean structure space of $RegX$. By Lemma 6.5.3 S_X is hyperseparable and $B(X)/M(X)$ is isomorphic to $C(S_X)$. Also by Lemma 6.5.3, X has no isolated points if, and only if, S_X has no isolated points. By Corollary 6.5.4, whenever X is a perfect Polish spaces, S_X is homeomorphic to $S_\mathbb{R}$.

It follows that if $2^\mathbb{N}$ is the Cantor space and X is any perfect Polish space then $B(X)/M(X)$ is isomorphic to $B(2^\mathbb{N})/M(2^\mathbb{N})$. We recall that this is the *Dixmier algebra* (see Sect. 4.2). Clearly its spectrum is homeomorphic to $S_\mathbb{R}$. Let Γ be the Dyadic Group $\oplus \mathbb{Z}_2$. Let η be the natural action of Γ on $2^\mathbb{N}$, that is, $\eta_\gamma(x) = \gamma + x$. (Here we identify $2^\mathbb{N}$ with $\prod \mathbb{Z}_2$). Then, as in Sect. 6.4, there is a corresponding action $\hat{\eta}$ on $S_\mathbb{R}$. We shall call $\hat{\eta}$ the *canonical action* of $\oplus \mathbb{Z}_2$ on $S_\mathbb{R}$. Let $\rho : S_\mathbb{R} \to 2^\mathbb{N}$ be the continuous surjection such that $f \mapsto f \circ \rho$ is the natural embedding of $C(2^\mathbb{N})$ into $B(2^\mathbb{N})/M(2^\mathbb{N}) = C(S_\mathbb{R})$. Let s_0 be a point in $S_\mathbb{R}$ such that $\rho(s_0)$ is the zero of the group $\Gamma \subset 2^\mathbb{N}$. Then, by Propositions 6.4.4 and 6.4.5, the orbit $\{\hat{\eta}_\gamma(s_0) : \gamma \in \Gamma\}$ is free and dense in $S_\mathbb{R}$. So the action $\hat{\eta}$ is ergodic.

Let $(S_\mathbb{R}, G, \varepsilon)$ be a dynamical system where the (countably infinite) group G has a non-degenerate action as homeomorphism of $S_\mathbb{R}$. Let this action be ergodic. We claim that this action is generically equivalent to the action of Γ described above. We first show that the G-action is equivalent to some ergodic action of Γ, the Dyadic Group, with a free, dense orbit $\Gamma[s_0]$.

Lemma 6.9.2 *Let $(S_\mathbb{R}, G, \varepsilon)$ be a dynamical system where the (countably infinite) group G has a non-degenerate action as homeomorphisms of $S_\mathbb{R}$. Let this action be ergodic. Let Γ be the Dyadic Group. Then the G-action is equivalent to some ergodic action of Γ, with a free, dense orbit $\Gamma[s_0]$.*

Proof Since $S_\mathbb{R}$ is hyperseparable and the G-action is ergodic, it follows from Proposition 6.5.5 that there exists a G-invariant, dense G_δ set $Y_1 \subset S_\mathbb{R}$ such that for each $y \in Y_1$, the orbit $G[y]$ is dense. By Proposition 6.5.10, there exists a G-invariant, dense G_δ-set $Y_2 \subset S_\mathbb{R}$, such that the action of G on Y_2 is pseudo free. Since $S_\mathbb{R}$ is a

Baire space, $Y_1 \cap Y_2$ is a dense G_δ set. Let s_0 be any point in $Y_1 \cap Y_2$. Then $G[s_0]$ is an orbit which is dense and pseudo free. It now follows from Theorem 6.7.2 that the action of G on $S_{\mathbb{R}}$ is equivalent to an action of Γ on $S_{\mathbb{R}}$, where $\Gamma[s_0]$ is a free, dense orbit. $\quad\square$

It follows from Lemma 6.9.2 and Exercise 6.6.7 that it suffices to prove our claim when G is the Dyadic Group and the orbit $G[s_0]$ is free and dense. So from now on we make these assumptions. Incidentally, since $g = g^{-1}$ for each element of the Dyadic Group, it follows that

$$\varepsilon'_g(f) = f \circ \varepsilon_{g^{-1}} = f \circ \varepsilon_g = h_g(f).$$

Let us recapitulate (see Corollary 6.5.4 some observations about $ProjC(S_{\mathbb{R}})$, the complete Boolean algebra of projections in $C(S_{\mathbb{R}})$. It contains a countable subalgebra B_0 such that each element of $ProjC(S_{\mathbb{R}})$ is the supremum of the (countable) set of elements it dominates; that is B_0 is order dense in $ProjC(S_{\mathbb{R}})$. To see this, just identify $C(S_{\mathbb{R}})$ with $B(2^{\mathbb{N}})/M(2^{\mathbb{N}})$ and take B_0 to be the countable Boolean algebra whose elements are χ_K, where K is a clopen subset of $2^{\mathbb{N}}$. Now let Λ be a countable group of homomorphisms of $C(S_{\mathbb{R}})$. Then the saturation of B_0 by Λ is a countable set which generates a countable Boolean algebra $B_1 \subset ProjC(S_{\mathbb{R}})$ where B_1 is Λ-invariant. Clearly B_1 is order dense in $ProjC(S_{\mathbb{R}})$. If B_1 has an atom then this would also be an atom of $ProjC(S_{\mathbb{R}})$ so B_1 is non-atomic.

Since B_1 is countably infinite, it is a quotient of the free Boolean algebra on $\aleph_0$ generators. The structure space of this free algebra is $2^{\mathbb{N}}$. By the duality between the category of Boolean algebras and the category of compact Hausdorff totally disconnected spaces, it follows that the structure space of B_1 is homeomorphic to F, a closed subspace of $2^{\mathbb{N}}$. Then F is a separable, compact, totally disconnected metric space. Also, since B_1 has no atoms, F has no isolated points, that is, F is perfect. But the Cantor set is the unique totally disconnected, perfect compact metric space [167, Corollary 30.5]. So F is homeomorphic to $2^{\mathbb{N}}$.

Thus there is a Boolean isomorphism π from $ProjC(2^{\mathbb{N}})$ onto $B_1 \subset C(S_{\mathbb{R}})$. We may think of π as a finitely additive measure which we extend, first to simple functions then to a linear map Π from $C(2^{\mathbb{N}})$ into $C(S_{\mathbb{R}})$. It is straightforward to check that Π is a $*$-isomorphism whose range, A_1, is a closed $*$-subalgebra of $C(S_{\mathbb{R}})$. Since A_1 is the closed linear span of B_1, A_1 is invariant under the action of Λ.

Alternatively we may establish directly that A_1 is a regular subalgebra of $C(S_{\mathbb{R}})$ and so $C(S_{\mathbb{R}})$ is the regular completion of A_1. Since A_1 is generated by its projections and is a separable Banach space, its spectrum F is a zero-dimensional compact metric space with no isolated points and so F is homeomorphic to $2^{\mathbb{N}}$.

We shall make use of the results of Exercise 6.4.9.

Theorem 6.9.3 *Let $S_{\mathbb{R}}$ be the spectrum of the Dixmier algebra. Let $(S_{\mathbb{R}}, G, \varepsilon)$ be a dynamical system where the (countably infinite) group G has a non-degenerate action as homeomorphisms of $S_{\mathbb{R}}$. Let the action ε be ergodic. Then this action is generically equivalent to the canonical action of $\bigoplus \mathbb{Z}_2$ on $S_{\mathbb{R}}$.*

Proof By Lemma 6.9.2 we can suppose that G is the Dyadic Group $\oplus \mathbb{Z}_2$ and that its action on $S_\mathbb{R}$ has a free, dense orbit. Let ε' be the dual action of G as automorphisms of $C(S_\mathbb{R})$. Using the observations made above, these automorphisms act on a subalgebra A_0 which is isomorphic to $C(2^\mathbb{N})$ (and where $B(2^\mathbb{N})/M(2^\mathbb{N})$ can be identified with $C(S_\mathbb{R})$).

Let h_g^0 be the restriction of $h_g = \varepsilon_g'$ to $A_0 = C(2^\mathbb{N})$. Then, by duality, there is an action $\beta : G \mapsto Homeo(2^\mathbb{N})$ such that $h_g^0(a) = a \circ \beta_g$ for each a in $C(2^\mathbb{N})$. By Lemma 6.4.2, h_g^0 has a *unique* extension to a homomorphism H_g of $B(2^\mathbb{N})/M(2^\mathbb{N})$. It follows from this uniqueness that H_g can be identified with ε_g'.

Let π be the quotient homomorphism of $B(2^\mathbb{N})$ onto $B(2^\mathbb{N})/M(2^\mathbb{N})$. Then, by Lemma 6.4.2, $H_g(\pi(f)) = \pi(f \circ \beta_g)$. That is $\varepsilon_g'(\pi(f)) = \pi(f \circ \beta_g)$. In the notation of that Lemma, $\widehat{\beta_g} = \varepsilon_g$.

It now follows from Proposition 6.4.4 that the action β has a dense orbit in the perfect Polish space $2^\mathbb{N}$. We can now apply Theorem 6.8.1.

As above, we use η for the natural action of Γ on $2^\mathbb{N}$. Then there exist dense G_δ sets $X_1 \subset 2^\mathbb{N}$ and $X_2 \subset 2^\mathbb{N}$ with a homeomorphism $\theta : X_2 \mapsto X_1$ with the following properties. First, X_1 is invariant under the action of β and X_2 is invariant under the action of η. Secondly, the action β, restricted to X_1 is strongly equivalent to the action $\gamma \mapsto \theta \eta_\gamma \theta^{-1}$ on X_1.

Fix γ and put $\alpha_\gamma = \theta \eta_\gamma \theta^{-1}$. Then, strong equivalence tells us that there exists a sequence of disjoint Borel sets $(K(j))$, where each $K(j)$ is a clopen subset of X_1 in the relative topology and

$$\bigcup_{j \geq 1} K(j) = X_1.$$

Also there exists a sequence $(g(j))$ in G, such that, for each j, $\alpha_\gamma|_{K(j)} = \beta_{g(j)}|_{K(j)}$.

We saw in Exercise 6.4.9 that we may identify $B(2^\mathbb{N})/M(2^\mathbb{N})$ with $B(X_1)/M(X_1)$. When π_1 is the quotient homomorphism of $B(X_1)$ onto $B(X_1)/M(X_1)$ then we can identify $\pi(f)$ with $\pi_1(Rf)$ where Rf is the restriction of f to X_1. To avoid unnecessary complications of notation we shall usually blur the distinction between f and Rf; and write π instead of π_1. This is analogous to measure theory where we do not distinguish between two functions which only differ on a set of measure zero, indeed if the functions are only defined almost everywhere this does not bother us. In the following we shall make free use of such identifications.

For each j and any $f \in B(X_1)$,

$$\chi_{K(j)} Rf \circ \alpha_\gamma = \chi_{K(j)} f \circ \beta_{g(j)}.$$

So

$$\pi(\chi_{K(j)}) \pi(Rf \circ \alpha_\gamma) = \pi(\chi_{K(j)}) \pi(f \circ \beta_{g(j)})$$
$$= \pi(\chi_{K(j)}) H_{g(j)}(\pi(f))$$
$$= \pi(\chi_{K(j)}) \varepsilon_{g(j)}'(\pi(f)).$$

For each j, $\pi(\chi_{K(j)})$ can be identified with a projection in $C(S_{\mathbb{R}})$ and so with $\chi_{C(j)}$ where $C(j)$ is a clopen set. Since π is a homomorphism it maps orthogonal projections to orthogonal projections. So $(C(j))$ is a sequence of disjoint clopen sets.

Since π is a σ-homomorphism onto $C(S_{\mathbb{R}})$, and $\bigcup_{j\geq 1} K(j)$ is a dense open subset of X_1, it follows that $\bigcup_{j\geq 1} C(j)$ is a dense open subset of $S_{\mathbb{R}}$.

From the above, we have $\chi_{C(j)}\pi(f \circ \alpha_\gamma) = \chi_{C(j)}\varepsilon'_{g(j)}(\pi(f)) = \chi_{C(j)}\pi(f) \circ \varepsilon_{g(j)}$.

Using Lemma 6.4.2, the action $\lambda \mapsto \alpha_\lambda$ on X_1 induces an action $\lambda \mapsto \widehat{\alpha_\lambda}$ on $S_{\mathbb{R}}$ where $\pi(f \circ \alpha_\lambda) = \pi(f) \circ \widehat{\alpha_\lambda}$.

So $\chi_{C(j)}\pi(f) \circ \widehat{\alpha_\gamma} = \chi_{C(j)}\pi(f) \circ \varepsilon_{g(j)}$.

It now follows that $\widehat{\alpha_\gamma}$ is strongly decomposable over $\bigcup_{j\geq 1} C(j)$ with respect to the action $g \mapsto \beta_g$. Similar arguments show that each $\widehat{\beta_g}$ is decomposable over S with respect to the action $\lambda \mapsto \widehat{\alpha_\lambda}$. So the actions $\hat{\alpha}$ and $\hat{\beta}$ are equivalent over S.

If we can show that the action $\hat{\alpha}$ is generically equivalent to the action $\hat{\eta}$ then we are done.

By Exercise 6.4.9 (c) the map $f \mapsto f \circ \theta$ induces an isomorphism of $B(X_1)/M(X_1)$ onto $B(X_2)/M(X_2)$. On identifying these algebras with $C(S_{\mathbb{R}})$ we see that there is an automorphism of $C(S_{\mathbb{R}})$, V such that the $\hat{\alpha}'$ and $\hat{\eta}'$ actions on $C(S_{\mathbb{R}})$ are conjugate. Hence the actions $\hat{\alpha}$ and $\hat{\eta}$ are conjugate. So they are generically equivalent. $\quad\square$

6.10 Summary and Preview

We claimed in Sect. 1.2 that when a countable group G acts as homeomorphisms of a compact extremally disconnected space S then we can construct a corresponding monotone complete C^*-algebra M_E, where M_E is determined by the orbit equivalence relation E for the G-action. We have seen that actions by different groups can give rise (modulo meagre sets) to the same orbit equivalence relation; hence to the same monotone complete C^*-algebra. We shall show how to construct algebras from orbit equivalence relations in the next chapter.

We shall see that ergodicity implies that the corresponding algebra is a factor, that is, has one-dimensional centre. We know, from Lemma 6.3.1, that if G acts ergodically on S, then each orbit is either dense or nowhere dense. It is possible for every orbit to be nowhere dense. (In particular this must hold when E is not separable). But the existence of a single (pseudo) free dense G-orbit in S has major implications. In particular the corresponding algebra M_E will be shown to be generated by an increasing sequence of finite dimensional matrix algebras i.e. "hyperfinite".

Because of their importance for constructing monotone complete C^*-algebras we shall, later on, give many examples of G-actions on compact extremally disconnected spaces.

When the action of G has a pseudo free, dense orbit in S then we have shown that the orbit equivalence relation (and hence the associated algebra) can be obtained (modulo meagre sets) from an action of $\bigoplus \mathbb{Z}_2$ with a free, dense orbit. So, for the purpose of constructing monotone complete C^*-algebras, it suffices to find free, dense group actions when the group is $\bigoplus \mathbb{Z}_2$. In essence, we shall show that such an action exists on each of 2^c compact, separable spaces. This gives rise to 2^c small hyperfinite factors, where each such factor has a different weight in the classification semigroup.

Chapter 7
Constructing Monotone Complete C^*-Algebras

We build monotone complete C^*-algebras from equivalence relations on topological spaces. This is applied to orbit equivalence relations associated with the action of a countable group G. In general, these algebras may be identified with monotone cross-product algebras arising from actions of G on commutative monotone complete C^*-algebras. Since different groups can give rise to the same orbit equivalence relation, this can be used to show that, apparently different monotone cross-product algebras, are in fact, isomorphic.

7.1 Monotone Complete C^*-Algebra of an Equivalence Relation

The idea of constructing a C^*-algebra or a von Neumann algebra from a groupoid has a long history and a vast literature; there is an excellent exposition in [162]. Here, instead of general groupoids, we use an equivalence relation with countable equivalence classes. Our aim is to construct monotone complete (monotone σ-complete) algebras by a modification of the approach used in [157]. We try to balance conciseness with putting in enough detail to convince the reader that this is an easy and transparent way to construct huge numbers of examples of monotone complete C^*-algebras which are factors. In other words these algebras have one-dimensional centres.

We shall also give a sketch of the close connections with monotone cross-product algebras.

In this section, X is a topological space which is homeomorphic to a G_δ-subset of a compact Hausdorff space. Every locally compact space and every Polish space (see Lemma 6.5.1) satisfies this condition. Clearly X is a completely regular, Baire space. We could work with more general topological spaces but this is not necessary for our applications.

© Springer-Verlag London 2015

181

K. Saitô, J.D.M. Wright, *Monotone Complete C*-algebras and Generic Dynamics*,
Springer Monographs in Mathematics, DOI 10.1007/978-1-4471-6775-4_7

As before, let $B(X)$ be the set of all bounded complex valued Borel functions on X. When equipped with the obvious algebraic operations and the supremum norm, it becomes a commutative C^*-algebra. Clearly $B(X)$ is monotone σ-complete.

In the following it would be easy to use a more general setting, where we do not assume a topology for X, replace the field of Borel sets with a σ-field $\mathcal{T}$, and use $\mathcal{T}$-measurable bijections instead of homeomorphisms. But we stick to a topological setting which is what we need later.

Let G be a countable group of homeomorphisms of X and let

$$E = \{(x, y) \in X \times X : \exists g \in G \text{ such that } y = g(x)\}.$$

Then E is the graph of the orbit equivalence relation on X arising from the action of G. We shall identify this equivalence relation with its graph. We know, from the work of Chap. 6, that the same orbit equivalence relation can arise from actions by different groups.

We see below that E is the union of countably many closed sets. Thus E is an F_σ-subset of $X \times X$. Each equivalence class for E is an orbit $G[x]$ and so countable.

Let us recall that for $A \subset X$, the *saturation* of A (by E) is

$$E[A] = \{y \in X : \exists \, x \in A \text{ such that } xEy\}$$
$$= \bigcup\{g[A] : g \in G\}.$$

Since each g is a homeomorphism, it follows that the saturation of a Borel set is also a Borel set.

A slightly different approach, which has many merits, is to "axiomatise" equivalence relations. More precisely, take as fundamental object, an equivalence relation E, whose graph is a Borel set and whose equivalence classes are countable. (Countable Borel equivalence relations and their relationship with von Neumann algebras were penetratingly analysed in [41] and [42].) See also the important contributions of Hjorth and Kechris [80] and [79]. When X is a Polish space then, by a theorem of Feldman and Moore [41] and [42], E is the orbit equivalence relation arising from the action of a countable group of Borel bijections of X. How far the Feldman-Moore theorem can be generalised when X is not a standard Borel space is unknown. Instead of pursuing this issue, we shall only consider orbit equivalence relations which arise from actions by countable groups of homeomorphisms. For all our applications this will suffice.

Exercise 7.1.1 Show that the separable space $\{0, 1\}^{\mathbb{R}}$ is not a standard Borel space.
Hint: When X is a standard Borel space, either it is countable or $\#X = c = \#\mathbb{R}$.

Definition 7.1.2 Let $\mathcal{I}$ be a σ-ideal of the Boolean algebra of Borel subsets of X with $X \notin \mathcal{I}$.

Definition 7.1.3 Let $B_{\mathcal{I}}$ be the set of all f in $B(X)$ such that $\{x \in X : f(x) \neq 0\}$ is in $\mathcal{I}$.

Then $B_{\mathcal{I}}$ is a σ-ideal of $B(X)$. (See Sect. 4.2.) Let q be the quotient homomorphism from $B(X)$ onto $B(X)/B_{\mathcal{I}}$.

Lemma 7.1.4 *Let $A \in \mathcal{I}$. Then $E[A] \in \mathcal{I}$ if, and only if, $g[A] \in \mathcal{I}$ for every $g \in G$.*

Proof For each $g \in G$, $g[A] \subset E[A]$. Since $\mathcal{I}$ is an ideal, if $E[A] \in \mathcal{I}$ then $g[A] \in \mathcal{I}$.

Conversely, if $g[A] \in \mathcal{I}$ for each g, then $E[A]$ is the union of countably many elements of the σ-ideal and hence in the ideal. $\qquad\qquad\qquad\qquad\qquad$ $\square$

In the following we require that the action of G maps the ideal $\mathcal{I}$ into itself. Equivalently, for any $A \in \mathcal{I}$, its saturation by E is again in $\mathcal{I}$. This is automatically satisfied if $\mathcal{I}$ is the ideal of meagre Borel sets but we do not wish to confine ourselves to this situation.

We show how orbit equivalence relations on X give rise to monotone complete C^*-algebras. A key point (see remarks above) is that these algebras are constructed from the equivalence relation without explicit mention of G. But in establishing the properties of these algebras, the existence of an underlying group is assumed. This construction (similar to a groupoid C^*-algebra) seems particularly natural and transparent. When applied to examples of countable groups acting ergodically on compact extremally disconnected spaces, this machinery produces huge numbers of small factors; factors which take 2^c different values in the classification semigroup.

We shall consider monotone cross-products in Sect. 7.2. We will indicate why, for countable groups acting on commutative algebras, the corresponding monotone cross-products are essentially the same as the algebras constructed from an equivalence relation.

An account of monotone cross-products for a countable group Γ acting as homomorphisms of a commutative monotone complete C^*-algebra A, was given in [132]. This was generalised to the situation where A is not commutative in [64].

We could work in greater generality (for example we could weaken the condition that the elements of G be homeomorphisms or consider more general groupoid constructions) but for ease and simplicity we avoid this. We are mainly interested in two situations. First, where X is an "exotic" space but $\mathcal{I}$ is only the ideal of meagre Borel subsets of X. Secondly, where X is just the Cantor space but $\mathcal{I}$ is an "exotic" ideal of the Borel sets.

For each $x \in X$ let $[x]$ be the equivalence class generated by x. Let $[X]$ be the set of all equivalence classes. Let $\ell^2([x])$ be the Hilbert space of all square summable, complex valued functions from $[x]$ to $\mathbb{C}$. For each $y \in [x]$ let $\delta_y \in \ell^2([x])$ be defined by

$$\delta_y(z) = 0 \text{ for } z \neq y; \; \delta_y(y) = 1.$$

Then $\{\delta_y : y \in [x]\}$ is an orthonormal basis for $\ell^2([x])$ which we shall call the *canonical basis* for $\ell^2([x])$. For each $x \in X$, $\mathcal{L}(\ell^2([x]))$ is the von Neumann algebra of all bounded operators on $\ell^2([x])$. (In the rest of the book we would denote the bounded operators on $\ell^2([x])$ by $L(\ell^2([x]))$ but because, in this section, we shall use "L" for something else, we use "$\mathcal{L}$" to avoid confusion.) We now form a direct product of these algebras by:

$$S = \prod_{[x] \in [X]} \mathcal{L}(\ell^2([x])).$$

Remark 7.1.5 One possible source of confusion is a clash of terminology. The "direct product" of von Neumann algebras is more usually called a "direct sum of von Neumann algebras" [157, 161]. But, in order to be consistent with our Chap. 3, we shall use "direct product".

The direct product of a family of C^*-algebras, as defined in [15] is

$$\prod_{\omega \in \Omega} A_\omega = \{(a_\omega) : \sup||a_\omega|| < \infty\}.$$

We observed in Chap. 3, that when each A_ω is monotone complete then so, also, is $\prod_{\omega \in \Omega} A_\omega$. When the A_ω are all von Neumann algebras then so is $\prod_{\omega \in \Omega} A_\omega$.

We see that S is a Type I von Neumann algebra, being a direct product of such algebras. It acts on the Hilbert space which is the direct sum of the spaces $\ell^2([x])$. It is of no independent interest but is a framework in which we embed an algebra of "Borel matrices" and then take a quotient, obtaining monotone complete C^*-algebras. To each operator F in S we can associate, uniquely, a function $f : E \mapsto \mathbb{C}$ as follows. First we decompose F as:

$$F = \prod_{[x] \in [X]} F_{[x]}.$$

Here each $F_{[x]}$ is a bounded operator on $\ell^2([x])$. Then F acts on the direct sum of these Hilbert spaces. Now recall that $(x, y) \in E$ precisely when $y \in [x]$. We now define $f : E \mapsto \mathbb{C}$ by

$$f(x, y) = < F_{[x]}\delta_y, \delta_x >.$$

When f is restricted to $[x] \times [x]$ then it becomes the matrix representation of $F_{[x]}$ with respect to the canonical orthonormal basis of $\ell^2([x])$. It follows that there is a bijection between operators in S and those functions $f : E \mapsto \mathbb{C}$ for which there is a constant k such that, for each $[x] \in [X]$, the restriction of f to $[x] \times [x]$ is the matrix of a bounded operator on $\ell^2([x])$ whose norm is bounded by k. Call such an f *matrix bounded*. For each matrix bounded f let $L(f)$ be the corresponding element of S. Then $f(x, y) = < L(f)\delta_y, \delta_x >$ for $y \in [x]$.

For any $\xi \in \ell^2([x])$ we have

$$||\xi||^2 = \sum_{y \in [x]} | < \xi, \delta_y > |^2.$$

So

$$\|L(f)^*\delta_x\|^2 = \sum_{y\in[x]} |< L(f)^*\delta_x, \delta_y >|^2 = \sum_{y\in[x]} |f(x,y)|^2.$$

On applying the Cauchy-Schwarz inequality, it follows that when f and h are matrix bounded functions from E to $\mathbb{C}$, then $\sum_{y\in[x]} f(x,y)h(y,z)$ is absolutely convergent.

Straightforward matrix manipulations give

$$L(f)L(h) = L(f \circ h)$$

where

$$f \circ h(x,z) = \sum_{y\in[x]} f(x,y)h(y,z).$$

Also $L(f)^* = L(f^*)$, where $f^*(x,y) = \overline{f(y,x)}$ for all $(x,y) \in E$.
Furthermore $f + h$ is matrix bounded and $L(f + h) = L(f) + L(h)$.
Let $\|f\| = \|L(f)\|$.

Exercise 7.1.6 Give a detailed proof that the matrix bounded functions on E, when equipped with the norm and algebraic operations described above, form a C^*-algebra and $f \mapsto L(f)$ is an (isometric) $*$-isomorphism of this algebra onto S.

Let Δ be the diagonal set $\{(x,x) : x \in X\}$. It is closed, because the topology of X is Hausdorff. It is an easy calculation to show that $L(\chi_\Delta)$ is the unit element of S. We note that χ_Δ is matrix bounded because it comes from the identity operator in S.

For each $g \in G$, the map $(x,y) \mapsto (x,g(y))$ is a homeomorphism of $X \times X$ onto itself. So this homeomorphism maps Δ onto a closed set $\{(x,g(x)) : x \in X\}$.

Let us recall that

$$E = \bigcup_{g\in G} \{(x,g(x)) : x \in X\}.$$

So E is the union of countably many closed sets; it is an F_σ- set. In particular, E is a Borel subset of $X \times X$.

Definition 7.1.7 Let $\mathcal{M}(E)$ be the set of all Borel measurable functions $f : E \mapsto \mathbb{C}$ which are matrix bounded.

Exercise 7.1.8 Let Y and Z be Hausdorff spaces. Let $f : Y \mapsto Z$ be a Borel map. Prove that $f^{-1}[T]$ is a Borel subset of Y whenever T is a Borel subset of Z.
Hint: Let $\mathcal{B} = \{T \in BorZ : f^{-1}[T] \in BorY\}$. Show that $\mathcal{B}$ is a σ-field containing the open sets of Z.

Exercise 7.1.9 Let Y, Z, W be Hausdorff spaces. Let $f : Y \mapsto Z$ and $g : Z \mapsto W$ be Borel maps. Show that $gf : Y \mapsto W$ is a Borel map.

Lemma 7.1.10 *The set* $M = \{L(f) : f \in \mathcal{M}(E)\}$ *is a* C^*-*subalgebra of* S *which is sequentially closed with respect to the weak operator topology of* S. *When equipped with the appropriate algebraic operations and norm,* $\mathcal{M}(E)$ *is a* C^*-*algebra isomorphic to* M.

Proof Let f and h be in $\mathcal{M}(E)$. Clearly $f + h$ is Borel measurable, and, as remarked above, matrix bounded. So $f + h$ is in $\mathcal{M}(E)$. Similarly, scalar multiples of f are in $\mathcal{M}(E)$.

Let θ be the homeomorphism of $X \times X$ onto itself defined by $\theta(x, y) = (y, x)$. Then its restriction to E is a homeomorphism of E onto itself. So

$$(x, y) \mapsto f(\theta(x, y))$$

is a Borel function. But f^* is the complex conjugate of this Borel function. So $f^* \in \mathcal{M}(E)$.

We now wish to show that $f \circ h$ is Borel measurable.

First let $g_1, g_2, \ldots$ be an enumeration of G, without repetitions. (When G is a finite group the same arguments work, but with finite sums instead of infinite series.) Let $\Delta_r = \{(x, g_r x) : x \in X\}$. Since, see above, Δ_r is a closed subset of $X \times X$ it is a Borel subset of E. Because G is not required to act freely, the sets (Δ_r) $(r = 1, 2 \ldots)$ cannot be assumed to be pairwise disjoint. So we define $D_1 = \Delta_1$ and for $r \geq 1$,

$$D_{r+1} = \Delta_{r+1} \setminus \bigcup_{j=1}^{r} \Delta_j.$$

Then each D_r is a Borel subset of E; these Borel sets are pairwise disjoint and their union is E. Let χ_{D_n} be the characteristic function of D_r.

Let $f_r = \chi_{D_r} f$. Then f_r is a Borel function on E.

Fix x. Let $y \in [x]$, then $\chi_{D_n}(x, y) = 0$ unless $y = g_n x$ and $g_n x \neq g_r x$ for any $r < n$.

Absolute convergence of a series of complex numbers implies the order of summation does not matter. So

$$f \circ h(x, z) = \sum_{y \in [x]} f(x, y) h(y, z)$$

$$= \sum_{n=1}^{\infty} \chi_{D_n}(x, g_n x) f(x, g_n x) h(g_n x, z)$$

$$= \sum_{n=1}^{\infty} f_n(x, g_n x) h(g_n x, z).$$

The pointwise limit of a sequence of Borel functions is a Borel function. So, to show that $f \circ h$ is Borel, it suffices to show that $(x, z) \mapsto f_n(x, g_n x) h(g_n x, z)$ is a Borel function on E for each fixed n. But $x \mapsto (x, g_n x)$ is a homeomorphism from X onto Δ_n. So $x \mapsto f(x, g_n x)$ is a Borel function on X. Thus $(x, z) \mapsto f_n(x, g_n x)$ is a Borel function on E. Also the map $(x, z) \mapsto (g_n x, z)$ is a homeomorphism from E into E. So $(x, z) \mapsto h(g_n x, z)$ is a Borel function on E. The pointwise product of two scalar valued Borel functions on E is a Borel function. It now follows that $f \circ h$ is Borel. So M is a $*$-subalgebra of S.

Let $(L(f_n))$ be a sequence in M which converges in the weak operator topology of S to $L(f)$. So, for any $x \in X$ and any $y \in [x]$,

$$< L(f_n)\delta_y, \delta_x > \to < L(f)\delta_y, \delta_x > .$$

Thus $f_n(x, y) \to f(x, y)$. So f is a Borel function. Since it is also matrix bounded (by the norm of $L(f)$), $f \in \mathcal{M}(E)$ and so $L(f) \in M$.

Since S is a von Neumann algebra, it is a Banach dual space. When the weak operator topology of S, and the weak*-topology of S are restricted to the (norm closed) unit ball of S, they coincide.   □

Lemma 7.1.11 *Let (f_n) be a sequence in the unit ball of $\mathcal{M}(E)$ which converges pointwise to f. Then $f \in \mathcal{M}(E)$ and $L(f_n)$ converges to $L(f)$ in the weak operator topology of S. Also f is in the unit ball of $\mathcal{M}(E)$.*

Proof The weak operator topology gives a compact Hausdorff topology on the norm closed unit ball of S.

Let $K_n = \{L(f_j) : j \geq n\}$ and let clK_n be its closure in the weak operator topology of S. Then, by the finite intersection property, there exists $T \in \bigcap_{n \in \mathbb{N}} clK_n$. If U is any open neighbourhood of T, then $U \cap K_n$ is non-empty for all n.

Fix (x, y) in E. Fix $\varepsilon > 0$. Let $U = \{S \in S : | < (S - T)\delta_y, \delta_x > | < \varepsilon\}$. Then we can find a subsequence $(f_{n(r)})(r = 1, 2 \ldots)$ for which $L(f_{n(r)}) \in U$ for each r. Thus $|f_{n(r)}(x, y) - < T\delta_y, \delta_x > | < \varepsilon$. So $|f(x, y) - < T\delta_y, \delta_x > | \leq \varepsilon$. Since this holds for all positive ε, we have $f(x, y) = < T\delta_y, \delta_x >$. So $T = L(f)$.

Let $\mathcal{T}$ be the locally convex topology of S generated by all seminorms of the form $V \mapsto | < V\delta_x, \delta_y > |$ as (x, y) ranges over E. This is a Hausdorff topology which is weaker than the weak operator topology. Hence it coincides with the weak operator topology on the unit ball, because the latter topology is compact. But $< L(f_n)\delta_y, \delta_x > \to f(x, y) = < L(f)\delta_y, \delta_x >$ for all $(x, y) \in E$.

It now follows that $L(f_n) \to L(f)$ in the weak operator topology of S. Since f is the pointwise limit of a sequence of Borel measurable functions, it too, is Borel measurable. So $f \in \mathcal{M}(E)$ and, since T is in the unit ball of S, f is in the unit ball of $\mathcal{M}(E)$.   □

Let p be the homeomorphism of Δ onto X, given by $p(x, x) = x$. So $B(X)$, the algebra of bounded Borel measurable functions on X, is isometrically $*$-isomorphic to $B(\Delta)$ under the map $h \mapsto h \circ p$.

For each $f \in \mathcal{M}(E)$ let Df be the function on E which vanishes off the diagonal, Δ, and is such that, for each $x \in X$,

$$Df(x, x) = f(x, x).$$

Then D is a linear idempotent map from $\mathcal{M}(E)$ onto an abelian subalgebra which we can identify with $B(\Delta)$, which can, in turn, be identified with $B(X)$. Let $\tilde{D}f$ be the function on X such that $\tilde{D}f(x) = Df(x, x)$ for all $x \in X$. We shall sometimes abuse our notation by using Df instead of $\tilde{D}f$.

Let $\pi : B(X) \mapsto \mathcal{M}(E)$ be defined by $\pi(h)(x, x) = h(x)$ for $x \in X$ and $\pi(h)(x, y) = 0$ for $x \neq y$. Then π is a $*$-isomorphism of $B(X)$ onto an abelian $*$-subalgebra of $\mathcal{M}(E)$, which can be identified with the range of D.

We have $\pi\tilde{D}f = Df$ for $f \in \mathcal{M}(E)$. Also, for $g \in B(X)$, $\tilde{D}\pi(g) = g$.

Lemma 7.1.12 *The map D is positive.*

Proof Each positive element of $\mathcal{M}(E)$ is of the form $f \circ f^*$. But

$$D(f \circ f^*)(x, x) = \sum_{y \in [x]} f(x, y) f^*(y, x) \qquad (\#)$$

$$= \sum_{y \in [x]} f(x, y)\overline{f(x, y)}$$

$$= \sum_{y \in [x]} |f(x, y)|^2 \geq 0. \qquad \square$$

Definition 7.1.13

Let $I_\mathcal{I} = \{f \in \mathcal{M}(E) : q\tilde{D}(f \circ f^*) = 0\}$

$$= \{f \in \mathcal{M}(E) : \exists A \in \mathcal{I} \text{ such that } \tilde{D}(f \circ f^*)(x) = 0 \text{ for } x \notin A\}.$$

Lemma 7.1.14 *The set $I_\mathcal{I}$ is a two-sided ideal of $\mathcal{M}(E)$.*

Proof In any C^*-algebra,

$$(a + b)(a + b)^* \leq 2(aa^* + bb^*).$$

So

$$D((f + g) \circ (f + g)^*) \leq 2D(f \circ f^*) + 2D(g \circ g^*).$$

From this it follows that if f and g are both in $I_\mathcal{I}$ then so also is $f + g$.

In any C^*-algebra,

$$fz(fz)^* = fzz^*f^* \le ||z||^2 ff^*.$$

From this it follows that $f \in I_{\mathcal{I}}$ and $z \in \mathcal{M}(E)$ implies that $f \circ z \in I_{\mathcal{I}}$.

Now suppose that $f \in I_{\mathcal{I}}$. Then for some $A \in \mathcal{I}$, $D(f \circ f^*)(x) = 0$ for $x \notin A$.

Since $E[A] \in \mathcal{I}$ we can suppose that $A = E[A]$. Hence if $x \notin E[A]$ then $[x] \cap E[A] = \varnothing$. For $x \notin E[A]$, we have

$$0 = D(f \circ f^*)(x) = \sum_{y \in [x]} |f(x, y)|^2.$$

Thus $f(x, y) = 0$ for xEy and $x \notin A$. Then, for $z \notin A$, we have $D(f^* \circ f)(z) = \sum_{y \in [z]} |f^*(z, y)|^2 = \sum_{y \in [z]} |f(y, z)|^2 = 0$. So $f^* \in I_{\mathcal{I}}$. So $I_{\mathcal{I}}$ is a two-sided ideal of $\mathcal{M}(E)$. □

Lemma 7.1.15 *If $y \in I_{\mathcal{I}}$ then $q\tilde{D}(y) = 0$. Furthermore $y \in I_{\mathcal{I}}$ if, and only if, $q\tilde{D}(y \circ a) = 0$ for all $a \in \mathcal{M}(E)$.*

Proof Let T be the (compact Hausdorff) structure space of the algebra $B(X)/B_{\mathcal{I}}$. By applying the Cauchy-Schwarz inequality we see that for x, y in $\mathcal{M}(E)$ and $t \in T$,

$$|q\tilde{D}(x^* \circ y)(t)| \le q\tilde{D}(x^* \circ x)(t)^{1/2} q\tilde{D}(y^* \circ y)(t)^{1/2}.$$

Let $x = 1$ and let $y \in I_{\mathcal{I}}$. Then $y^* \in I_{\mathcal{I}}$. So $q\tilde{D}(y^* \circ y) = 0$. From the above inequality it follows that $q\tilde{D}(y) = 0$.

Because $I_{\mathcal{I}}$ is an ideal, if $y \in I_{\mathcal{I}}$ then $y \circ a$ is in the ideal for each $a \in \mathcal{M}(E)$. It now follows from the above that $q\tilde{D}(y \circ a) = 0$. Conversely, if $q\tilde{D}(y \circ a) = 0$ for all $a \in \mathcal{M}(E)$ then, on putting $a = y^*$ we see that $y \in I_{\mathcal{I}}$. □

Lemma 7.1.16 *$L[I_{\mathcal{I}}]$ is a (two-sided) ideal of $L[\mathcal{M}(E)]$ which is sequentially closed in the weak operator topology of S.*

Proof Let (f_r) be a sequence in $I_{\mathcal{I}}$ such that $(L(f_r))$ is a sequence which converges in the weak operator topology to an element T of S. Then it follows from the Uniform Boundedness Theorem that the sequence is bounded in norm. By Lemma 7.1.10, there exists $f \in \mathcal{M}(E)$ such that $L(f) = T$ where $L(f_r) \to L(f)$ in the weak operator topology. So $f_r \to f$ pointwise. Hence $\tilde{D}(f_r) \to \tilde{D}(f)$ pointwise. For each r there exists $A_r \in \mathcal{I}$ such that $x \notin A_r$ implies $\tilde{D}(f_r)(x) = 0$. Since $\mathcal{I}$ is a Boolean σ-ideal of the Boolean algebra of Borel subsets of X, $\cup\{A_r : r = 1, 2 \ldots\}$ is in $\mathcal{I}$. Hence $q\tilde{D}(f) = 0$.

For any $a \in \mathcal{M}(E)$, $(f_r \circ a)$ is a sequence in $I_{\mathcal{I}}$ such that $(L(f_r \circ a))$ converges in the weak operator topology to $L(f)L(a)$. So, as in the preceding paragraph, $q\tilde{D}(f \circ a) = 0$. By appealing to Lemma 7.1.15 we see that $f \in I_{\mathcal{I}}$. Hence $L[I_{\mathcal{I}}]$ is sequentially closed in the weak operator topology of S. □

Corollary 7.1.17 *The algebra $\mathcal{M}(E)$ is monotone σ-complete and $I_{\mathcal{I}}$ is a σ-ideal.*

Proof Each norm bounded monotone increasing sequence in $L(\mathcal{M}(E))$ converges in the strong operator topology to an element T of $\mathcal{S}$. By Lemma 7.1.10, $T \in L(\mathcal{M}(E))$. Then $T = L(f)$ for some $f \in \mathcal{M}(E)$. Hence $L(\mathcal{M}(E))$ (and its isomorphic image, $\mathcal{M}(E)$) are monotone σ-complete. It now follows from Lemma 7.1.16, that $I_{\mathcal{I}}$ is a σ-ideal.   □

Definition 7.1.18 Let Q be the quotient map from $\mathcal{M}(E)$ onto $\mathcal{M}(E)/I_{\mathcal{I}}$.

Proposition 7.1.19 *The algebra $\mathcal{M}(E)/I_{\mathcal{I}}$ is monotone σ-complete. There exists a positive, faithful, σ-normal, conditional expectation $\hat{D}$ from $\mathcal{M}(E)/I_{\mathcal{I}}$ onto a commutative σ-subalgebra, which is isomorphic to $B(X)/B_{\mathcal{I}}$. Furthermore, if there exists a strictly positive linear functional on $B(X)/B_{\mathcal{I}}$, then $\mathcal{M}(E)/I_{\mathcal{I}}$ is monotone complete and $\hat{D}$ is normal.*

Proof By Proposition 2.2.21 and Corollary 7.1.17, the quotient algebra $\mathcal{M}(E)/I_{\mathcal{I}}$ is monotone σ-complete.

Let $g \in B(X)$. Then, as remarked before Lemma 7.1.12, $\tilde{D}\pi(g) = g$.

Now $\pi(g) \in I_{\mathcal{I}}$ if, and only if, $q\tilde{D}(\pi(g) \circ \pi(g)^*) = 0$.

But $q\tilde{D}(\pi(g) \circ \pi(g)^*) = q\tilde{D}(\pi(|g|^2)) = q(|g|^2)$.

So $\pi(g) \in I_{\mathcal{I}}$ if and only if $|g|^2 \in B_{\mathcal{I}}$ i.e. if and only if g vanishes off some set $A \in \mathcal{I}$ i.e. if and only if $g \in B_{\mathcal{I}}$.

So π induces an isomorphism from $B(X)/B_{\mathcal{I}}$ onto $D[\mathcal{M}(E)]/I_{\mathcal{I}}$.

Let $h \in I_{\mathcal{I}}$. Then, by Lemma 7.1.15, $q\tilde{D}(h) = 0$. That is, $\tilde{D}(h) \in B_{\mathcal{I}}$. So $\pi\tilde{D}(h) \in I_{\mathcal{I}}$. But $\pi\tilde{D}(h) = Dh$.

So $h \in I_{\mathcal{I}}$ implies $QDh = 0$. It now follows that we can define $\hat{D}$ on $\mathcal{M}(E)/I_{\mathcal{I}}$ by $\hat{D}(f + I_{\mathcal{I}}) = QDf$.

It is clear that $\hat{D}$ is a positive linear map which is faithful. Its range is an abelian subalgebra of $\mathcal{M}(E)/I_{\mathcal{I}}$. This subalgebra is $D[\mathcal{M}(E)]/I_{\mathcal{I}}$ which, as we have seen above, is isomorphic to $B(X)/B_{\mathcal{I}}$. We shall denote $D[\mathcal{M}(E)]/I_{\mathcal{I}}$ by A and call it the diagonal algebra. Furthermore, $\hat{D}$ is idempotent, so it is a conditional expectation.

Let (f_n) be a sequence in $\mathcal{M}(E)$ such that (Qf_n) is an upper bounded monotone increasing sequence in $\mathcal{M}(E)/I_{\mathcal{I}}$. Then, by Corollary 2.2.18, we may assume that (f_n) is an upper bounded, monotone increasing sequence in $\mathcal{M}(E)$. Let Lf be the limit of (Lf_n) in the weak operator topology. (Since the sequence is monotone, Lf is also its limit in the strong operator topology.) By Lemma 7.1.10, $f \in \mathcal{M}(E)$, and $f_n(x, x) \to f(x, x)$ for all $x \in X$. Thus $Df_n \to Df$ pointwise on X. Also, since D is positive, (Df_n) is monotone increasing. Since Q is a σ-homomorphism, QDf is the least upper bound of (QDf_n). Since $\hat{D}(f + I_{\mathcal{I}}) = QDf$ it now follows that $\hat{D}$ is σ-normal.

If μ is a strictly positive functional on $B(X)/B_{\mathcal{I}}$ then $\mu\hat{D}$ is a strictly positive linear functional on $\mathcal{M}(E)/I_{\mathcal{I}}$. It then follows from Theorem 2.1.14 that $\mathcal{M}(E)/I_{\mathcal{I}}$ is monotone complete. Furthermore, if Λ is a downward directed subset of the self-adjoint part of $\mathcal{M}(E)/I_{\mathcal{I}}$, with 0 as its greatest lower bound, then there exists a monotone decreasing sequence (x_n), with each x_n in Λ, and $\bigwedge x_n = 0$. It now follows from the σ-normality of $\hat{D}$ that 0 is the infimum of $\{\hat{D}(x) : x \in \Lambda\}$. Hence $\hat{D}$ is normal. $\qquad\qquad$ □

We now make additional assumptions about the action of G and use this to construct a natural unitary representation of G. We give some technical results which give an analogue of Mercer-Bures convergence, see [101] and [23]. This will be useful later when we wish to approximate elements of $\mathcal{M}(E)/I_{\mathcal{I}}$ by finite dimensional subalgebras.

For the rest of this section we suppose that the action of G on X is free on each orbit i.e. for each $x \in X$, x is not a fixed point of g, where $g \in G$, unless g is the identity element of G.

For each $g \in G$, let $\Delta_g = \{(x, gx) : x \in X\}$. Then the Δ_g are pairwise disjoint and $E = \cup_{g \in G} \Delta_g$.

For each $g \in G$, let $u_g : E \mapsto \{0, 1\}$ be the characteristic function of $\Delta_{g^{-1}}$. As we pointed out earlier, χ_Δ is the unit element of $\mathcal{M}(E)$, so, in this notation, u_ι is the unit element of $\mathcal{M}(E)$.

The function u_g on E is matrix bounded. To show this, we shall define an operator $T \in S$ such that $< T\delta_y, \delta_x > = u_g(x, y)$ for $(x, y) \in E$.

For each $x \in X$, let $\lambda_g([x]) \in \mathcal{L}(\ell^2([x]))$ be defined by

$$\lambda_g([x])\underline{\xi}(y) = \underline{\xi}(g^{-1}y) \text{ for each } y \in [x] \text{ and } \underline{\xi} \in \ell^2([x]).$$

Then $\lambda_g([x])$ is a unitary operator on $\ell^2([x])$ for each $g \in G$. Let $V_g \in S$ be $V_g = \oplus_{[x]} \lambda_g([x])$. Clearly $V_g \in S$. For each $(x, y) \in E$, we have

$$< V_g \delta_y, \delta_x > = < \lambda_g([x])\delta_y, \delta_x > = \chi_{\Delta_{g^{-1}}}(x, y).$$

So, u_g is matrix bounded.

For each $(x, y) \in E$ we have:

$$u_g \circ u_h(x, y) = \sum_{k \in G} u_g(x, kx)u_h(kx, y).$$

But $u_g(x, kx) \neq 0$, only if $k = g^{-1}$ and $u_h(g^{-1}x, y) \neq 0$ only if $y = h^{-1}g^{-1}x = (gh)^{-1}x$. So $u_g \circ u_h = u_{gh}$.

Also $u_g^*(x, y) = \overline{u_g(y, x)} = u_g(y, x)$. But $u_g(y, x) \neq 0$ only if $x = g^{-1}y$, that is, only if $y = gx$. So $u_g^*(x, y) = u_{g^{-1}}(x, y)$. It follows that $g \mapsto u_g$ is a unitary representation of G in $\mathcal{M}(E)$.

Let f be any element of $\mathcal{M}(E)$. Then

$$f \circ u_g(x, y) = \sum_{z \in [x]} f(x, z)u_g(z, y) = f(x, gy). \qquad\qquad \text{(i)}$$

So, for each $x \in X$,

$$D(f \circ u_g)(x, x) = f(x, gx).$$

Then

$$D(f \circ u_g) \circ u_{g^{-1}}(x, y) = \sum_{z \in [x]} D(f \circ u_g)(x, z) u_{g^{-1}}(z, y)$$

$$= D(f \circ u_g)(x, x) u_{g^{-1}}(x, y)$$

$$= f(x, gx) \chi_{\Delta_g}(x, y).$$

So

$$D(f \circ u_g) \circ u_{g^{-1}}(x, y) = \begin{cases} f(x, y) & \text{if } (x, y) \in \Delta_g \\ 0 & \text{if } (x, y) \notin \Delta_g. \end{cases} \qquad \text{(ii)}$$

The identity (#), used in Lemma 7.1.12, can be re-written as

$$D(f \circ f^*)(x, x) = \sum_{g \in G} |f(x, gx)|^2 \qquad \text{(iii)}$$

$$= \sum_{g \in G} |D(f \circ u_g)(x, x)|^2$$

$$= \sum_{g \in G} |\tilde{D}(f \circ u_g)(x)|^2.$$

Let F be any finite subset of G. Let $f_F = \sum_{g \in F} D(f \circ u_g) \circ u_{g^{-1}}$. Then, using (ii),

$$(f - f_F)(x, y) = \begin{cases} 0 & \text{if } (x, y) \in \Delta_g \text{ and } g \in F \\ f(x, y) & \text{if } (x, y) \in \Delta_g \text{ and } g \notin F. \end{cases}$$

We now replace f by $f - f_F$ in (iii) and get:

$$D((f - f_F) \circ (f - f_F)^*)(x, x) = \sum_{g \in G \backslash F} |D(f \circ u_g)(x, x)|^2 = \sum_{g \in G \backslash F} |\tilde{D}(f \circ u_g)(x)|^2. \qquad \text{(iv)}$$

Now let $(F_n)(n = 1, 2 \ldots)$ be any strictly increasing sequence of finite subsets of G whose union is G. Write f_n for f_{F_n}. Then
$D((f - f_n) \circ (f - f_n)^*)(x, x)$ decreases monotonically to 0 as $n \to \infty$.
Since Q is a σ-homomorphism,

$$\bigwedge Q D((f - f_n) \circ (f - f_n)^*) = 0. \qquad \text{(v)}$$

For each $g \in G$, let $U_g = Qu_g$. Since Q is a $*$-homomorphism onto $\mathcal{M}(E)/I_{\mathcal{I}}$, U_g is a unitary and $g \mapsto U_g$ is a unitary representation of G in $\mathcal{M}(E)/I_{\mathcal{I}}$. On applying the preceding paragraph we get:

Proposition 7.1.20 *Let* $z \in \mathcal{M}(E)/I_{\mathcal{I}}$. *Let* $(F(n))(n = 1, 2 \ldots)$ *be a strictly increasing sequence of finite subsets of* G *whose union is* G. *Let* $z_n = \sum_{g \in F(n)} \hat{D}(zU_g)U_{g^{-1}}$. *Then*

$$\bigwedge \hat{D}((z - z_n)(z - z_n)^*) = 0.$$

Corollary 7.1.21 *Let* $z \in \mathcal{M}(E)/I_{\mathcal{I}}$ *such that* $\hat{D}(zU_g) = 0$ *for each* g. *Then* $z = 0$.

Proof This follows from Proposition 7.1.20 because $z_n = 0$ for every n.   □

Lemma 7.1.22 *For each* $f \in \mathcal{M}(E)$ *and each* $g \in G$, $u_g^* \circ f \circ u_g(x, y) = f(gx, gy)$. *In particular,* f *vanishes off* Δ *if, and only if,* $u_g^* \circ f \circ u_g$ *vanishes off* Δ.

Proof Let $h \in \mathcal{M}(E)$. Then, applying identity (i) we get $(u_g^* \circ h)(a, b) = (h^* \circ u_g)^*(a, b) = \overline{(h^* \circ u_g)(b, a)} = \overline{h^*(b, ga)} = h(ga, b)$.

Let $h = f \circ u_g$. Then $u_g^* \circ f \circ u_g(x, y) = f \circ u_g(gx, y) = f(gx, gy)$.   □

Corollary 7.1.23 *For each* $a \in A$, *the diagonal algebra, and for each* $g \in G$, $U_g a U_g^*$ *is in* A.

7.1.1 Induced Actions

Let K be a compact Hausdorff space. Let ϕ be a homeomorphism of K onto itself. Then, see the discussion following Lemma 6.4.1 in Sect. 6.4, the induced automorphism on $C(K)$ is defined by

$$h_\phi(f) = f \circ \phi.$$

Every $*$-automorphism of $C(K)$ arises in this way from a homeomorphism of K.

We recall that $Homeo(K)$ is the group of homeomorphisms of K onto itself. Then the map $\theta \mapsto h_\theta$ is a bijective map from $HomeoK$ onto $Aut(C(K))$, the group of $*$-automorphisms. In general this is not a group isomorphism but an anti-isomorphism.

Let us recall that for any group Γ, the opposite group, Γ^{op}, is the same underlying set as Γ but with a new group operation defined by $x \times y = yx$. Also Γ^{op} and Γ are isomorphic groups, the map $g \mapsto g^{-1}$ gives an isomorphism. So, in the preceding paragraph, $g \mapsto h_g$ is a group isomorphism of $Homeo(K)^{op}$ into $Aut(C(K))$. Since $Homeo(K)^{op}$ is isomorphic to $Homeo(K)$, this is not of major significance. But we define $h^g = h_{g^{-1}}$ for each $g \in Homeo(K)$. Then $g \mapsto h^g$ is a group isomorphism of $Homeo(K)$ onto the automorphism group of $C(K)$.

Let $\gamma \mapsto \alpha_\gamma$ be an action of a group Γ on K. Then the induced action of Γ as automorphisms of $C(K)$ is $\gamma \mapsto \alpha^\gamma$ where

$$\alpha^\gamma(f) = f \circ \alpha_{\gamma^{-1}}.$$

Conversely, given an action $\gamma \mapsto \alpha^\gamma$ of Γ as automorphisms of $C(K)$ this can be identified with the dual of an action $\gamma \mapsto \alpha_\gamma$, of the group Γ as homeomorphisms of K.

There is a potential clash of terminology which arises as follows. Consider the category whose objects are the compact Hausdorff spaces and whose arrows are the continuous maps between them. By Gelfand-Naimark theory, this is dual to the category whose objects are the commutative, unital C^*-algebras and whose arrows are the $*$-homomorphisms. Take a compact Hausdorff X and a homeomorphism θ of X. It might seem natural to describe the automorphism of $C(X)$ induced by θ as "dual" to θ. But this could cause confusion. Because, when we consider actions of abelian groups, we also refer to actions of the dual group. So we prefer to use "induced".

Lemma 7.1.24 *Let T be a compact, totally disconnected space. Let θ be a homeomorphism of T onto T. Let h_θ be the automorphism of $C(T)$ induced by θ. Let K be a non-empty clopen set such that, for each clopen $Q \subset T$, $(h_\theta(\chi_Q) - \chi_Q)\chi_K = 0$. Then $\theta(t) = t$ for each $t \in K$. In other words, $h_\theta(f) = f$ for each $f \in \chi_K C(T)$.*

Proof Let us assume that $t_0 \in K$ such that $\theta(t_0) \neq t_0$. By total disconnectedness, there exists a clopen set Q with $\theta(t_0) \in Q$ and $t_0 \notin Q$. We have $h_\theta(\chi_Q) = \chi_{\theta^{-1}[Q]}$. So

$$\theta^{-1}[Q] \cap K = Q \cap K.$$

But t_0 is an element of $\theta^{-1}[Q] \cap K$ and $t_0 \notin Q$. This is a contradiction. $\square$

Let C be a monotone (σ-)complete C^*-algebra. We call an automorphism α (of C) *properly outer* if there does not exist a non-zero α-invariant projection $e \in C$ such that α restricts to an inner automorphism of eCe. When C is commutative and α is properly outer, then for an α-invariant projection e, if α restricts to the identity on eCe, then $e = 0$. Let H be a group and let $g \mapsto \alpha_g$ be an injective group homomorphism $\alpha : H \mapsto \mathrm{Aut}C$. That is, α is an action of H on C.

When, for each $h \in H \setminus \{\iota\}$, α_h is properly outer, we call α a *free action*.

We recall that, on putting $A = \hat{D}[\mathcal{M}(E)/I_{\mathcal{I}}]$, we have $U_g A U_g^* = A$ for each $g \in G$. Then we can define the $*$-automorphism λ_g of A by $\lambda_g(a) = U_g a U_g^*$ for $a \in A$ and $g \in G$. Then $g \mapsto \lambda_g$ is a group homomorphism of G into $\mathrm{Aut}A$.

Proposition 7.1.25 *When the action λ of G is free on A then $\hat{D}[\mathcal{M}(E)/I_{\mathcal{I}}]$ is a maximal abelian $*$-subalgebra of $\mathcal{M}(E)/I_{\mathcal{I}}$.*

Proof Let z commute with each element of $\hat{D}[\mathcal{M}(E)/I_{\mathcal{I}}]$. Let $a = \hat{D}(z)$. We shall show that $z = a$. By Corollary 7.1.21 it will suffice to show that $\hat{D}((z - a)U_g) = 0$ for each $g \in G$. We remark that $\hat{D}(aU_g) = a\hat{D}(U_g) = 0$ for each $g \neq \iota$.

So it is enough to show that if $g \in G$ and g is not the identity element of G then $\hat{D}(zU_g) = 0$.

We have, for each $b \in \hat{D}[\mathcal{M}(E)/I_I]$, $bz = zb$. So

$$b\hat{D}(zU_g) = \hat{D}(bzU_g) = \hat{D}(zbU_g)$$

$$= \hat{D}(zU_g U_g^* bU_g)$$

$$= \hat{D}(zU_g)U_g^* bU_g$$

$$= U_g^* bU_g\hat{D}(zU_g).$$

This implies that $(\lambda_{g^{-1}}(b) - b)\hat{D}(zU_g) = 0$. For shortness put $c = \hat{D}(zU_g)$.

Assume that $c \neq 0$. Then, by spectral theory, there exists a non-zero projection e and a strictly positive real number δ such that $\delta e \leq cc^*$. Then

$$0 \leq \delta(\lambda_{g^{-1}}(b) - b)e(\lambda_{g^{-1}}(b) - b)^*$$

$$\leq (\lambda_{g^{-1}}(b) - b)cc^*(\lambda_{g^{-1}}(b) - b)^* = 0.$$

So, $(\lambda_{g^{-1}}(b) - b)e = 0$ for each b in the range of $\hat{D}$. It now follows from Lemma 7.1.24 that $\lambda_{g^{-1}}(ea) = ea$ for each a in the range of $\hat{D}$. But this contradicts the freeness of the action of G. So $\hat{D}(zU_g) = 0$. It now follows that z is in $\hat{D}[\mathcal{M}(E)/I_I]$. Hence $\hat{D}[\mathcal{M}(E)/I_I]$ is a maximal abelian $*$-subalgebra of $\mathcal{M}(E)/I_I$. □

Exercise 7.1.26 Let G act as a group of homeomorphisms of a compact zero-dimensional space Z. Then this induces a dual action $g \mapsto h^g$ as automorphisms of $C(Z)$. Suppose that $g \mapsto h^g$ is a free action, that is, h^g is a properly outer automorphism whenever $g \neq \iota$. Show that there exists a dense, G_δ-subset X, where X is G-invariant and $\{g(x) : g \in G\}$ is a free orbit for every $x \in X$. (Hint see [157].) Could we strengthen this by showing that $G[z]$ is a free orbit for *every* point of Z? To see that this is not always possible, put $Z = 2^{\mathbb{Z}}$ and let s be the shift operation on $2^{\mathbb{Z}}$. Now let $\mathbb{Z}$ act on $2^{\mathbb{Z}}$ by $n \mapsto s^n$. This action has a fixed point.

Remark 7.1.27 We have said nothing about the centre of $\mathcal{M}(E)/I_I$. Clearly it is contained in each maximal abelian $*$-subalgebra. Let c be a central projection. Then

$$U_g c U_g^* = c$$

for each g. From this it can be deduced that c corresponds to a Borel subset of X, C, where C is G-invariant. That is $C = E[C]$. It now follows that $\mathcal{M}(E)/I_I$ has one-dimensional centre precisely when each G-invariant Borel subset of X is in I or is the compliment of a set in I. Now specialise to the situation where I is the Boolean ideal of meagre Borel sets. Each Borel subset of X has the Baire property i.e. differs from an open set only by a meagre set. It follows that when the action of G on X is generically ergodic then $\mathcal{M}(E)/I_I$ has one-dimensional centre.

A unitary w in a C^*-algebra M is said to normalise a $*$-subalgebra A if $wAw^* = A$. For future reference we define the *normaliser subalgebra* of $\mathcal{M}(E)/I_{\mathcal{I}}$ to be the smallest monotone complete $*$-subalgebra of $\mathcal{M}(E)/I_{\mathcal{I}}$ which contains every unitary which normalises the diagonal subalgebra. It follows from Corollary 7.1.23 that each U_g is a normalising unitary for the diagonal algebra, $A = \hat{D}[\mathcal{M}(E)/I_{\mathcal{I}}]$. Since each element of A is a finite linear combination of unitaries in A, it follows immediately that A is contained in the normaliser subalgebra.

7.2   Introduction to Cross-Product Algebras

Cross-product constructions are of importance in C^*-algebra theory and in the theory of von Neumann algebras.

A detailed account of monotone cross-products for a countable group Γ acting as homomorphisms of a commutative monotone complete C^*-algebra A is given in Saitô [132]; they were considered earlier by Takenouchi [159]. Eventually this was generalised to the situation where A is not commutative by Hamana [64], who gives a meticulous exposition.

Here we shall give a brief outline and sketch connections with the algebras constructed in Sect. 7.1.

First let us recall some familiar facts. Let A be a unital C^*-algebra. Let α be an automorphism of A. If there exists a unitary $u \in A$ such that, for each $z \in A$, $\alpha(z) = uzu^*$ then α is said to be an *inner automorphism*. When no such unitary exists in A then α is an *outer* automorphism.

Let G be a countable group and let $g \mapsto \beta_g$ be an homomorphism of G into the group of all automorphisms of A. Intuitively, a cross-product algebra, for this action of G, is a larger C^*-algebra, B, in which A is embedded as a subalgebra and where each β_g is induced by a unitary in B. More precisely, there is an injective $*$-homomorphism $j : A \mapsto B$, and a group homomorphism $g \mapsto U_g$, (from G into the group of unitaries in B), such that, for each $z \in A$, $j\beta_g(z) = U_g j(z) U_g^*$. So when we identify A with its image $j[A]$ in B, although β_g need not be an inner automorphism of A it can be extended to an inner automorphism of the larger algebra B. We also require that B is "in some sense" generated by $j[A]$ and the collection of unitaries U_g. When A is a monotone complete C^*-algebra, we can always construct a B which is monotone complete.

For the purposes of this chapter we shall only consider the situation where A is commutative. So for the rest of this section, A shall be a monotone complete commutative C^*-algebra. Hence $A \simeq C(S)$ where S is a compact, Hausdorff, extremally disconnected space. We shall outline below properties of the monotone cross-product of A by the action of countable groups. It turns out that they can be identified with algebras already constructed in Sect. 7.1. *In fact the constructions in Sect. 7.1 can be used to give an alternative definition of these cross-products.* Historically, cross-products came first. One advantage of the Sect. 7.1 construction is that it makes clear why some, apparently different, cross-product algebras are, in fact, isomorphic.

We shall suppose for the rest of this section that S has no isolated points. Then, as remarked in Sect. 6.2, any dense subset Y has no isolated points. We shall also require that $C(S)$ possesses a faithful state. We have seen that when S has a countable dense set then $C(S)$ has a faithful state. The converse is false (consider $L^\infty[0, 1]$.)

We now use Sect. 7.1 to relate monotone cross-products to the monotone complete C^*-algebra of an orbit equivalence relation.

Suppose that G is a countably infinite group of homeomorphisms of S, where the action of G has a free, dense orbit. We shall show that the corresponding monotone cross-product is isomorphic to one obtained by an action of $\bigoplus \mathbb{Z}_2$.

For the convenience of the reader we recall some points from earlier chapters.

Let X be any dense subset of S. Then we recall that S can be identified with the Stone-Czech compactification of X. So if $f : X \mapsto \mathbb{C}$ is a bounded continuous function then it has a unique extension to a continuous function $\hat{f} : S \mapsto \mathbb{C}$. It follows that $f \mapsto \hat{f}$ is an isometric $*$-isomorphism of $C_b(X)$, the algebra of bounded continuous functions on X, onto $C(S)$. Similarly, as remarked earlier, any homeomorphism θ from X onto X has a unique extension to a homeomorphism $\hat{\theta}$ from S onto S. We may abuse our notation by using θ instead of $\hat{\theta}$ i.e. using the same symbol for a homeomorphism of X and for its unique extension to a homeomorphism of S. Slightly more generally, when X_1 and X_2 are dense subsets of S, if there exists an homeomorphism of X_1 onto X_2 then it has a unique extension to a homeomorphism of S onto itself.

Following Corollary 7.1.23 we discussed induced actions. Let G be a countable group and $g \mapsto \alpha_g$ an action of G as homeomorphisms of S then $g \mapsto \alpha^g$ is the dual action of G as homomorphisms of $C(S)$. Here $\alpha^g(f) = f \circ \alpha_{g^{-1}}$.

WARNING We also use $g \mapsto \lambda_g$ to indicate an action of G as automorphisms of an algebra. In other words, we prefer to indicate group actions by using subscripts unless we wish to emphasise duality between a G-action on a space and the induced action as automorphisms.

Let h be a $*$-automorphism of A. Recall that h is said to be *properly outer* if, for each non-zero h-invariant projection p, the restriction of h to pA is not the identity.

Exercise 7.2.1 Let θ be a homeomorphism of S onto itself. Let h_θ be the corresponding automorphism of $C(S)$. Let F be the fixed point set $\{s \in S : \theta(s) = s\}$. Show that h_θ is properly outer if, and only if, the closed set F has empty interior.

Let Γ be a subgroup of $Aut(A)$ such that every element, except the identity, is properly outer. Then recall that the action of Γ on A is said to be *free*.

Let G be a countable group of homeomorphisms of S. Then, by applying Exercise 7.2.1, if $g \mapsto \alpha^g$ is a free action of G on A then there exists a dense G_δ-set $Y \subset S$, where Y is G-invariant, such that, whenever $g \in G$ is not the identity, then α_g has no fixed points in Y. In other words, for each $y \in Y$, $G[y]$ is a free orbit.

Conversely, we have:

Lemma 7.2.2 *Let X be a dense subset of S, where X is G-invariant. Let $g \mapsto \alpha_g$ be an action of G as homeomorphisms of S. Suppose $G[x]$ is a free orbit for each $x \in X$. Then $g \mapsto \alpha^g$ is a free action of G on $C(S)$.*

Proof Let $g \in G$, such that, α^g is not properly outer. So, for some non-empty clopen set K, and each $f \in C(S)$,

$$\chi_K f = \alpha^g(\chi_K f) = (\chi_K \circ g^{-1})(f \circ g^{-1}) = \chi_{g[K]}(f \circ g^{-1}).$$

In particular, $K = g[K]$. Suppose that $x_1 \in K$ with $g(x_1) \neq x_1$. Then we can find a continuous function f which takes the value 1 at $g(x_1)$ and 0 at x_1. But, from the equation above, this implies $0 = 1$. So $g(x) = x$ for each $x \in K$. Because X is dense in S and K is a non-empty open set, there exists $y \in X \cap K$. So y is a fixed point of g and $G[y]$ is a free orbit. This is only possible if g is the identity element of G. Hence the action $g \mapsto \alpha^g$ is free. $\square$

In the following, G is a countably infinite group of homeomorphisms of S and Y is a dense G_δ-subset of S, where Y is G-invariant. Let $g \mapsto \beta^g$ be the corresponding action of G as automorphisms of A. We use $M(C(S), G)$ to denote the associated (Takenouchi) monotone cross-product. (A formal definition of the Takenouchi cross-product can be found in Saitô [132] or [159].) We hope the discussion below will make it clear why these algebras can be obtained in an alternative way by using Sect. 7.1. We shall sketch a description of the monotone cross-product below. We hope this "bird's eye view" will be helpful.

The key fact is that, provided the G-action is free, the monotone cross-product algebra can be identified with the monotone complete C^*-algebra arising from the G-orbit equivalence relation. The end part of Sect. 7.1 already makes this plausible.

Before saying more about the monotone cross-product, we outline some properties of monotone complete tensor products.

(Comment: An alternative, equivalent, approach avoiding the tensor product, is to use the theory of Kaplansky-Hilbert modules [90–92, 166, 169].)

For the rest of this section, H is a separable Hilbert space and H_1 is an arbitrary Hilbert space. Let us fix an orthonormal basis for H. Then, with respect to this basis, each $V \in L(H_1)\overline{\otimes}L(H)$ has a unique representation as a matrix (V_{ij}), where each V_{ij} is in $L(H_1)$. Let M be a von Neumann subalgebra of $L(H_1)$. Then the elements of $M\overline{\otimes}L(H)$ are those V for which each V_{ij} is in M. Let T be any set and $Bnd(T)$ the commutative von Neumann algebra of all bounded functions on T. (It may be thought of as acting on the Hilbert space $\ell^2(T)$.)

Then, it can be shown that $Bnd(T)\overline{\otimes}L(H)$ is a von Neumann subalgebra of $L(\ell^2(T))\overline{\otimes}L(H)$. It can be identified with the algebra of all matrices $[m_{ij}]$ over $Bnd(T)$ for which $t \mapsto [m_{ij}(t)]$ is a norm bounded function over T.

We denote the commutative C^*-algebra of bounded, complex valued Borel measurable functions on Y by $B(Y)$. Then the product $B(Y)\tilde{\otimes}L(H)$ may be defined as the monotone σ-closure of $B(Y) \otimes_{\min} L(H)$ inside $Bnd(Y)\overline{\otimes}L(H)$. The elements of $B(Y)\tilde{\otimes}L(H)$ correspond to the matrices $[b_{ij}]$ where each $b_{ij} \in B(Y)$ and $y \mapsto [b_{ij}(y)]$ is a norm bounded function from Y into $L(H)$. See also [120].

Let H be the Hilbert space $\ell^2(G)$ with the canonical orthonormal basis $\{\delta_g : g \in G\}$, where $\delta_g(h) = \delta_{g,h}$ for $g, h \in G$. Let $M^\sigma(B(Y), G)$ be the subalgebra of $B(Y)\tilde{\otimes}L(H)$, consisting of those elements of the tensor product which have a matrix representation over $B(Y)$ of the form $[a_{\gamma,\sigma}]$ $(\gamma \in G, \sigma \in G)$ where $a_{\gamma\tau,\sigma\tau}(y) = a_{\gamma,\sigma}(\tau y)$ for all $y \in Y$ and all γ, σ, τ in G. Let E be the orbit equivalence relation on Y arising from G, that is

$$E = \{(y, gy) : y \in Y, g \in G\}.$$

Lemma 7.2.3 *Assume that each $g \in G$ has no fixed points in Y unless g is the identity element. Then $M^\sigma(B(Y), G)$ is naturally isomorphic to $\mathcal{M}(E)$.*

Proof (Sketch) The correspondence between these two algebras is given as follows. Let $f \in \mathcal{M}(E)$. For each σ, γ in G and, for all $y \in Y$, let $a_{\gamma,\sigma}(y) = f(\gamma y, \sigma y)$.

Then $a_{\gamma,\sigma}$ is in $B(Y)$. Also the norm of $[a_{\gamma,\sigma}(y)]$ is uniformly bounded for $y \in Y$. So $[a_{\gamma,\sigma}]$ is in $B(Y)\tilde{\otimes}L(H)$. Also

$$a_{\gamma\tau,\sigma\tau}(y) = f(\gamma\tau y, \sigma\tau y) = a_{\gamma,\sigma}(\tau y).$$

It now follows that $[a_{\gamma,\sigma}]$ is in $M^\sigma(B(Y), G)$.

Conversely, let $[a_{\gamma,\sigma}]$ be in $M^\sigma(B(Y), G)$. We now use the hypothesis of freeness for the action of G on Y, to deduce that $(\{(y, \tau y) : y \in Y\})(\tau \in G)$ is a countable family of pairwise disjoint closed subsets of E. So we can now define, unambiguously, a function $f : E \mapsto \mathbb{C}$ by $f(y, \tau y) = a_{\iota,\tau}(y)$. This is a bounded Borel function on E. It follows from the definition of $M^\sigma(B(Y), G)$ that $a_{\gamma,\sigma}(y) = a_{\iota,\sigma\gamma^{-1}}(\gamma y) = f(\gamma y, \sigma y)$ for all σ, γ in G and all y in Y. From this it follows that f is in $\mathcal{M}(E)$.

Since Y is a dense G_δ-subset of S, $S\backslash Y$ is a meagre Borel subset of S. Let π_1 be the canonical σ-homomorphism of $B(S)$ onto $C(S)$ such that $\pi_1(f) = f$ for each $f \in C(S)$; the kernel of π_1 is the σ-ideal of Borel functions which vanish off a meagre set. Then, when $h \in B(S)$ and h vanishes on Y then $\pi_1(h) = 0$. Now, given g in $B(Y)$, it can be extended to a function $g_1 \in B(S)$. For example let g_1 take the value 0 on $S\backslash Y$. Let g_2 be any other extension. Then $\pi_1(g_1) = \pi_1(g_2)$. So we can define $\pi(g) = \pi_1(g_1)$.

Then π is the canonical quotient homomorphism from $B(Y)$ onto $C(S)$, whose kernel is the ideal of Borel functions which vanish off a meagre set. Also π is a σ-homomorphism satisfying $\pi(f|_Y) = f$ for each $f \in C(S)$. In fact, Exercise 6.4.9 (a) and (b) together with Corollary 4.2.8 tell us that there exists a surjective σ-normal $*$-homomorphism π from $B(Y)$ onto $C(S)$ such that $\pi(f \circ \gamma) = \pi(f) \circ \gamma$ for all $f \in B(Y)$ and $\gamma \in G$. $\square$

We consider the monotone tensor product $C(S)\overline{\otimes}L(H)$. For our purposes here, it is enough to observe that each element of the monotone tensor product $C(S)\overline{\otimes}L(H)$ has a representation as a matrix over $C(S)$. (The product is not straightforward. But it turns out that there exists a σ-homomorphism Π from $B(Y)\tilde{\otimes}L(H)$, where the product *is* straightforward, onto $C(S)\overline{\otimes}L(H)$; with $\Pi([b_{\gamma,\sigma}]) = [\pi(b_{\gamma,\sigma})]$.)

The monotone cross-product can be defined to be the subalgebra of $C(S) \overline{\otimes} L(H)$ corresponding to matrices $[a_{\gamma,\sigma}]$ over $C(S)$ for which $\beta^{\tau^{-1}}(a_{\gamma,\sigma}) = a_{\gamma,\sigma} \circ \beta_\tau = a_{\gamma\tau,\sigma\tau}$ for all γ, σ, τ in G. Equivalently, $a_{\gamma\tau,\sigma\tau}(s) = a_{\gamma,\sigma}(\tau s)$ for all γ, σ, τ in G and $s \in S$. From this it can be shown that the homomorphism Π maps $M^\sigma(B(Y), G)$ onto $M(C(S), G)$.

The diagonal subalgebra of $M(C(S), G)$ consists of those matrices $[a_{\gamma,\sigma}]$ which vanish off the diagonal i.e. $a_{\gamma,\sigma} = 0$ for $\gamma \neq \sigma$. Also, $\beta^{\tau^{-1}}(a_{\iota,\iota}) = a_{\iota,\iota} \circ \beta_\tau = a_{\tau,\tau}$ for each $\tau \in G$. It follows that we can define an isomorphism from A onto the diagonal of $M(C(S), G)$ by $j(a) = Diag(\ldots, \beta^{\tau^{-1}}(a), \ldots)$. We find, see Proposition 7.1.25, the freeness of the action of G implies that the diagonal algebra of $M(C(S), G)$ is a maximal abelian $*$-subalgebra of $M(C(S), G)$.

Then, by Lemma 3.3 [157], we have:

Lemma 7.2.4 *Let E be the graph of the relation of orbit equivalence given by G acting on Y. Then there exists a σ-normal homomorphism δ from $\mathcal{M}(E)$ onto $M(C(S), \beta, G)$. The kernel of δ is*

$$J = \{z \in \mathcal{M}(E) : D(zz^*) \text{ vanishes off a meagre subset of } Y\}.$$

Furthermore, δ maps the diagonal subalgebra of $\mathcal{M}(E)$ onto the diagonal subalgebra of $M(C(S), \beta, G)$. In particular, δ induces an isomorphism of $\mathcal{M}(E)/J$ onto $M(C(S), G)$.

(*Comment*: Instead of using tensor products to construct cross-products, as above, we may use Kaplansky-Hilbert modules instead.)

Let $C(S) \times_\beta G$ be the smallest monotone closed $*$-subalgebra of $M(C(S), G)$ which contains the diagonal and each unitary which implements the β-action of G. It will sometimes be convenient to call $C(S) \times_\beta G$ the "small" monotone cross-product. It turns out that $C(S) \times_\beta G$ does not depend on G, only on the orbit equivalence relation. This is not at all obvious but this will be established in Sect. 7.3. We will show that when w is a unitary in $M(C(S), G)$ such that w normalises the diagonal then w is in $C(S) \times_\beta G$. So, the isomorphism of $M(C(S), G)$ onto $\mathcal{M}(E)/J$ maps $C(S) \times_\beta G$ onto the normaliser subalgebra of $\mathcal{M}(E)/J$.

Does the small monotone cross product equal the "big" monotone cross-product? Equivalently, is $\mathcal{M}(E)/J$ equal to its normaliser subalgebra? This is unknown, but we can approximate each element of $M(C(S), G)$ by a sequence in $C(S) \times_\beta G$, in the following precise sense:

Lemma 7.2.5 *Let $z \in M(C(S), G)$. Then there exists a sequence (z_n) in $C(S) \times_\beta G$ such that the sequence $(D(z - z_n)^*(z - z_n))$ is monotone decreasing and*

$$\bigwedge_{n=1}^{\infty} D((z - z_n)^*(z - z_n)) = 0.$$

Proof This follows from Proposition 7.1.20.   □

Theorem 7.2.6 *Let $G_j (j = 1, 2)$ be countable, infinite groups of homeomorphisms of S. Let $g \mapsto \beta_j^g$ be the corresponding action of G_j as automorphisms of $C(S)$. Let Y be a dense G_δ-subset of S such that $G_j[Y] = Y$ and G_j acts freely on Y. Let E_j be the orbit equivalence relation on Y arising from the action of G_j. Suppose that $E_1 = E_2$. Then there exists an isomorphism of $M(C(S), G_1)$ onto $M(C(S), G_2)$ which maps the diagonal algebra of $M(C(S), G_1)$ onto the diagonal algebra of $M(C(S), G_2)$.*

Proof The first part is a straight forward application of Lemma 7.2.4. Let $E = E_1 = E_2$. Then both algebras are isomorphic to $\mathcal{M}(E)/J$.   □

As a consequence of a theorem which will be proved in Sect. 7.3, the isomorphism in Theorem 7.2.6 maps $C(S) \times_{\beta_1} G_1$ onto $C(S) \times_{\beta_2} G_2$.

In the next result, we require S to be separable.

Corollary 7.2.7 *Let G be a countable, infinite group of homeomorphisms of S. Suppose, for some $s_0 \in S$, $G[s_0]$ is a free, dense orbit. Then there exists an isomorphism ϕ of $\bigoplus \mathbb{Z}_2$ into $Homeo(S)$, such that there exists an isomorphism of $M(C(S), G)$ onto $M(C(S), \bigoplus \mathbb{Z}_2)$ which maps the diagonal algebra of $M(C(S), G)$ onto the diagonal algebra of $M(C(S), \bigoplus \mathbb{Z}_2)$.*

Proof By Lemma 6.4.7 G acts freely on a dense G_δ set containing $G[s_0]$. By Theorem 6.7.2, there exists an isomorphism ϕ from $\bigoplus \mathbb{Z}_2$ into $Homeo(S)$ such that there exists a dense G_δ set Y in S with the following properties. First, Y is invariant under the action of both G and $\bigoplus \mathbb{Z}_2$. Secondly the induced orbit equivalence relations coincide on Y. Theorem 7.2.6 then gives the result.   □

Remark 7.2.8 By Lemma 6.3.4, when G has a dense orbit in S then the action on S is such that the only invariant clopen set is empty or the whole space. Then by Remark 7.1.27, this implies that the algebra $M(C(S), G)$ is a monotone complete factor.

Lemma 7.2.9 *Let $\mathcal{B}$ be a Boolean σ-algebra. Let (p_n) be a sequence in $\mathcal{B}$ which σ-generates $\mathcal{B}$, that is, $\mathcal{B}$ is the smallest σ-subalgebra of $\mathcal{B}$ which contains each p_n. Let Γ be a group of automorphisms of $\mathcal{B}$. Let Γ be the union of an increasing sequence of finite subgroups (Γ_n). Then we can find an increasing sequence of finite Boolean algebras $(\mathcal{B}_n)$ where each $\mathcal{B}_n$ is invariant under the action of Γ_n and $\cup \mathcal{B}_n$ is a Boolean algebra which σ-generates $\mathcal{B}$.*

Proof For any natural number k the free Boolean algebra on k generators has 2^k elements. So a Boolean algebra with k generators, being a quotient of the corresponding free algebra, has a finite number of elements.

We proceed inductively. Let $\mathcal{B}_1$ be the subalgebra generated by $\{g(p_1) : g \in \Gamma_1\}$. Then $\mathcal{B}_1$ is finite and Γ_1-invariant. Suppose we have constructed $\mathcal{B}_1, \mathcal{B}_2, \ldots, \mathcal{B}_n$. Then $\mathcal{B}_n \cup \{p_{n+1}\}$ is a finite set. So its saturation by the finite group Γ_{n+1} is again a finite set. So the Boolean algebra this generates, call it $\mathcal{B}_{n+1}$, is finite. Clearly $\mathcal{B}_n \subset \mathcal{B}_{n+1}$ and $\mathcal{B}_{n+1}$ is invariant under the action of Γ_{n+1}.   □

A commutative monotone complete C^*-algebra is *countably* σ-*generated* if its Boolean algebra of projections is σ-generated by a countable subset.

Proposition 7.2.10 *Let the Boolean algebra of projections in $C(S)$ be countably σ-generated by (p_n). Let Γ be a group of automorphisms of $C(S)$. Let Γ be the union of an increasing sequence of finite subgroups (Γ_n). Let $g \mapsto u_g$ be the unitary representation of Γ in $M(C(S), \Gamma)$ which implements the action of Γ on the diagonal algebra A. Let π be the canonical isomorphism of $C(S)$ onto A. Then the C^*-algebra generated by $\{u_g : g \in \Gamma\} \cup \{\pi(p_n) : n = 1, 2, \ldots\}$ is the closure of an increasing sequence of finite dimensional subalgebras.*

Proof Let $\mathcal{B}$ be the complete Boolean algebra of all projections in A. By Lemma 7.2.9, we can find an increasing sequence of finite Boolean algebras of projections $(\mathcal{B}_n)$ where $\mathcal{B}$ is σ-generated by $\cup \mathcal{B}_n$ and each $\mathcal{B}_n$ is invariant under the action of Γ_n.

Let A_n be the (complex) linear span of $\mathcal{B}_n$. Then A_n is a finite dimensional $*$-subalgebra of A. Also, for $g \in \Gamma_n$, $u_g A_n u_g^* = A_n$.

Now let B_n be the linear span of $\{bu_g : g \in \Gamma_n \text{ and } b \in A_n\}$. Then B_n is a finite dimensional $*$-subalgebra. Clearly (B_n) is an increasing sequence and $\cup_{n=1}^{\infty} B_n$ is a $*$-subalgebra generated by $\{u_g : g \in \Gamma\} \cup \{\pi(p_n) : n = 1, 2, \ldots\}$. $\qquad\square$

7.3 The Normaliser Algebra

In this section M is a monotone complete C^*-algebra with a maximal abelian $*$-subalgebra A and $D : M \mapsto A$ a positive, linear, idempotent map of M onto A. It follows from a theorem of Tomiyama [163], see Definition III.3.3 and Theorem III.3.4 of [161] that D is a conditional expectation. That is,

$$D(azb) = a(Dz)b$$

for each $z \in M$ and every a, b in A.

The monotone complete algebras considered in Sects. 7.1 and 7.2 satisfy these conditions when the diagonal algebra, A, is maximal abelian; and A is maximal abelian when the action $g \mapsto \beta^g$ is free.

We recall that a unitary w in M is a *normaliser* of A if $wAw^* = A$. It is clear that the normalisers of A form a subgroup of the unitaries in M. We use $\mathcal{N}(A, M)$ to denote this normaliser subgroup. Let $M_{\mathcal{N}}$ be the smallest monotone closed $*$-subalgebra of M which contains $\mathcal{N}(A, M)$. Then $M_{\mathcal{N}}$ is said to be the *normaliser subalgebra* of M.

Let G be a countable group. Let $g \mapsto u_g$ be a unitary representation of G in $\mathcal{N}(A, M)$. Let $\lambda_g(a) = u_g a u_g^*$.

Let α be an automorphism of A. We recall that α is *properly outer* if, for each non-zero α-invariant projection $e \in A$, the restriction of α to eA is not the identity map. We further recall that the action $g \mapsto \lambda_g$ is *free* provided, for each g other than the identity, λ_g is properly outer.

Theorem 7.3.1 *Let M_0 be the smallest monotone closed subalgebra of M which contains $A \cup \{u_g : g \in G\}$. We suppose that:*

(i) *The action $g \mapsto \lambda_g$ is free.*
(ii) *For each $z \in M$, if $D(zu_g) = 0$ for every $g \in G$ then $z = 0$.*

Then M_0 contains every unitary in M which normalises A, that is, $M_0 = M_{\mathcal{N}}$.

Proof Let w be a unitary in M which normalises A. Let σ be the automorphism of A induced by w. Then, for each $a \in A$, we have $waw^* = \sigma(a)$. So $wa = \sigma(a)w$. Hence, for each g, we have

$$D(wau_g) = D(\sigma(a)wu_g).$$

But D is a conditional expectation. So

$$D(wu_g)u_g^* au_g = D(wau_g)$$
$$= D(\sigma(a)wu_g)$$
$$= \sigma(a)D(wu_g).$$

Because A is abelian, it follows that $(\sigma(a) - \lambda_g^{-1}(a))D(wu_g) = 0$.

Let p_g be the range projection of $D(wu_g)D(wu_g)^*$ in A. Equivalently this is the left projection of $D(wu_g)$. So, for every $a \in A$,

$$(\sigma(a) - \lambda_g^{-1}(a))p_g = 0. \tag{#}$$

Fix g and h with $g \neq h$, and let e be the projection $p_g p_h$. Then we have, for each $a \in A$,

$$(\lambda_h^{-1}(a) - \lambda_g^{-1}(a))e = (\sigma(a) - \lambda_g^{-1}(a))p_g p_h - (\sigma(a) - \lambda_h^{-1}(a))p_h p_g = 0.$$

Let b be any element of A and let $a = \lambda_g(b)$. Then $(\lambda_{h^{-1}g}(b) - b)e = 0$. If $e \neq 0$, then by (i) it follows that $h^{-1}g$ is the identity element of G. But this implies $g = h$, which is a contradiction. So $0 = e = p_g p_h$. So $\{p_g : g \in G\}$ is a (countable) family of orthogonal projections.

Let q be a projection in A which is orthogonal to each p_g. Then $qD(wu_g) = qp_gD(wu_g) = 0$. So $D(qwu_g) = 0$ for each $g \in G$. Hence, by applying hypothesis (ii), $qw = 0$. But $ww^* = 1$. So $q = 0$. Thus $\sum p_g = 1$. This convergence is in A. But since A is monotone (σ-)closed in M_0 this convergence is also in M_0, and so in M.

From (#) we see that

$$(\lambda_g\sigma(a) - a)\lambda_g(p_g) = 0.$$

We define q_g to be the projection $\lambda_g(p_g)$. Then

$$(a - \lambda_g\sigma(a))q_g = 0. \tag{##}$$

By arguing in a similar fashion to the above, we find that $\{q_g : g \in G\}$ is a family of orthogonal projections in A with $\sum q_g = 1$.

(Suppose that $pq_g = 0$ for all $g \in G$. Then we have

$$0 = p\lambda_g(p_g)\lambda_g(D(wu_g)) = p\lambda_g(D(wu_g))$$

$$= \lambda_g(D(wu_g))p = \lambda_g[D(wu_g)\lambda_g^{-1}(p)] = \lambda_g[D(wpu_g)]$$

and so we have $D(wpu_g) = 0$ for all g, which implies that $wp = 0$, that is, $p = 0$.)

For each $g \in G$, let $v_g = u_g p_g$. Then v_g is in M_0 and is a partial isometry with $v_g v_g^* = q_g$ and $v_g^* v_g = p_g$.

We wish to show that there exists a unitary v in M_0 such that

$$q_g v = v_g \text{ and } vp_g = v_g p_g = u_g p_g.$$

We could do this by applying the General Additivity of Equivalence for AW^*-algebras, see page 129 [13]. But we prefer to use the order convergence results of Chap. 2. Since $\{v_g\}_{g \in G}$ and $\{q_g\}_{g \in G}$ are ℓ^2-summable over G and M is monotone (σ-)complete, $v = \sum_{g \in G} v_g$ exists and satisfies the required conditions. (See Lemma 2.1.38 and Proposition 8.3.3.) (We again note that M_0 is monotone closed in M and so v is convergent in M and belongs to M_0.) From (#), for each $a \in A$,

$$\sigma(a)p_g = u_g^* a u_g p_g = p_g v^* a v p_g$$

$$= v^* v p_g v^* a v p_g$$

$$= v^* u_g p_g u_g^* a v p_g$$

$$= v^* a v p_g.$$

So $(\sigma(a) - v^* av)p_g = 0$. Let $y = \sigma(a) - v^* av$. Then $y^* y p_g = 0$.

So the range projection of $y^* y$ (the right projection of y) is orthogonal to p_g for each g.

Since A is monotone closed in M and $\sum_{g \in G} p_g = 1$ in A_{sa}, $\sum_{g \in G} p_g = 1$ in M_{sa}. Hence the right projection of y is 0. So $y = 0$. It now follows that $waw^* = vav^*$ for each $a \in A$. Then $v^* w$ commutes with each element of A. Since A is maximal abelian in M it follows that $v^* w$ is in A. Since v is in M_0, it now follows that w is in M_0.   □

We note that the above theorem does not require the action $g \mapsto \lambda_g$ to be ergodic. It is applicable to the algebras considered in Sects. 7.1 and 7.2.

Definition 7.3.2 Let $C(S) \times_\beta G$ be the smallest monotone closed $*$-subalgebra of $M(C(S), G)$ which contains the diagonal and each unitary which implements the β-action of G. It will sometimes be convenient to call $C(S) \times_\beta G$ the *"small"* *monotone cross product.*

It follows from Theorem 7.3.1 that $C(S) \times_\beta G$ is the normaliser subalgebra of $M(C(S), G)$.

It turns out that $C(S) \times_\beta G$ does not depend on the group G, only on the orbit equivalence relation. More precisely, using the same assumptions as in Theorem 7.2.6, we have the following:

Corollary 7.3.3 *Let $G_j(j = 1, 2)$ be countable, infinite groups of homeomorphisms of S. Let Y be a dense G_δ-subset of S such that $G_j[Y] = Y$ and G_j acts freely on Y. Let E_j be the orbit equivalence relation on Y arising from the action β_j of G_j. Suppose that $E_1 = E_2$. Then there is an isomorphism of $M(C(S), G_1)$ onto $M(C(S), G_2)$ which maps $C(S) \times_{\beta_1} G_1$ onto $C(S) \times_{\beta_2} G_2$.*

Proof This is a direct consequence of Theorems 7.2.6 and 7.3.1. More explicitly, when w is a unitary in $M(C(S), G_j)$ such that w normalises the diagonal then w is in $C(S) \times_{\beta_j} G_j$. So, the isomorphism of $M(C(S), G_j)$ onto $\mathcal{M}(E_j)/J$ maps $C(S) \times_{\beta_j} G_j$ onto the normaliser subalgebra of $\mathcal{M}(E_j)/J$, where $E_1 = E_2$. $\square$

7.4 Free Dense Actions of the Dyadic Group

We have said a great deal about G-actions with a free dense orbit and the algebras associated with them. It is incumbent on us to provide examples. We do this in this section. We have seen that when constructing monotone complete algebras from the action of a countably infinite group G on an extremally disconnected space S, what matters is the orbit equivalence relation induced on S. When the action of G has a free, dense orbit in S then we have shown that the orbit equivalence relation (and hence the associated algebras) can be obtained from an action of $\bigoplus \mathbb{Z}_2$ with a free, dense orbit. So, when searching for free, dense group actions, it suffices to find them when the group is $\bigoplus \mathbb{Z}_2$.

In this section we construct such actions of $\bigoplus \mathbb{Z}_2$. As an application, we will find 2^c hyperfinite factors which take 2^c different values in the classification semigroup that was constructed in Chap. 3.

We begin with some purely algebraic considerations before introducing topologies and continuity. We will end up with a huge number of examples.

We use $F(S)$ to denote the collection of all finite subsets of a set S. Here we shall regard the empty set, the set with no elements, as a finite set. We use $\mathbb{N}$ to be the set of natural numbers, excluding 0.

Let $C = \{f_{\mathbf{k}} : \mathbf{k} \in F(\mathbb{N})\}$ be a countable set where $\mathbf{k} \mapsto f_{\mathbf{k}}$ is a bijection. For each $n \in \mathbb{N}$ let σ_n be defined on C by

$$\sigma_n(f_{\mathbf{k}}) = \begin{cases} f_{\mathbf{k} \setminus \{n\}} & \text{if } n \in \mathbf{k} \\ f_{\mathbf{k} \cup \{n\}} & \text{if } n \notin \mathbf{k}. \end{cases}$$

Lemma 7.4.1   **(i)** *For each n, σ_n is a bijection of C onto C, and $\sigma_n\sigma_n = id$, where id is the identity map on C.*

(ii) *When $m \neq n$ then $\sigma_m\sigma_n = \sigma_n\sigma_m$.*

Proof   (i) It is clear that $\sigma_n\sigma_n = id$ and hence σ_n is a bijection.

(ii) Fix $f_\mathbf{k}$. Then we need to show $\sigma_m\sigma_n(f_\mathbf{k}) = \sigma_n\sigma_m(f_\mathbf{k})$. This is a straightforward calculation, considering separately the four cases when $\mathbf{k}$ contains neither m nor n, contains both m and n, contains m but not n and contains n but not m.

     We recall that $\bigoplus \mathbb{Z}_2 = \oplus_{n\in\mathbb{N}}\mathbb{Z}_2$, the countable commutative discrete group obtained by the countable restricted direct sum of $\mathbb{Z}_2$. For each $n \in \mathbb{N}$, let g_n, be the element defined by $g_n(m) = \delta_{m,n}$ for all $m \in \mathbb{N}$. Then $\{g_n : n \in \mathbb{N}\}$ is a set of generators of $\bigoplus \mathbb{Z}_2$.

     Take any $g \in \bigoplus \mathbb{Z}_2$ then g has a unique representation as $g = g_{n_1} + \cdots + g_{n_p}$ where $1 \leq n_1 < \cdots < n_p$ or g is the zero. Let us define

$$\varepsilon_g = \sigma_{n_1}\sigma_{n_2}\ldots\sigma_{n_p}.$$

Here we adopt the notational convention that $\sigma_{n_1}\sigma_{n_2}\ldots\sigma_{n_p}$ denotes the identity map of C onto itself when $\{n_1,\ldots,n_p\} = \emptyset$.

     Then $g \mapsto \varepsilon_g$ is a group homomorphism of $\bigoplus \mathbb{Z}_2$ into the group of bijections of C onto C. It will follow from Lemma 7.4.2 (ii) that this homomorphism is injective.   □

Lemma 7.4.2   **(i)** *The set C is an orbit. More precisely*

$$C = \{\varepsilon_g(f_\emptyset) : g \in \bigoplus \mathbb{Z}_2\}$$
$$= \{\sigma_{n_1}\sigma_{n_2}\ldots\sigma_{n_p}(f_\emptyset) : \{n_1, n_2 \cdots, n_p\} \in F(\mathbb{N})\}.$$

(ii) *For each $\mathbf{k} \in F(\mathbb{N})$*

$$\sigma_{n_1}\sigma_{n_2}\ldots\sigma_{n_p}(f_\mathbf{k}) = f_\mathbf{k}$$

only if $\sigma_{n_1}\sigma_{n_2}\ldots\sigma_{n_p} = id$.

Proof   (i) Let $\mathbf{k} = \{n_1, \cdots, n_p\}$ where $n_i \neq n_j$ for $i \neq j$. Then $\sigma_{n_1}\sigma_{n_2}\ldots\sigma_{n_p}(f_\emptyset) = f_\mathbf{k}$.

(ii) Assume this is false. Then, for some $\mathbf{k} \in F(\mathbb{N})$ we have $\sigma_{n_1}\sigma_{n_2}\ldots\sigma_{n_p}(f_\mathbf{k}) = f_\mathbf{k}$ where $\sigma_{n_1}\sigma_{n_2}\ldots\sigma_{n_p}$ is not the identity map. So we may assume, without loss of generality, that $\{n_1, n_2, \ldots, n_p\} = \mathbf{m}$ is a non-empty set of p natural numbers.

     First consider the case where $\mathbf{k}$ is the empty set. Then $\sigma_{n_1}\sigma_{n_2}\ldots\sigma_{n_p}(f_\emptyset) = f_\emptyset$. So $f_\mathbf{m} = f_\emptyset$. But this is not possible because the map $\mathbf{k} \mapsto f_\mathbf{k}$ is injective.

So $\mathbf{k}$ cannot be the empty set; let $\mathbf{k} = \{m_1, m_2, \ldots, m_q\}$. Then $\sigma_{m_1}\sigma_{m_2} \ldots \sigma_{m_q}(f_\emptyset) = f_{\mathbf{k}}$. Hence

$$\sigma_{n_1}\sigma_{n_2} \ldots \sigma_{n_p}\sigma_{m_1}\sigma_{m_2} \ldots \sigma_{m_q}(f_\emptyset) = \sigma_{m_1}\sigma_{m_2} \ldots \sigma_{m_q}(f_\emptyset).$$

On using the fact that the σ_j are idempotent and mutually commutative, we find that $\sigma_{n_1}\sigma_{n_2} \ldots \sigma_{n_p}(f_\emptyset) = f_\emptyset$. But, from the above argument, this is impossible. So (ii) is proved.   □

Let us recall that the "Big Cantor Space" $\{0, 1\}^{\mathbb{R}}$, is compact, totally disconnected and separable but not metrisable or second countable. Also each compact, separable, totally disconnected space is homeomorphic to a subspace of $\{0, 1\}^{\mathbb{R}}$. Let C be a countable subset of $\{0, 1\}^{\mathbb{R}}$ then clC, the closure of C, is a compact separable, totally disconnected space. This implies that C is completely regular and hence has a Stone-Czech compactification βC.

We saw in Chap. 4, that the regular σ-completion of $C(clC)$ is monotone complete and can be identified with $B^\infty(clC)/M(clC)$. Let $\widehat{clC}$ be the spectrum (maximal ideal space) of $B^\infty(clC)/M(clC)$. Then this may be identified with the Stone space of the complete Boolean algebra of regular open subsets of clC. By varying C in a carefully controlled way, we exhibited 2^c essentially different extremally disconnected spaces in the form $\widehat{clC}$.

For each of these spaces $\widehat{clC}$ we shall construct an action of $\bigoplus \mathbb{Z}_2$ with a free dense orbit.

We need to begin by recalling some notions from Chap. 4. A pair $(T, \mathbf{O})$ is said to be *feasible* if it satisfies the following conditions:

(i) T is a set of cardinality $c = 2^{\aleph_0}$; $\mathbf{O} = (O_n)(n = 1, 2 \ldots)$ is an infinite sequence of non-empty subsets of T, with $O_m \neq O_n$ whenever $m \neq n$.

(ii) Let M be a finite subset of T and $t \in T \backslash M$. For each natural number m there exists $n > m$ such that $t \in O_n$ and $O_n \cap M = \emptyset$.

In other words $\{n \in \mathbb{N} : t \in O_n \text{ and } O_n \cap M = \emptyset\}$ is an infinite set.

An example satisfying these conditions can be obtained by putting $T = 2^{\mathbb{N}}$, the Cantor space and letting $\mathbf{O}$ be an enumeration (without repetitions) of the (countable) collection of all non-empty clopen subsets.

For the rest of this section $(T, \mathbf{O})$ will be a fixed but arbitrary feasible pair.

Let $(T, \mathbf{O})$ be a feasible pair and let R be a subset of T. Then R is said to be *admissible* if

(i) R is a subset of T, with $\#R = \#(T \backslash R) = c$.

(ii) O_n is not a subset of R for any natural number n.

Return to the example where T is the Cantor space and $\mathbf{O}$ an enumeration of the non-empty clopen subsets. Then, whenever $R \subset 2^{\mathbb{N}}$ is nowhere dense and of cardinality c, R is admissible.

Throughout this section the feasible pair is kept fixed and the existence of at least one admissible set is assumed. For the moment, R is a fixed admissible subset of T. Later on we shall vary R.

Since $F(\mathbb{N}) \times F(T)$ has cardinality c, we can identify the Big Cantor space with $2^{F(\mathbb{N}) \times F(T)}$. For each $\mathbf{k} \in F(\mathbb{N})$, let $f_{\mathbf{k}} \in 2^{F(\mathbb{N}) \times F(T)}$ be the characteristic function of the set

$$\{(\mathbf{l}, L) : L \in F(T \backslash R), \mathbf{l} \subset \mathbf{k} \text{ and } O_n \cap L = \varnothing \text{ whenever } n \in \mathbf{k} \text{ and } n \notin \mathbf{l}\}.$$

As in Chap. 4, let $N(t) = \{n \in \mathbb{N} : t \in O_n\}$. By feasibility, this set is infinite for each $t \in T$. It is immediate that $f_{\mathbf{k}}(\mathbf{l}, L) = 1$ precisely when

$$L \in F(T \backslash R), \mathbf{l} \subset \mathbf{k} \text{ and, for each } t \in L, \ N(t) \cap (\mathbf{k} \backslash \mathbf{l}) = \varnothing.$$

Let X_R be the countable set $\{f_{\mathbf{k}} : \mathbf{k} \in F(\mathbb{N})\}$. Let K_R be the closure of X_R in the Big Cantor space. Then K_R is a (separable) compact Hausdorff totally disconnected space with respect to the relative topology induced by the product topology of the Big Cantor space. We always suppose X_R to be equipped with the relative topology induced by K_R.

Let $C = X_R$. If the map $\mathbf{k} \mapsto f_{\mathbf{k}}$ is an injection then we can define σ_n on X_R as before.

Lemma 7.4.3 *Let $f_{\mathbf{k}} = f_{\mathbf{m}}$. Then $\mathbf{k} = \mathbf{m}$.*

Proof By definition, $f_{\mathbf{k}}(\mathbf{l}, \varnothing) = 1$ precisely when $\mathbf{l} \subset \mathbf{k}$. Since $f_{\mathbf{k}}(\mathbf{m}, \varnothing) = f_{\mathbf{m}}(\mathbf{m}, \varnothing) = 1$ it follows that $\mathbf{m} \subset \mathbf{k}$. Similarly, $\mathbf{k} \subset \mathbf{m}$. Hence $\mathbf{m} = \mathbf{k}$.   □

For each $(\mathbf{k}, K) \in F(\mathbb{N}) \times F(T)$ let $E_{(\mathbf{k},K)} = \{x \in K_R : x(\mathbf{k}, K) = 1\}$. The definition of the product topology of the Big Cantor space implies that $E_{(\mathbf{k},K)}$ and its compliment $E_{(\mathbf{k},K)}^c$ are clopen subsets of K_R. It also follows from the definition of the product topology that finite intersections of such clopen sets form a base for the topology of K_R. Hence their intersections with X_R give a base for the relative topology of X_R. But we saw in Chap. 4 that, in fact,

$$\{E_{(\mathbf{k},K)} \cap X_R : \mathbf{k} \in F(\mathbb{N}), K \in F(T \backslash R)\}$$

is a base for the topology of X_R. Also $E_{(\mathbf{k},K)} = \varnothing$ unless $K \subset T \backslash R$.

Since each $E_{(\mathbf{k},K)}$ is clopen, it follows from Lemma 6.2.1 that $E_{(\mathbf{k},K)}$ is the closure of $E_{(\mathbf{k},K)} \cap X_R$.

To slightly simplify our notation, we shall write $E(\mathbf{k}, K)$ for $E_{(\mathbf{k},K)} \cap X_R$ and E_n for $E_{(\{n\},\varnothing)} \cap X_R$. Also E_n^c is the compliment of E_n in X_R, which is, $E_{(\{n\},\varnothing)}^c \cap X_R$. We shall see, below, that $\{f_{\mathbf{h}} : n \notin \mathbf{h}\} = E_n^c$, equivalently, $E_n = \{f_{\mathbf{h}} : n \in \mathbf{h}\}$.

When G is a subset of X_R we denote its closure in βX_R by clG. When G is a clopen subset of X_R then clG is a clopen subset of βX_R. So the closure of E_n in βX_R is clE_n, whereas its closure in K_R is $E_{(\{n\},\varnothing)}$.

We need to show that each σ_n is continuous on X_R. Since σ_n is equal to its inverse, this implies that σ_n is a homeomorphism of X_R onto itself.

Our first step to establish continuity of σ_n is the following:

Lemma 7.4.4 *We have* $E_n = \{f_\mathbf{k} : n \in \mathbf{k}\}$ *and* $E_n^c = \{f_\mathbf{m} : n \notin \mathbf{m}\}$. *Also* σ_n *interchanges* E_n *and* E_n^c. *Furthermore, for* $m \neq n$, σ_m *maps* E_n *onto* E_n *and* E_n^c *onto* E_n^c.

Proof By definition $f_\mathbf{k}(\{n\}, \emptyset) = 1$ if, and only if $\{n\} \subset \mathbf{k}$. So $f_\mathbf{m} \in E_n^c$ precisely when $n \notin \mathbf{m}$.

For $f_\mathbf{k} \in E_n$ we have $\sigma_n(f_\mathbf{k}) = f_{\mathbf{k}\setminus\{n\}}$. So σ_n maps E_n onto E_n^c. Similarly, it maps E_n^c onto E_n.

When $m \neq n$, consider $f_\mathbf{k} \in E_n$. Then $n \in \mathbf{k}$. So $n \in \mathbf{k} \cup \{m\}$ and $n \in \mathbf{k}\setminus\{m\}$. Thus $\sigma_m(f_\mathbf{k})$ is in E_n. i.e. $\sigma_m[E_n] \subset E_n$.

Since σ_m is idempotent, we get $\sigma_m[E_n] = E_n$. Similarly $\sigma_m[E_n^c] = E_n^c$. □

Lemma 7.4.5 *The map* $\sigma_n : X_R \mapsto X_R$ *is continuous.*

Proof It suffices to show that $\sigma_n^{-1}[E(\mathbf{l}, L)]$ is open when $L \subset T\setminus R$.

Let $f_\mathbf{h}$ be in $\sigma_n^{-1}[E(\mathbf{l}, L)]$. We shall find, U, an open neighbourhood of $f_\mathbf{h}$ such that $\sigma_n[U] \subset E(\mathbf{l}, L)$.

We need to consider three possibilities.

(1) First suppose that $n \in \mathbf{h}$, that is $f_\mathbf{h} \in E_n$. Then $f_{\mathbf{h}\setminus\{n\}} = \sigma_n(f_\mathbf{h})$, which is in $E(\mathbf{l}, L)$. So $\mathbf{l} \subset \mathbf{h}\setminus\{n\}$ which implies $n \notin \mathbf{l}$. Also $N(t) \cap ((\mathbf{h}\setminus\{n\})\setminus\mathbf{l}) = \emptyset$ for all $t \in L$. It follows that $\mathbf{l} \cup \{n\} \subset \mathbf{h}$ and, for all $t \in L$, $N(t) \cap (\mathbf{h}\setminus(\mathbf{l} \cup \{n\})) = \emptyset$. Hence $f_\mathbf{h} \in E_n \cap E(\mathbf{l} \cup \{n\}, L)$.

   Let $f_\mathbf{k} \in E(\mathbf{l} \cup \{n\}, L)$. Then $\mathbf{l} \cup \{n\} \subset \mathbf{k}$. Also, for $t \in L$, $N(t) \cap (\mathbf{k}\setminus(\mathbf{l} \cup \{n\})) = \emptyset$.

   Hence $\mathbf{l} \subset \mathbf{k}\setminus\{n\}$ and, for $t \in L$, $N(t) \cap ((\mathbf{k}\setminus\{n\})\setminus\mathbf{l}) = \emptyset$. This implies $\sigma_n(f_\mathbf{k}) = f_{\mathbf{k}\setminus\{n\}} \in E(\mathbf{l}, L)$. Thus $E(\mathbf{l} \cup \{n\}, L)$ is a clopen set, which is a neighbourhood of $f_\mathbf{h}$ and a subset of $\sigma_n^{-1}[E(\mathbf{l}, L)]$.

(2) Now suppose $n \notin \mathbf{h}$. Then $f_{\mathbf{h}\cup\{n\}} = \sigma_n(f_\mathbf{h})$ which is in $E(\mathbf{l}, L)$. This gives (a) $\mathbf{h} \cup \{n\}$ contains $\mathbf{l}$. (b) For all $t \in L$, $N(t) \cap ((\mathbf{h} \cup \{n\})\setminus\mathbf{l}) = \emptyset$. (c) $f_\mathbf{h} \in E_n^c$.

   Suppose, additionally, that $n \in \mathbf{l}$. Then $(\mathbf{h} \cup \{n\})\setminus\mathbf{l} = \mathbf{h}\setminus(\mathbf{l}\setminus\{n\})$. So, for all $t \in L$, $N(t) \cap (\mathbf{h}\setminus(\mathbf{l}\setminus\{n\})) = \emptyset$. So $f_\mathbf{h}$ is in $E(\mathbf{l}\setminus\{n\}, L)$. Hence, by (c) $f_\mathbf{h}$ is in $E_n^c \cap E(\mathbf{l}\setminus\{n\}, L)$.

   Now let $f_\mathbf{k} \in E_n^c \cap E(\mathbf{l}\setminus\{n\}, L)$. Since $f_\mathbf{k} \in E_n^c$, it follows that $n \notin \mathbf{k}$. So $f_{\mathbf{k}\cup\{n\}} = \sigma_n(f_\mathbf{k})$.

   Since $f_\mathbf{k} \in E(\mathbf{l}\setminus\{n\}, L)$, we have $\mathbf{l}\setminus\{n\} \subset \mathbf{k}$. So $\mathbf{l} \subset \mathbf{k}\cup\{n\}$. Also $(\mathbf{k}\cup\{n\})\setminus\mathbf{l} = \mathbf{k}\setminus(\mathbf{l}\setminus\{n\})$. So, for any $t \in L, N(t) \cap ((\mathbf{k} \cup \{n\})\setminus\mathbf{l}) = \emptyset$. Thus $f_{\mathbf{k}\cup\{n\}} \in E(\mathbf{l}, L)$. That is, $\sigma_n(f_\mathbf{k}) \in E(\mathbf{l}, L)$.

(3) We now suppose that $n \notin \mathbf{h}$ and $n \notin \mathbf{l}$. As in (2), statements (a), (b) and (c) hold.

   Note $\mathbf{h}\setminus\mathbf{l} = (\mathbf{h} \cup \{n\})\setminus(\mathbf{l} \cup \{n\})$. It follows from (b) that $N(t) \cap (\mathbf{h}\setminus\mathbf{l}) = \emptyset$ for each $t \in L$. Hence $f_\mathbf{h} \in E(\mathbf{l}, L) \cap E_n^c$.

We also observe that, because $n \notin \mathbf{l}$, (b) implies that

$$(d) \; \{n\} \cap N(t) = \emptyset \text{ for each } t \in L.$$

Now let $f_\mathbf{k} \in E(\mathbf{l}, L) \cap E_n^c$. Then $n \notin \mathbf{k}$. So $\sigma_n(f_\mathbf{k}) = f_{\mathbf{k} \cup \{n\}}$.

Also $\mathbf{l} \subset \mathbf{k}$ and $N(t) \cap (\mathbf{k} \backslash \mathbf{l}) = \emptyset$ for any $t \in L$. It now follows from (d) that $((\mathbf{k} \cup \{n\}) \backslash \mathbf{l}) \cap N(t) = \emptyset$ whenever $t \in L$. Hence $f_{\mathbf{k} \cup \{n\}} \in E(\mathbf{l}, L)$. Thus $E(\mathbf{l}, L) \cap E_n^c$ is a clopen neighbourhood of $f_\mathbf{h}$ and it is a subset of $\sigma_n[E(\mathbf{l}, L)]$.

It follows from (1), (2) and (3) that every point of $\sigma_n[E(\mathbf{l}, L)]$ has an open neighbourhood contained in $\sigma_n[E(\mathbf{l}, L)]$. In other words, the set is open.   □

We recall that X_R is completely regular because it is a subspace of the compact Hausdorff space K_R. Let βX_R be its Stone-Cech compactification. Then each continuous function $f : X_R \mapsto X_R$ has a unique extension to a continuous function F from βX_R to βX_R. When f is a homeomorphism, then by considering the extension of f^{-1} it follows that F is a homeomorphism of βX_R. In particular, each σ_n has a unique extension to a homeomorphism of βX_R. We abuse our notation by also denoting this extension by σ_n.

Let us recall from Sect. 6.4, that when θ is in $Homeo(\beta X_R)$ then it induces a homomorphism h_θ of $C(\beta X_R)$ by $h_\theta(f) = f \circ \theta$. It also induces an automorphism of $B^\infty(\beta X_R)/M_0(\beta X_R)$ by $H_\theta([F]) = [F \circ \theta]$. Then H_θ is the unique automorphism of $B^\infty(\beta X_R)/M_0(\beta X_R)$ which extends h_θ. Let S_R be the (extremally disconnected) structure space of $B^\infty(\beta X_R)/M_0(\beta X_R)$; this algebra can then be identified with $C(S_R)$. Then H_θ corresponds to $\hat{\theta}$, an homeomorphism of S_R. Then $\theta \mapsto h_\theta$ is a group anti-isomorphism of $Homeo(\beta X_R)$ onto $AutC(\beta X_R)$; $h_\theta \mapsto H_\theta$ is an isomorphism of $AutC(\beta X_R)$ into $AutC(S_R)$. Also $H_\theta \mapsto \hat{\theta}$ is a group anti-isomorphism of $AutC(S_R)$ into $Homeo(S_R)$. When G is an Abelian subgroup of $Homeo(\beta X_R)$ it follows that $\theta \mapsto H_\theta$ and $\theta \mapsto \hat{\theta}$ are group isomorphisms of G into $AutC(S_R)$ and $Homeo(S_R)$, respectively.

We recall that $g \mapsto \varepsilon_g$ is an injective group homomorphism of $\bigoplus \mathbb{Z}_2$ into the group of bijections of C onto C. By taking the natural bijection from C onto X_R, and by applying Lemmas 7.4.2 and 7.4.5, we may regard ε_* as an injective group homomorphism of $\bigoplus \mathbb{Z}_2$ into $Homeo(X_R)$. Since each homeomorphism of X_R onto itself has a unique extension to a homeomorphism of βX_R onto itself, we may identify ε_* with an injective group homomorphism of $\bigoplus \mathbb{Z}_2$ into the group $Homeo(\beta X_R)$. This induces a group isomorphism, $g \mapsto \hat{\varepsilon}^g$ from $\bigoplus \mathbb{Z}_2$ into $AutC(S_R)$ by putting $\hat{\varepsilon}^g = H_{\varepsilon_g}$. The corresponding isomorphism, $g \mapsto \hat{\varepsilon}_g$, from $\bigoplus \mathbb{Z}_2$ into $Homeo(S_R)$, is defined by

$$\hat{\varepsilon}_g = \widehat{\varepsilon_g} \text{ for each } g \in \bigoplus \mathbb{Z}_2.$$

As in Sect. 6.4, ρ is the continuous surjection from S_R onto βX_R which is dual to the natural injection from $C(\beta X_R)$ into $B^\infty(\beta X_R)/M_0(\beta X_R) \simeq C(S_R)$. Let $s_0 \in S_R$ such that $\rho(s_0) = f_\emptyset$.

Theorem 7.4.6 *Let $g \mapsto \widehat{\varepsilon}_g$ be the representation of $\bigoplus \mathbb{Z}_2$, as homeomorphisms of S_R, as defined above. Then the orbit $\{\widehat{\varepsilon}_g(s_0) : g \in \bigoplus \mathbb{Z}_2\}$ is a free, dense orbit in S_R. Furthermore, there exists Y, a G-invariant, dense G_δ-subset of S_R, with $s_0 \in Y$, such that the action $\widehat{\varepsilon}$ is free on Y.*

Proof By Lemma 7.4.2 (i), $X_R = \{\varepsilon_g(f_\emptyset) : g \in \bigoplus \mathbb{Z}_2\}$. By Proposition 6.4.4 this implies the orbit $\{\widehat{\varepsilon}_g(s_0) : g \in \bigoplus \mathbb{Z}_2\}$ is dense in S_R.

By Lemma 7.4.2 (ii) $\{\varepsilon_g(f_\emptyset) : g \in \bigoplus \mathbb{Z}_2\}$ is a free orbit. This theorem now follows from Lemmas 6.7.1 and 6.5.9. $\qquad\qquad\qquad\qquad\qquad\qquad\qquad\square$

Corollary 7.4.7 *The group isomorphism, $g \mapsto \widehat{\varepsilon}^g$, from $\bigoplus \mathbb{Z}_2$ into $AutC(S_R)$ is free and ergodic.*

We shall see below that we can now obtain some additional information about this action of $\bigoplus \mathbb{Z}_2$ as automorphisms of $C(S_R)$. This will enable us to construct huge numbers of small wild factors which are hyperfinite, that is, generated by an increasing sequence of matrix algebras.

We have seen that, for each natural number n, σ_n is a homeomorphism of X_R onto itself with the following properties. First, $\sigma_n = \sigma_n^{-1}$. Secondly, $\sigma_n[E_n] = E_n^c$ and, for $m \neq n$, we have $\sigma_n[E_m] = E_m$. (This notation was introduced just before Lemma 7.4.4, above.)

We have seen that σ_n has a unique extension to a homeomorphism of βX_R, which we again denote by σ_n. Then

$\sigma_n[clE_n] = clE_n^c$ and, for $m \neq n$, we have $\sigma_n[clE_m] = clE_m$.

We define $e_n \in C(\beta X_R)$ as the characteristic function of the clopen set $clE(\{n\}, \emptyset) = clE_n$.

Using the above notation, $\widehat{\varepsilon}^{\sigma_n}$ is the $*$-automorphism of $B^\infty(\beta X_R)/M_0(\beta X_R) \simeq C(S_R)$ induced by σ_n. We have

$$\widehat{\varepsilon}^{\sigma_n}(e_n) = 1 - e_n \text{ and, for } m \neq n, \ \widehat{\varepsilon}^{\sigma_n}(e_m) = e_m.$$

Since $Reg(\beta X_R) \approx Reg(X_R) \approx Reg(K_R)$, we can apply Proposition 4.3.15 and Corollary 4.3.16 to show that the smallest monotone σ-closed $*$-subalgebra of $B^\infty(\beta X_R)/M_0(\beta X_R)$ which contains $\{e_n : n = 1, 2, \dots\}$ is $B^\infty(\beta X_R)/M_0(\beta X_R)$ itself. We shall see that the (norm-closed) $*$-algebra generated by $\{e_n : n = 1, 2 \dots\}$ is naturally isomorphic to $C(2^\mathbb{N})$.

When $S \subset \mathbb{N}$ we use η_S to denote the element of $2^\mathbb{N}$ which takes the value 1 when $n \in S$ and 0 otherwise. Let G_n be the clopen set $\{\eta_S \in 2^\mathbb{N} : n \in S\}$. These clopen sets generate the (countable) Boolean algebra of clopen subsets of $2^\mathbb{N}$. An application of the Stone-Weierstrass Theorem shows that the $*$-subalgebra of $C(2^\mathbb{N})$, containing each χ_{G_n} is dense in $C(2^\mathbb{N})$.

Lemma 7.4.8 *There exists an isometric isomorphism, π_0, from $C(2^\mathbb{N})$ into $C(\beta X_R)$ such that $\pi_0(\chi_{G_n}) = e_n$.*

Proof As in Sect. 4.3, we define a map Γ from the Big Cantor space, $2^{F(\mathbb{N}) \times F(T)}$, onto the classical Cantor space, $2^{\mathbb{N}}$, by $\Gamma(\mathbf{x})(n) = \mathbf{x}((\{n\}, \emptyset))$. Put $J = \{(\{n\}, \emptyset) : n = 1, 2, \ldots\}$. Then we identify $2^{\mathbb{N}}$ with 2^J. So Γ may be regarded as a restriction map and, by definition of the topology for product spaces, it is continuous.

From the definition of $f_{\mathbf{k}}$, we see that $f_{\mathbf{k}}(\{n\}, \emptyset) = 1$ precisely when $n \in \mathbf{k}$. So $\Gamma f_{\mathbf{k}} = \eta_{\mathbf{k}}$. Hence

$$\Gamma[E_n] = \{\eta_{\mathbf{k}} : n \in \mathbf{k} \text{ and } \mathbf{k} \in F(\mathbb{N})\}.$$

By the basic property of the Stone-Czech compactification, the natural embedding of X_R into K_R factors through βX_R. So there exists a continuous surjection ϕ from βX_R onto K_R which restricts to the identity map on X_R. Then $\Gamma\phi$ maps clE_n onto G_n and clE_n^c onto G_n^c. For $f \in C(2^{\mathbb{N}})$ let $\pi_0(f) = f \circ \Gamma\phi$. Then π_0 is the required isometric isomorphism into $C(\beta X_R) \subset B^{\infty}(\beta X_R)/M_0(\beta X_R)$. $\qquad\square$

Let $\hat{\varepsilon}$ be the action of the Dyadic Group on $C(S_R)$ considered above. Let M_R be the corresponding monotone cross-product algebra. So there exists an isomorphism π_R from $C(S_R)$ onto the diagonal subalgebra of M_R and a group representation $g \mapsto u_g$ of the Dyadic Group in the unitary group of M_R such that $u_g \pi_R(a) u_g^* = \pi_R(\hat{\varepsilon}^g(a))$. Since each element of the Dyadic Group is its own inverse, we see that each u_g is self-adjoint. Since the Dyadic Group is abelian, $u_g u_h = u_h u_g$ for each g and h.

As before, let g_n be the nth term in the standard sequence of generators of $\bigoplus \mathbb{Z}_2$ that is, g_n takes the value 1 in the nth coordinate and 0 elsewhere. We abuse our notation by writing "u_n" for the unitary u_{g_n} and "e_n" for the projection $\pi_R(e_n)$ in the diagonal subalgebra of M_R. We then have:

$$u_n e_n u_n = 1 - e_n \text{ and, for } m \neq n, \ u_n e_m u_n = e_m.$$

Let $A_R = \pi_R[C(S_R)]$ be the diagonal algebra of M_R. We recall that the Boolean σ-subalgebra of the projections of A_R, generated by $\{e_n : n = 1, 2, \ldots\}$, contains all the projections of A_R.

Let us also recall, see [121, p205], that the Fermion algebra is the most basic of the uniformly hyperfinite C^*-algebras (Glimm algebras). It is generated as a C^*-algebra by an increasing sequence of full matrix algebras of dimensions $2^n \times 2^n$ ($n = 1, 2, \ldots$) and is unital and simple.

For any projection p we define $p^{(0)} = p$ and $p^{(1)} = 1 - p$.

For each choice of n and for each choice of $\alpha_1, \ldots, \alpha_n$ from $\mathbb{Z}_2$, it follows from Lemma 7.4.8 that the product $e_1^{\alpha_1} e_2^{\alpha_2} \ldots e_n^{\alpha_n}$ is neither 1 nor 0. In the terminology of [180], (e_n) is a sequence of (mutually commutative) independent projections. The following Lemma follows from the proof of Proposition 2.1 [180]. But for the reader's convenience we give a different, self-contained argument. In particular, for each n, $\{u_j : j = 1, 2, \ldots, n\} \cup \{e_j : j = 1, 2, \ldots, n\}$ generates a subalgebra isomorphic to the algebra of all $2^n \times 2^n$ complex matrices.

Lemma 7.4.9 *Let $\mathcal{F}$ be the Fermion algebra. Then there exists an isomorphism Π from $\mathcal{F}$ onto the smallest norm closed $*$-subalgebra of M_R which contains $\{u_n : n = 1, 2, \ldots\}$ and $\{e_n : n = 1, 2, \ldots\}$. This isomorphism takes the diagonal of $\mathcal{F}$ onto the smallest closed abelian $*$-subalgebra containing $\{e_n : n = 1, 2, \ldots\}$.*

Proof Let B be the C^*-subalgebra generated by $\{u_n : n = 1, 2, \cdots\}$ and $\{e_n : n = 1, 2, \cdots\}$. For each n, let $f_{11}^{(n)} = e_n$, $f_{12}^{(n)} = e_n u_n$, $f_{21}^{(n)} = u_n e_n$ and $f_{22}^{(n)} = 1 - e_n$. Then for each n, $\{f_{ij}^{(n)}\}_{1 \leq i, j \leq 2}$ is a system of 2×2 matrix units in B such that $f_{ij}^{(n)} f_{kl}^{(n)} = f_{kl}^{(n)} f_{ij}^{(m)}$ for all m, n in $\mathbb{N}$ and i, j, k, l in $\{1, 2\}$ and such that the commuting family $[\{f_{ij}^{(n)} : i, j = 1, 2\} : n = 1, 2, \cdots]$ generates the norm closed subalgebra B. More precisely, if B_n is the C^*-subalgebra of B generated by $[\{f_{ij}^{(n)} : i, j = 1, 2\} : n = 1, 2, \cdots, n]$ for each n, then we have

$$B_1 \subset B_2 \subset \cdots \subset B_n \subset B_{n+1} \subset \cdots \subset B$$

such that $B_n = C^*(u_1, \cdots, u_n; e_1, \cdots, e_n)$, the C^*-subalgebra generated by $\{u_1, \cdots, u_n; e_1, \cdots, e_n\}$ for each n and B is the norm closure of the $*$-subalgebra $\bigcup_{n \geq 1} B_n$ in M_R. Here we used the fact that $u_n = e_n u_n + u_n e_n = f_{12}^{(n)} + f_{21}^{(n)}$ for each n.

On the other hand, there exists an inductive system

$$M_2 \xrightarrow{\pi_1} M_{2^2} \xrightarrow{\pi_2} \cdots \xrightarrow{\pi_{n-1}} M_{2^n} \xrightarrow{\pi_n} M_{2^{n+1}} \xrightarrow{\pi_{n+1}} \cdots,$$

with the following properties. First $M_{2^n} = M_2(\mathbb{C}) \otimes \cdots \otimes M_2(\mathbb{C})$ (n-times). Secondly,

$$\pi_n : M_{2^n} \longmapsto M_{2^{n+1}}$$

where for each n, and each $a \in M_{2^n}$,

$$\pi_n(a) = a \otimes 1_{M_2(\mathbb{C})}.$$

Third, $\mathcal{F}$ is the C^*-algebra obtained by the C^*-completion of the inductive limit $*$-subalgebra $F_\infty = \bigcup_{n \geq 1} \iota_n(M_{2^n})$, where ι_n is the canonical $*$-isomorphism from M_{2^n} into F_∞. (When $a_1, \cdots, a_n \in M_2(\mathbb{C})$ we sometimes write

$$\iota_n(a_1 \otimes \cdots \otimes a_n) = a_1 \otimes \cdots \otimes a_n \otimes 1 \otimes \cdots.$$

For details see [121, p205] or [162, p83, Vol III]. Let $\{e_{ij}\}_{1 \leq i, j \leq 2}$ be the standard system of matrix units for $M_2(\mathbb{C})$. Since each M_{2^n} is $*$-isomorphic to the simple C^*-algebra $M_{2^n}(\mathbb{C})$, there exists a unique $*$-isomorphism θ_n from M_{2^n} onto B_n satisfying $\theta_n(e_{i_1 j_1} \otimes \cdots e_{i_n j_n}) = f_{i_1 j_1}^{(1)} \cdots f_{i_n j_n}^{(n)}$ for each $i_k, j_k \in \{1, 2\}$ with $1 \leq k \leq n$ and $\theta_{n+1}(x \otimes e_{ij}) = \theta_n(x) f_{ij}^{(n+1)}$ for each $x \in M_{2^n}$ and $1 \leq i, j \leq 2$, that is, we have $\theta_{n+1} \circ \pi_n = \theta_n$ for each n. Hence there exists a unique $*$-homomorphism θ from F_∞ onto $\bigcup_{n \geq 1} B_n$ satisfying $\theta \circ \iota_n = \theta_n$ for each n. Since each θ_n is an isometry,

θ extends uniquely to a $*$-isomorphism Π from $\mathcal{F}$ onto B which maps the diagonal of $\mathcal{F}$ onto the smallest norm closed $*$-subalgebra containing $\{e_n : n = 1, 2, \cdots\}$. This completes the proof. $\qquad\square$

Definition 7.4.10 Let B_R be the smallest monotone σ-closed $*$-subalgebra of M_R which contains $\Pi[\mathcal{F}]$.

Proposition 7.4.11 *The algebra B_R is a monotone complete factor which contains A_R as a maximal abelian $*$-subalgebra. There exists a faithful normal conditional expectation from B_R onto A_R. The state space of B_R is separable. The factor B_R is wild and of Type III. It is also a small C^*-algebra.*

Proof Let D_R be the faithful normal conditional expectation from M_R onto A_R. The maximal ideal space of A_R can be identified with the separable space S_R. Then, by Corollary 4.2.21, there exists a faithful state ϕ on A_R. Hence ϕD_R is a faithful state on M_R and restricts to a faithful state on B_R. So, by Theorem 2.1.14, B_R is monotone complete. Let D be the restriction of D_R to B_R then D is a faithful and normal conditional expectation from B_R onto A_R.

Since each e_n is in B_R it follows that A_R is a $*$-subalgebra of B_R. Since it is maximal abelian in M_R it must be a maximal abelian $*$-subalgebra of B_R. So the centre of B_R is a subalgebra of A_R. Each u_n is in B_R and so each central projection of B_R commutes with each u_n. Since the action $\hat{\varepsilon}$ of the Dyadic Group is ergodic (see Corollary 7.4.7), it follows that the only projections in A_R which commute with every u_n are 0 and 1. So B_R is a (monotone complete) factor.

The state space of every unital C^*-subalgebra of M_R is a surjective image of the state space of M_R, which is separable. So the state space of B_R is separable. Equivalently, B_R is almost separably representable.

Since B_R contains a maximal abelian $*$-subalgebra which is not a von Neumann algebra it is a wild factor. Also M_R is almost separably representable, hence it possesses a strictly positive state (see the proof of Lemma 3.1.1) and so is a Type III factor [176], see also [131]. In Sects. 5.8 and 5.9, it is shown that a monotone complete factor is a small C^*-algebra whenever it has a separable state space. It now follows immediately that the factor is a small C^*-algebra. $\qquad\square$

Alternatively we may use results from Theorem 5.9.9 on small C^*-algebras to argue as follows.

Since $C(S_R)$ is small and $\ell^2(\oplus Z_2)$ is a separable Hilbert space, $C(S_R)\overline{\otimes}L(\ell^2(\oplus Z_2))$ is small. Any unital C*-subalgebra of a small C*-algebra is also small. From the identification of M_R as a monotone cross product

$$M_R \subset C(S_R)\overline{\otimes}L(\ell^2(\oplus Z_2)).$$

So M_R and B_R are both small. Hence they both possess a faithful state and, as above, must be wild Type III factors.

Proposition 7.4.12 *The homomorphism* Π *extends to a* σ-*homomorphism* Π^∞ *from* $\mathcal{F}^\infty$, *the Pedersen-Baire envelope of the Fermion algebra, onto* B_R. *Let* J_R *be the kernel of* Π^∞ *then* $\mathcal{F}^\infty/J_R$ *is isomorphic to* B_R.

Proof Let $\tilde{\Pi} : \mathcal{F}^\infty \mapsto B_R^\infty$ be the canonical σ-normal extension of Π (see [121, p60]). By Theorem 5.4.5, there exists a surjective σ-normal unital $*$-homomorphism $q : B_R^\infty \mapsto B_R$. Let $\Pi^\infty = q \circ \tilde{\Pi}$. Then this Π^∞ satisfies our requirements. $\qquad\square$

7.5 Approximately Finite Dimensional Algebras

We could proceed in greater generality, but for ease and simplicity, we shall only consider monotone complete C^*-algebra which possess a faithful state. Every almost separably representable algebra has this property and hence so does every small C^*-algebra.

Ever since the pioneering work of Connes [30], the notion of a hyperfinite or approximately finite dimensional von Neumann algebra has been well understood. The situation for monotone complete C^*-algebras is much more obscure, with many questions unanswered. We have to make a number of delicate distinctions which are unnecessary for von Neumann algebras.

Definition 7.5.1 Let B be a monotone complete C^*-algebra with a faithful state. Then B is said to be *approximately finite dimensional* if there exists an increasing sequence of finite dimensional $*$-subalgebras (F_n) such that the smallest monotone closed subalgebra of B which contains $\cup_{n=1}^\infty F_n$ is B itself.

Definition 7.5.2 Let B be approximately finite dimensional. If we can take each F_n to be a full matrix algebra, then B is said to be *strongly hyperfinite*.

Definition 7.5.3 Let M be a monotone complete C^*-algebra with a faithful state. We call M *nearly approximately finite dimensional* (with respect to B and D) if it satisfies the following conditions:

(i) M contains a monotone closed subalgebra B, where B is approximately finite dimensional.

(ii) There exists a linear map $D : M \mapsto B$ which is completely positive, unital, faithful and normal.

(iii) For each $z \in M$, there exists a sequence $(z_n)(n = 1, 2 \ldots)$ in B, such that

$$D((z - z_n)(z - z_n)^*) \geq D((z - z_{n+1})(z - z_{n+1})^*)$$

for each n, and

$$\bigwedge_{n=1}^\infty D((z - z_n)(z - z_n)^*) = 0.$$

Suppose M, B and D satisfy the above conditions. In the notation of Chap. 3, $M \precsim B$ because $D : M \mapsto B$ is faithful and normal. Also $B \precsim M$ because B is monotone closed in $\dot{M}$. So B and M are equivalent.

Definition 7.5.4 Let M be a monotone complete C^*-algebra with a faithful state. We call M *nearly hyperfinite* if it contains a monotone closed subalgebra B such that (i) M is nearly approximately finite dimensional with respect to B and (ii) B is strongly hyperfinite.

Theorem 7.5.5 *Let S be a compact Hausdorff extremally disconnected space. Let G be a countably infinite group and $g \mapsto \beta^g$ be a free action of G as automorphisms of $C(S)$. Let $M(C(S), G)$ be the corresponding monotone cross-product; let $\pi[C(S)]$ be the diagonal subalgebra; let $D : M(C(S), G) \mapsto \pi[C(S)]$ be the diagonal map. Let $g \mapsto u_g$ be a unitary representation of G in $M(C(S), G)$ such that $\beta^g(a) = u_g a u_g^*$ for each $a \in \pi[C(S)]$. Let B be the monotone closure of the $*$-algebra generated by $\pi[C(S)] \cup \{u_g : g \in G\}$. If B is approximately finite dimensional then $M(C(S), G)$ is nearly approximately finite dimensional.*

Proof By Theorem 7.3.1, B is the normaliser subalgebra of $M(C(S), G)$.

By Lemma 7.2.5, $M(C(S), G)$ satisfies condition (iii) of Definition 7.5, with respect to B and the diagonal map D. It follows immediately that $M(C(S), G)$ is nearly approximately finite dimensional whenever B is approximately finite dimensional.   □

In Chap. 3 we constructed a weight semigroup, $\mathcal{W}$, which classified monotone complete C^*-algebras (of bounded cardinality). In particular, given algebras C_1 and C_2, they are equivalent (as defined in Chap. 3) precisely when their values in the weight semigroup, wC_1 and wC_2, are the same.

Let $(T, \mathbf{O})$ be a feasible pair, see Definition 4.3.3. Let $\mathcal{R}$ be the collection of all admissible subsets of T. For each $R \in \mathcal{R}$ let $A_R = C(S_R) = B^\infty(K_R)/M_0(K_R)$. (See Remark 6.4.8 and Corollary 4.2.24.) Then, by Corollary 4.3.25, we can find $\mathcal{R}_0 \subset \mathcal{R}$ such that $\#\mathcal{R}_0 = 2^c$, where $c = 2^{\aleph_0}$ with the following property. Whenever R_1 and R_2 are distinct elements of $\mathcal{R}_0$ then A_{R_1} is not equivalent to A_{R_2} that is, $wA_{R_1} \neq wA_{R_2}$.

Theorem 7.5.6 *There exists a family of monotone complete C^*-algebras, $(B_\lambda, \lambda \in \Lambda)$ with the following properties: Each B_λ is a wild, strongly hyperfinite, Type III factor. Each B_λ is a small C^*-algebra. Each B_λ is a quotient of the Pedersen-Baire envelope of the Fermion algebra. The cardinality of Λ is 2^c, where $c = 2^{\aleph_0}$. When $\lambda \neq \mu$ then B_λ and B_μ take different values in the classification semigroup $\mathcal{W}$; in particular, they cannot be isomorphic.*

Proof First we put $\Lambda = \mathcal{R}_0$. For each $R \in \mathcal{R}_0$ we have a faithful normal conditional expectation from B_R onto the maximal abelian $*$-subalgebra A_R. We use the partial ordering defined in Chap. 3. Since A_R is a monotone closed subalgebra of B_R and B_R is a monotone closed subalgebra of M_R we obtain $A_R \precsim B_R \precsim M_R$. Since D_R is a faithful normal positive map from M_R onto A_R, we have $M_R \precsim A_R$. Hence

$A_R \sim B_R \sim M_R$. By using the classification weight semigroup $\mathcal{W}$, we get $wA_R = wB_R = wM_R$. Since, for $R_1 \neq R_2$, we have $wA_{R_1} \neq wA_{R_2}$ this implies $wB_{R_1} \neq wB_{R_2}$.

The only item left to prove is that each factor B_R is strongly hyperfinite. But the Fermion algebra is isomorphic to $\Pi[\mathcal{F}]$. So $\Pi[\mathcal{F}]$ is the closure of an increasing sequence of full matrix algebras. It now follows that B_R is strongly hyperfinite. $\quad\square$

Corollary 7.5.7 *For each orbit equivalence relation $E(R)$, corresponding to R, the orbit equivalence factor $M_{E(R)}$ is nearly hyperfinite.*

Theorem 7.5.8 *Let S be a separable compact Hausdorff extremally disconnected space. Let G be a countably infinite group of homeomorphisms of S with a free, dense orbit. Let E be the orbit equivalence engendered by G and M_E the corresponding monotone complete factor. Suppose that the Boolean algebra of projections of $C(S)$ is countably generated. Then M_E is nearly approximately finite dimensional.*

Proof By Corollary 7.2.7, we may identify M_E with $M(C(S), \bigoplus \mathbb{Z}_2)$. In other words we can assume that $G = \bigoplus \mathbb{Z}_2$. We can further assume that $g \mapsto \beta_g$, the action of G as homeomorphisms of S, has a free dense orbit. Hence the corresponding action $g \mapsto \beta^g$, as automorphisms of $C(S)$, is free and ergodic. Let π be the isomorphism from $C(S)$ onto the diagonal. Let $g \mapsto u_g$ be a unitary representation of $\bigoplus \mathbb{Z}_2$ such that $u_g \pi(a) u_g^* = \pi(\beta^g(a))$ for each $a \in C(S)$.

Let (p_n) be a sequence of projections in $C(S)$ which σ-generate $C(S)$. By Proposition 7.2.10, the C^*-algebra, B_0, generated by $\{p_n : n = 1, 2, \ldots\} \cup \{u_n : n = 1, 2, \ldots\}$ is the closure of the union of an increasing sequence of finite dimensional subalgebras. Let B be the smallest monotone σ-closed subalgebra containing B_0. (B is the normaliser subalgebra.) Then B is monotone closed (because M_E has a faithful state) and approximately finite dimensional; B can be identified with $C(S) \times_\beta G$. Hence M_E is nearly approximately finite dimensional. $\quad\square$

Let B be a monotone complete factor which is almost separably representable (or, equivalently, small). When B is a von Neumann algebra, being approximately finite dimensional is equivalent to being injective. Surprisingly, this does not hold for wild factors.

There exists a wild factor which is approximately finite dimensional but *NOT* injective. This follows by applying a deep result of Hjorth and Kechris [80] (see also [43, 79, 165]). We shall present details of this elsewhere.

This is very different from what the theory of von Neumann algebras would lead us to expect.

Marczewski asked if the Banach-Tarski paradox could be strengthened so that the paradoxical pieces had the Baire property. It was proved in [181] that either the (hyperfinite) Takenouchi-Dyer factor was not injective or the Marczewski problem had a positive solution. Using completely different methods, in a tour de force, a positive solution to the Marczewski problem was established in [35].

7.5.1   Open Problems

In Theorem 7.5.8, the hypotheses allow us to deduce that we can approximate factors by increasing sequences of finite dimensional subalgebras but in the 2^c examples constructed in Sect. 7.4, we can do better. We can approximate by sequences of full matrix algebras (strongly hyperfinite). Let M be a monotone complete factor which is approximately finite dimensional. Is M strongly hyperfinite? Experience with von Neumann algebras would suggest a positive answer but for wild factors this is unknown.

Is M_E equal to its normaliser subalgebra? Equivalently is the "small" monotone cross-product equal to the "big" monotone cross-product?

Are nearly approximately finite dimensional algebras, in fact, approximately finite dimensional?

Chapter 8
Envelopes, Completions and AW^*-Algebras

An AW^*-algebra is a (unital) C^*-algebra where each maximal abelian $*$-subalgebra is monotone complete. We sketch a short introduction to AW^*-algebras in Sects. 8.2 and 8.3. This does not depend on any results subsequent to Chap. 2. In Sect. 8.1 we give a brief account of injective envelopes of C^*-algebras and regular completions of arbitrary C^*-algebras.

8.1 Injective Envelopes and Regular Completions

Injective envelopes of commutative C^*-algebras have been well understood for many years. Hamana made the very important advance of showing that an arbitrary C^*-algebra always has an injective envelope and always has a regular completion. See [59–61]. For "small" C^*-algebras, the regular completion coincides with the regular σ-completion, discussed earlier.

8.1.1 Completely Positive Linear Maps and Choi-Effros Algebras

The following proposition is in [38] (see Lemma 6.1.2, page 105).

Proposition 8.1.1 *Let B be a C^*-algebra and let $\Phi : B \longmapsto B$ be any completely positive idempotent linear map. Then, we have*

$$\Phi(\Phi(x)\Phi(y)) = \Phi(\Phi(x)y) = \Phi(x\Phi(y))$$

for all x, $y \in B$.

© Springer-Verlag London 2015

K. Saitô, J.D.M. Wright, *Monotone Complete C*-algebras and Generic Dynamics*,
Springer Monographs in Mathematics, DOI 10.1007/978-1-4471-6775-4_8

Corollary 8.1.2 *For any a, b and c in $\Phi(B)$,*

$$\Phi(\Phi(ab)c) = \Phi(a\Phi(bc)).$$

Proof Because Φ is idempotent, $\Phi(a) = a$ and $\Phi(c) = c$. So we find that

$$\begin{aligned}
\Phi(\Phi(ab)c) &= \Phi(\Phi(ab)\Phi(c)) \\
&= \Phi(ab\Phi(c)) \\
&= \Phi(abc) \\
&= \Phi(\Phi(a)bc) \\
&= \Phi(\Phi(a)\Phi(bc)) = \Phi(a\Phi(bc)).
\end{aligned}$$

$\square$

Next, we shall define a new multiplication $\circ$ on $\Phi(B)$ by $x \circ y = \Phi(xy)$ for all x, $y \in \Phi(B)$. Then Corollary 8.1.2 tells us that $(x \circ y) \circ z = x \circ (y \circ z)$ for all y, $z \in \Phi(B)$ and so $(\Phi(B), \circ)$ becomes an associative algebra with a $*$-operation.

Corollary 8.1.3 *Let B be a unital C^*-algebra and let Φ be a unital completely positive idempotent linear map from B into itself. Then $(\Phi(B), \circ)$ becomes a unital C^*-algebra with the unit 1 of B. Let A be a unital C^*-subalgebra of B and $\Phi|_A = \iota_A$. Then A is a unital C^*-subalgebra of $(\Phi(B), \circ)$ with respect to the given multiplication of A.*

Proof Since Φ is norm-continuous and idempotent, it is easy to check that $\Phi(B)$ is a norm-closed $\star$-subspace of B. Since $\|\Phi\| = 1$, $\|x \circ y\| \leq \|x\| \cdot \|y\|$ for all x, $y \in \Phi(B)$. We shall show that $\|x\|^2 = \|x^* \circ x\|$ for all $x \in \Phi(B)$. Note that $\Phi(\Phi(x)) = \Phi(x)$ for all $x \in B$. If $x \in \Phi(B)$, then by the Kadison-Schwarz inequality (see Lemma 5.8.1), $x^*x = \Phi(x^*)\Phi(x) \leq \Phi(x^*x) \leq \|x^*x\| \cdot 1_B$ and so we have $\|x\|^2 = \|x^* \circ x\|$. Since $1 \circ x = \Phi(1 \cdot x) = \Phi(x) = x = x \circ 1$ for all $x \in \Phi(B)$, 1 is a unit of $\Phi(B)$. If $a, b \in A$, then $a \circ b = \Phi(ab) = ab$ and so A is a $*$-subalgebra of $(\Phi(B), \circ)$. $\square$

Corollary 8.1.4 *For any $x \in \Phi(B)_{sa}$, $x \geq 0$ in B if, and only if, $x \geq 0$ in $(\Phi(B), \circ)$.*

Proof Let us recall that $\rho_B(x)$ is the resolvent set, that is, $\{\lambda \in \mathbb{C} : \lambda \notin \sigma_B(x)\}$.

If $x \geq 0$ in $(\Phi(B), \circ)$, then there is an $a \in \Phi(B)$ such that $x = a^* \circ a = \Phi(a^*a) \geq 0$ in B. On the other hand, we know that when $x \in \Phi(B)$ and $\lambda \in \rho_B(x)$, one has $y \in B$ such that $(\lambda \cdot 1 - x)y = y(\lambda \cdot 1 - x) = 1$, which implies, by applying Φ, that $\Phi((\lambda \cdot 1 - x)y) = \Phi(y(\lambda \cdot 1 - x)) = 1$ and so it follows that

$$\begin{aligned}
1 &= \Phi((\lambda \cdot 1 - x)y) \\
&= \Phi(\Phi(\lambda \cdot 1 - x)y) \\
&= \Phi(\Phi(\lambda \cdot 1 - x)\Phi(y))
\end{aligned}$$

$$= \Phi((\lambda \cdot 1 - x)\Phi(y))$$
$$= (\lambda \cdot 1 - x) \circ \Phi(y)$$

and similarly, we get that $1 = \Phi(y) \circ (\lambda \cdot 1 - x)$. So it follows that $\lambda \in \rho_{(\Phi(\mathcal{B}),\circ)}(x)$. Hence $\sigma_{(\Phi(B),\circ)}(x) \subset \sigma_B(x)$, that is, if $x \geq 0$ in B, then, $x \geq 0$ in $(\Phi(B), \circ)$. $\quad\square$

Corollary 8.1.5 *If B is monotone complete, then $\Phi(B)$ is also monotone complete with respect to the order inherited from B. Moreover, the C^*-algebra $(\Phi(B), \circ)$ is monotone complete.*

Proof If (a_γ) is any norm bounded increasing net in $\Phi(B)_{sa}$, one has $a_0 = \sup a_\gamma \in B$ in B_{sa}, because B is monotone complete. We shall show that $\Phi(a_0) = \sup a_\gamma$ in $\Phi(B)_{sa} = \Phi(B_{sa})$. Since $a_0 \geq a_\gamma$ for all γ, $\Phi(a_0) \geq a_\gamma$ for all γ in $\Phi(B)$. On the other hand, if $c \in \Phi(B)$ satisfies that $c \geq a_\gamma$ in $\Phi(B)$, so is in B. Hence $c \geq a_0$ in B and $c = \Phi(c) \geq \Phi(a_0)$ in $\Phi(B)$ follows and so $\Phi(a_0) = \sup_{\Phi(B)} a_\gamma$. Hence $\Phi(B)$ is monotone complete. By Corollary 8.1.4, the above (a_γ) in $(\Phi(B), \circ)_{sa}$ satisfies $a_\gamma \uparrow \Phi(a_0)$ in $(\Phi(B), \circ)_{sa}$. $\quad\square$

Completely Positive Idempotent Maps on a C^*-Algebra

Definition 8.1.6 Let B be a unital C^*-algebra and A be a C^*-subalgebra of B with unit 1_B. A seminorm p on B_{sa} is called an A-seminorm if

  (i) $p(x) = \|x\|$ if $x \in A_{sa}$,
  (ii) $p(u^*xu) = p(x)$ if $x \in B_{sa}$ and $u \in A$ is a unitary element,
  (iii) $p(x) \leq \|x\|$ if $x \in B_{sa}$.

We partially order the A-seminorms on B_{sa}, by

$$p_1 \leq p_2 \text{ if, and only if, } p_1(x) \leq p_2(x) \text{ for all } x \in B_{sa}.$$

An A-seminorm p on B_{sa} is said to be *minimal* if, whenever q is an A-seminorm on B_{sa} with $q \leq p$, then $q = p$.

Lemma 8.1.7 *Let p be any A-seminorm on B_{sa}. Then there exists a minimal A-seminorm p_0 on B_{sa} such that $p_0 \leq p$.*

Proof Let $P = \{q : q \text{ is an } A\text{-seminorm such that } q \leq p\}$. Since $p \in P$, it follows that $P \neq \emptyset$. We shall see that each chain in $(P, \leq)$ has a lower bound in P and so, by Zorn's Lemma, P has a minimal element. Take any chain L of P. Let

$$q_0(x) = \inf\{q(x) : q \in L\} \text{ for each } x \in B_{sa}.$$

Since $\|x\| \geq p(x) \geq q(x)$ for all $q \in L$ and for all $x \in B_{sa}$, we get that $\|x\| \geq q_0(x)$ for all $x \in B_{sa}$. If $x \in A_{sa}$, $\|x\| = p(x) = q(x)$ for all $q \in L$ and so $\|x\| = q_0(x)$ for all $x \in A_{sa}$. When $u \in A$ with u a unitary and $x \in B_{sa}$,

$$q_0(uxu^*) = \inf\{q(uxu^*) : q \in L\}$$
$$= \inf\{q(x) : q \in L\} = q_0(x).$$

Since L is a chain, for any x, $y \in B_{sa}$ and for all q, $q' \in L$, $q_0(x+y) \leq q(x) + q'(y)$ which implies that $q_0(x+y) \leq q_0(x) + q_0(y)$ for all x, $y \in B_{sa}$. Since it is easy to get that $q_0(\alpha x) = |\alpha| q_0(x)$ for all $\alpha \in \mathbb{R}$, $x \in B_{sa}$, q_0 is an A-seminorm such that $q_0 \leq p$. So $q_0 \in P$ and q_0 is a lower bound of L in P. So P has a minimal element p_0.

It now follows that p_0 is minimal in the set of all A-seminorms on B_{sa}. $\square$

Let B be a unital C^*-algebra and let A be a unital C^*-subalgebra of B. Let p be an A-seminorm on B_{sa}. Let us recall that when ϕ is a state of A then there is a corresponding Gelfand-Naimark-Segal representation (π_ϕ, H_ϕ). Let η_ϕ be the canonical embedding of A into H_ϕ such that $\{\eta_\phi(a) : a \in A\}$ is dense in H_ϕ. By the Hahn-Banach theorem, for each state ϕ of A, there exists a state ψ_ϕ of B such that $\psi_\phi(x) \leq p(x)$ for all $x \in B_{sa}$ and $\psi_\phi |_A = \phi$.

Let S be a set of states of A. Let $\bigoplus_{\phi \in S} H_\phi$ be the (orthogonal) direct sum of the Hilbert spaces H_ϕ. Then let $\sum_{\phi \in S}^{\oplus} \pi_\phi$ be the representation of A defined for each $x \in A$ and $\xi \in \bigoplus_{\phi \in S} H_\phi$ with $\xi = \oplus_{\phi \in S} \xi_\phi$ by

$$(\sum_{\phi \in S}^{\oplus} \pi_\phi)(x)\xi = \bigoplus_{\phi \in S} \pi_\phi(x)\xi_\phi.$$

Now let K be the state space of A. Then ∂K, the set of extreme points of K, is the set of pure states of A.

Let S be a subset of ∂K, such that $(\pi, H) = (\sum_{\phi \in S}^{\oplus} \pi_\phi, \bigoplus_{\phi \in S} H_\phi)$ is an injective representation of A. We may, for example, take $S = \partial K$.

Let $(\tilde{\pi}, \tilde{H}) = (\sum_{\phi \in S}^{\oplus} \pi_{\psi_\phi}, \bigoplus_{\phi \in S} H_{\psi_\phi})$, where we may assume that $\tilde{H} \supset H$. Let E be the orthogonal projection onto H. Then, $E \in \tilde{\pi}[A]'$. Define a completely positive map Ψ from B into $L(\tilde{H})$ by $\Psi(x) = E\tilde{\pi}(x)E$ for $x \in B$. Then we assert:

Lemma 8.1.8 *The set* $\Psi(B) \subset EL(\tilde{H})E$ *and* $\|\Psi(x)\| \leq p(x)$ *for all* $x \in B_{sa}$.

Proof We let E be the orthogonal projection from $\bigoplus_{\phi \in S} H_{\psi_\phi} (= \tilde{H})$ onto the closed subspace $\bigoplus_{\phi \in S} H_\phi = H$. Take any $x \in B_{sa}$. Since vectors of the form $\xi = \bigoplus_{\phi \in S} \eta_{\psi_\phi}(a_\phi)$ are norm dense in $E\tilde{H} (= H)$, by the definition of the norm of $E\tilde{\pi}(x)E$, for every positive real number ε, there exists $\xi = \bigoplus_{\phi \in S} \eta_{\psi_\phi}(a_\phi)$ with $\|\xi\| \leq 1$ such that

$$\|\Psi(x)\| < \varepsilon + |(\tilde{\pi}(x)\xi, \xi)|$$

$$= \varepsilon + \left| \sum_{\phi \in S} (\tilde{\pi}(x)(\bigoplus_{\phi \in S} \eta_{\psi_\phi}(a_\phi)), \bigoplus_{\phi \in S} \eta_{\psi_\phi}(a_\phi)) \right|$$

$$= \varepsilon + \left| \sum_{\phi \in S} (\pi_{\psi_\phi}(x) \eta_{\psi_\phi}(a_\phi), \eta_{\psi_\phi}(a_\phi)) \right|$$

$$= \epsilon + \left| \sum_{\phi \in S} (\tilde{\pi}(x) \eta_{\psi_\phi}(a_\phi), \eta_{\psi_\phi}(a_\phi)) \right|.$$

Since $\tilde{\pi}(A)$ acts *irreducibly* on $\eta_{\psi_\phi}(A)$, there exists, for each ϕ, a unitary u_ϕ in A (see [161] page 92, Theorem II.4.18) such that

$$\frac{1}{\|\eta_{\psi_\phi}(a_\phi)\|} \cdot \eta_{\psi_\phi}(a_\phi) = \eta_{\psi_\phi}(u_\phi)(= \pi_{\psi_\phi}(u_\phi)\eta_{\psi_\phi}(1)).$$

So, it follows that

$$\|\Psi(x)\| \le \epsilon + \sum_{\phi \in S} \|\eta_{\psi_\phi}(a_\phi)\|^2 |(\pi_{\psi_\phi}(x) \eta_{\psi_\phi}(u_\phi), \eta_{\psi_\phi}(u_\phi)|$$

$$= \epsilon + \sum_{\phi \in S} \|\eta_{\psi_\phi}(a_\phi)\|^2 |\psi_\phi(u_\phi^* x u_\phi)| \le \epsilon + \sum_{\phi \in S} \|\eta_{\psi_\phi}(a_\phi)\|^2 p(u_\phi^* x u_\phi)$$

$$= \epsilon + p(x) \sum_{\phi \in S} \|\eta_{\psi_\phi}(a_\phi)\|^2 = p(x)\|\xi\|^2 + \epsilon \le p(x) + \epsilon,$$

that is, $\|\Psi(x)\| \le p(x)$ for all $x \in B_{sa}$. □

In the following, as before, A is a unital C^*–subalgebra of B and p is an A–seminorm on B.

Proposition 8.1.9 *Let D be any injective C^*-algebra . Let Φ be any completely positive contraction map from A into D. Then, there exists a completely positive map $\hat{\Phi}$ from B into D such that $\hat{\Phi}|_A = \Phi$ and such that $\|\hat{\Phi}(x)\| \le p(x)$ for all $x \in B_{sa}$.*

Proof For each $a \in A$, let $\tilde{\Phi}(\tilde{\pi}(a)E) = \Phi(a)$. Then, $\tilde{\Phi}$ is well defined, because $\tilde{\pi}(a)E = 0$ implies that $a = 0$. Moreover, $\tilde{\Phi}(1) = \Phi(1)$ and hence $\|\tilde{\Phi}\| \le 1$. Since D is injective, there is a completely positive map Φ' from $E(L(\tilde{H}))E$ into D such that $\Phi'|_{\tilde{\pi}(A)E} = \tilde{\Phi}$. Moreover, $\|\Phi'\| = \|\Phi'(E)\| = \|\tilde{\Phi}(E)\| = \|\tilde{\Phi}\| \le 1$. Let $\hat{\Phi} = \Phi' \circ \Psi$. Then, $\hat{\Phi}$ satisfies all the requirements. □

Let us specialise Proposition 8.1.9 by requiring B to be injective and putting $B = D$. Let i_A be the identity map of A into B. Put $\Phi = i_A$. Let $\hat{\Phi} = \Psi$. Then Ψ is a completely positive map from B into B such that

$$\|\Psi(x)\| \le p(x) \text{ for all } x \in B_{sa}.$$

Also $\Psi(a) = a$ for all a in A.

Lemma 8.1.10 *Let A be a unital C^*-algebra and let B be any* injective *C^*-algebra which contains A as a unital C^*-subalgebra. (For example, when A acts on a Hilbert space H as a unital C^*-algebra, we may put $B = L(H)$. In particular, we may let H be the universal representation space.) Let p_0 be a minimal A-seminorm on B. Then there exists a completely positive map Ψ from B into B, such that $\Psi|_A = \iota_A$ and $\|\Psi(x)\| \le p_0(x)$ for all $x \in B_{sa}$. Then, Ψ is an* idempotent *A-projection.*

Proof Define $q_1(x) = \|\Psi(x)\|$ and $q_2(x) = \limsup_{n\to\infty} \|\frac{(\Psi+\cdots+\Psi^n)(x)}{n}\|$ for every $x \in B_{sa}$. Since A is in the multiplicative domain of Ψ, we have $\Psi(uxu^*) = u\Psi(x)u^*$ for $x \in B_{sa}$ and a unitary u in A by Lemma 5.8.1 (3). Then q_1 and q_2 are A-seminorms on B_{sa} such that $q_1 \le p_0$ and $q_2 \le p_0$. Since p_0 is minimal, $q_1 = p_0 = q_2$. So for every $x \in B_{sa}$,

$$\|\Psi(x) - \Psi^2(x)\| = \|\Psi(x - \Psi(x))\| = \limsup_{n\to\infty} \left\|\frac{(\Psi + \cdots + \Psi^n)(x - \Psi(x))}{n}\right\| = 0.$$

So $\Psi^2 = \Psi$ follows. $\qquad\qquad\qquad\qquad\qquad\qquad\qquad\qquad\qquad\qquad\qquad\square$

8.1.2 Constructing Regular Completions of Unital C^*-Algebras

Let A be any unital C^*-algebra. In this subsection we shall show that A has a regular completion. As pointed out in Remark 5.6.12 the proof of Theorem 5.6.11 gives the uniqueness of the regular completion.

Let (V, j_A) be the Dedekind completion of the order unit vector space A_{sa}. Let us recall that V is injective, by Proposition 2.3.9, and j_A is an isometric order isomorphic embedding.

The strategy for constructing a regular completion, $\hat{A}$, of A goes as follows. Consider the monotone closure $\hat{A}_{sa}$ of $j_A(A_{sa})$ in V and define a *multiplication* on $\hat{A} = \hat{A}_{sa} + i\hat{A}_{sa}$ in such a way that $\hat{A}$ becomes a monotone complete C^*-algebra where A is embedded as an order dense unital C^*-subalgebra.

To do this, take *any injective C^*-algebra B* such that A is a unital C^*-subalgebra of B. Since V is injective as a real Banach space, there exists an extension ϕ, a linear map from B_{sa} into V, such that $\|\phi\| = \|j_A\| = 1$ and $\phi|_{A_{sa}} = j_A$. So the set

$$S(B_{sa}, V) = \{\phi \in L(B_{sa}, V) : \|\phi\| = \|j_A\|, \ \phi|_{A_{sa}} = j_A\}$$

is not empty. (Here $L(B_{sa}, V)$ is the set of real linear maps from B_{sa} to V.) Since each $\phi \in S(B_{sa}, V)$ is of norm one and unital, ϕ is a positive map. (See the first two paragraphs of the proof of Lemma 5.8.7.)

Let us define a seminorm p on B_{sa} by

$$p(x) = \sup\{\|\phi(x)\| : \phi \in S(B_{sa}, V)\} \text{ for any } x \in B_{sa}.$$

Lemma 8.1.11 *The seminorm p is an A-seminorm on B_{sa}.*

Proof Since j_A is an isometry, if $x \in A_{sa}$, then $p(x) = \|x\|$. Recall that $\|\phi\| = \|j_A\| = 1$ and $p(x) \leq \|x\|$ follows for all $x \in B_{sa}$. We only have to show that $p(uxu^*) = p(x)$ for all $x \in B_{sa}$ and for all unitaries $u \in A$. Let T_u be the isometric map from B_{sa} onto B_{sa} defined by $T_u x = uxu^*$ for all $x \in B_{sa}$. Then $T_u(A_{sa}) = A_{sa}$. Since (V, j_A) is unique, $j_A \circ T_u \circ j_A^{-1}$ can be extended to a linear isometry S_u from V onto V such that $\|S_u\| = \|T_u\|$ and $S_u \big|_{j_A(A_{sa})} = j_A \circ T_u \circ j_A^{-1}$. So $S_{u^*} \circ \phi \circ T_u \in S(B_{sa}, V)$ for all unitaries $u \in A$ and $\phi \in S(B_{sa}, V)$. So $\|\phi(u^*xu)\| \leq p(x)$ for all $\phi \in S(B_{sa}, V)$, whence $p(u^*xu) \leq p(x)$ for all unitaries $u \in A$.

On replacing x by uxu^* we have $p(x) \leq p(uxu^*)$. On replacing u by u^*, $p(x) \leq p(u^*xu)$. So $p(u^*xu) = p(x)$ for all $x \in B_{sa}$ and all unitaries $u \in A$. $\qquad\square$

Corollary 8.1.12 *Let p_0 be a minimal A-seminorm majorised by p. There is a completely positive linear map $\Psi : B \longmapsto B$ such that $\|\Psi(x)\| \leq p_0(x)$ for all $x \in B_{sa}$ and $\Psi \big|_{A_{sa}} = \iota_{A_{sa}}$. Also Ψ is an idempotent A-projection on B satisfying*

$$\|\Psi(x)\| = \|x\| = \sup\{\|\phi(x)\| : \phi \in S(B_{sa}, V)\} \text{ for all } x \in \Psi(B_{sa}).$$

Proof By Lemma 8.1.10 there is a completely positive linear map $\Psi : B \mapsto B$ such that $\|\Psi(x)\| \leq p_0(x)$ for all $x \in B_{sa}$ and $\Psi \big|_{A_{sa}} = \iota_{A_{sa}}$. Also Ψ is an idempotent A-projection on B satisfying

$$\|\Psi(x)\| \leq \sup\{\|\phi(x)\| : \phi \in S(B_{sa}, V)\} \leq \|x\|$$

for all $x \in B_{sa}$. When $\Psi(B_{sa}) \ni x$, we have $x = \Psi(x)$. So

$$\|\Psi(x)\| = \|x\| = \sup\{\|\phi(x)\| : \phi \in S(B_{sa}, V)\} \text{ for all } x \in \Psi(B_{sa}). \qquad\square$$

Lemma 8.1.13 *Fix $x \in B_{sa}$. Then$\{\phi(x) : \phi \in S(B_{sa}, V)\} =$*

$$\{v \in V : \sup_V\{j_A(a) : a \in A_{sa}, a \leq x\} \leq v \leq \inf_V\{j_A(b) : b \in A_{sa}, b \geq x\}\}.$$

Proof When $x \in A$ this is trivial so we shall suppose that $x \notin A$.

Let ϕ be any element of $S(B_{sa}, V)$. Then, when $a \leq x \leq b$, and a, b are in A_{sa},

$$j_A(a) = \phi(a) \leq \phi(x) \leq j_A(b).$$

So

$$\sup_V\{j_A(a) : a \in A_{sa}, a \leq x\} \leq \phi(x) \leq \inf_V\{j_A(b) : b \in A_{sa}, b \geq x\}.$$

Conversely, keeping x fixed, let $v \in V$ such that

$$\sup_V \{ j_A(a) : a \in A_{sa}, a \le x \} \le v \le \inf_V \{ j_A(b) : b \in A_{sa}, b \ge x \}.$$

We claim that there exists $\phi \in S(B_{sa}, V)$ such that $v = \phi(x)$. First we observe, for all $a \in A_{sa}$,

$$-\|x + a\| \cdot 1 - a \le x \le \|x + a\| \cdot 1 - a.$$

So

$$-\|x + a\| \cdot 1 \le v + j_A(a) \le \|x + a\| \cdot 1. \tag{i}$$

Let $r \in \mathbb{R}$. For $r > 0$ it follows immediately from (i) that

$$-\|rx + a\| \cdot 1 \le rv + j_A(a) \le \|rx + a\| \cdot 1. \tag{ii}$$

In (ii) we replace a by $-a$, and observe that $\|rx - a\| = \|(-r)x + a\|$. We multiply the resulting inequality by -1. This gives (ii) with r replaced by $-r$. So (ii) is valid for all real r.

So, without ambiguity, we can define $\psi : A_{sa} + \mathbb{R}x \mapsto V$ by $\psi(rx + a) = rv + j_A(a)$. From (ii) we have $\|\psi\| \le 1$. So ψ has a unique extension to a bounded operator on $A_{sa} + \mathbb{R}x$. (Since ψ is bounded it obviously extends to the norm closure of this space in B_{sa}. In fact, because $x \notin A_{sa}$, it is straight forward to show that $A_{sa} + \mathbb{R}x$ is a real Banach space.) Because A is unital, $\psi(1) = 1$. So $\|\psi\| = 1 = \|j_A\|$. Also $\psi|_{A_{sa}} = j_A$. Since V is injective, ψ has an extension to some ϕ in $S(B_{sa}, V)$. $\qquad\square$

Let $W = \{ x \in \Psi(B_{sa}) : \phi(x) = \psi(x) \text{ for all } \phi \text{ and } \psi \in S(B_{sa}, V) \}$. Clearly W is a norm-closed subspace of $\Psi(B_{sa}) = (\Psi(B), \circ)_{sa}$ containing A_{sa}.

Lemma 8.1.14 *The (real) Banach space W is monotone closed in $(\Psi(B), \circ)_{sa}$ and A_{sa} is a regular subspace of W.*

Proof When $x \in W$, by Lemma 8.1.13 and the definition of W, for each $\phi \in S(B_{sa}, V)$,

$$\sup_V \{ j_A(a) : a \in A_{sa}, a \le x \} = \phi(x) = \inf_V \{ j_A(a) : a \in A_{sa}, a \ge x \}.$$

Recall that, for any $z \in \Psi(B_{sa})$, $\|z\| = \sup\{ \|\phi(z)\| : \phi \in S(B_{sa}, V) \}$ by Corollary 8.1.12. When $y \in \Psi(B_{sa})$, if $\phi(y) \ge 0$ for all $\phi \in S(B_{sa}, V)$, then we have $y \ge 0$. To show this, we may assume $\|y\| \le 1$. Then we have $0 \le \phi(y) \le 1$ and so $\|\phi(1 - y)\| = \|1 - \phi(y)\| \le 1$ for all $\phi \in S(B_{sa}, V)$, whence $\|1 - y\| \le 1$. Hence it follows that $y \ge 0$. So, $\phi|_W$ is a bipositive linear map from W into V.

We shall now show that $x = \sup\{a \in A_{sa} : a \le x\}$ in $\Psi(B_{sa})$. To see this, take $y \in \Psi(B_{sa})$ such that $a \le y$ for all $a \in A_{sa}$ with $a \le x$. Since ϕ is positive, $j_A(a) \le \phi(y)$ for all such $a \in A_{sa}$. So, $\phi(x) \le \phi(y)$ for all $\phi \in S(B_{sa}, V)$. Since $x - y \in \Psi(B_{sa})$, $x \le y$. So $x = \sup\{a \in A_{sa} : a \le x\}$ in $(\Psi(B), \circ)_{sa}$ by Corollary·8.1.4.

Next we shall show that W is monotone closed in $\Psi(B_{sa})$. Let (x_λ) be any increasing net in W with $x_\lambda \uparrow x$ in $\Psi(B_{sa})$ for some $x \in \Psi(B_{sa})$. Take any $\phi \in S(B_{sa}, V)$. Since $(\phi(x_\lambda))$ is increasing and bounded by $\|x\|1$ in V, there exists $v \in V$ such that $\phi(x_\lambda) \uparrow v$ in V. Since $\phi(x_\lambda) \le \phi(x)$ for all λ, we have $v \le \phi(x)$. On the other hand, if we take $a \in A_{sa}$ with $j_A(a) \ge v$, then we have $\phi(x_\lambda) \le j_A(a)$, which implies that $\phi(x_\lambda - a) \le 0$. So, the bipositivity of ϕ tells us that $x_\lambda \le a$ for all λ and so $x \le a$, that is, $\phi(x) \le j_A(a)$ for all such $a \in A_{sa}$ with $v \le j_A(a)$. Since $v = \inf\{j_A(a) : a \in A_{sa}, j_A(a) \ge v\}$ in V, it follows that $\phi(x) \le v$, that is, $v = \phi(x)$ for all $\phi \in S(B_{sa}, V)$. Since $\phi(x_\lambda) = \psi(x_\lambda)$ for all λ and $\phi, \psi \in S(B_{sa}, V)$, we have $x \in W$. This also implies that $(x_\lambda) \uparrow x$ in $(\Psi(B), \circ)$ by Corollary 8.1.4 again.   □

Theorem 8.1.15 *Let A be a unital C^*-algebra and let $\hat{A}_{sa}$ be the smallest monotone closed subset of $(\Psi(B), \circ)_{sa}(= \Psi(B_{sa}))$ that contains A_{sa}. The subspace $\hat{A} = \hat{A}_{sa} + i\hat{A}_{sa}$ is a monotone closed unital C^*-subalgebra of $(\Psi(B), \circ)$ that contains A and $(\hat{A}, \iota)$ is a regular completion of A. Moreover, for any $x \in \hat{A}_{sa}$,*

$$x = \sup\{a \in A_{sa} : a \le x\} \text{ in } (\Psi(B), \circ)_{sa}.$$

Proof By Corollary 8.1.3 A is a unital $*$-subalgebra of $(\Psi(B), \circ)$. By Corollary 8.1.5 $(\Psi(B), \circ)$ is monotone complete. In Theorem 5.3.7 put $B_1 = B_2 = (\Psi(B), \circ)$ and put $\Phi = \iota$. Then $\hat{A}$ is a unital monotone closed $*$-subalgebra that contains A. Also $\hat{A}_{sa}$ is the self-adjoint part of $\hat{A}$ and is a subspace of W. So $(\hat{A}, \iota)$ is a regular completion of A. In particular, for any $x \in \hat{A}_{sa}$,

$$x = \sup\{a \in A_{sa} : a \le x\} \text{ in } (\Psi(B), \circ)_{sa}.   □$$

InjectiveEnvelopes of Unital C^*-Algebras

Let B be a unital C^*-algebra which, for the moment, is not required to be injective. Let Φ be a unital completely positive idempotent linear map from B into itself. So, by Corollary 8.1.3, when $\Phi(B)$ is equipped with the product $\Phi(x) \circ \Phi(y) = \Phi(\Phi(x)\Phi(y))$, it is a unital C^*-algebra.

Lemma 8.1.16 *Let C be a unital C^*-algebra or an operator system on some Hilbert space and let $\Gamma : C \mapsto (\Phi(B), \circ)$ be a unital completely positive linear map. Then $\Gamma : C \mapsto B$ is also a unital completely positive map.*

Proof Take any $n \in \mathbb{N}$ and consider a map $\Phi_n : M_n(B) \mapsto M_n(B)$ defined by $\Phi_n((a_{ij})) = (\Phi(a_{ij}))$ for each $(a_{ij}) \in M_n(B)$. Then Φ_n is a completely positive idempotent map and for any $\mathbf{a} = (a_{ij})$ and $\mathbf{b} = (b_{ij})$ in $M_n(B)$, we have

$$\mathbf{a} \circ \mathbf{b} = \Phi_n(\mathbf{ab}) = \left(\sum_{k=1}^n \Phi(a_{ik} b_{kj}) \right)_{1 \le i,j \le n} = \left(\sum_{k=1}^n a_{ik} \circ b_{kj} \right)_{1 \le i,j \le n}$$

and so $M_n((\Phi(B), \circ)) = (\Phi_n(M_n(B)), \circ)$ which is an operator subsystem of $M_n(B)$.

So, using Definition 5.8.5, we only have to check that Γ is a positive linear map from C into B. Take any $T \in \mathcal{C}$ and suppose that $T \ge 0$. Then $\Gamma(T)$ is positive in $(\Phi(B), \circ)$ and so by Corollary 8.1.4, $\Gamma(T)$ is positive in B. So, Γ is a positive linear map from C into B.   □

Lemma 8.1.17 *Let B be an injective C^*-algebra and let Φ be a completely positive idempotent map from B into itself. Then the C^*-algebra $(\Phi(B), \circ)$ is injective.*

Proof Take any unital C^*-algebra D and any operator subsystem C of D. Let Ψ be any completely positive contractive map from C into $(\Phi(B), \circ)$.

For each n, consider the map

$$\Phi_n : M_n(B) \longmapsto M_n((\Phi(B), \circ))(= (\Phi_n(M_n(B)), \circ)).$$

It is unital and positive and its range is an operator subsystem of $M_n(B)$. So the map $\Phi : B \mapsto (\Phi(B), \circ)$ is completely positive and $\iota : (\Phi(B), \circ) \mapsto B$ is a completely positive embedding. See Corollary 8.1.4.

By Lemma 8.1.16, Ψ is a completely positive contractive map from C into B. Recall that B is injective. There is a completely positive contraction Ψ' from D into B such that $\Psi'|C = \iota \circ \Psi$. Let $\tilde{\Psi} = \Phi \circ \Psi'$, then $\tilde{\Psi}$ is a completely positive map from D into $(\Phi(B), \circ)$ such that $\tilde{\Psi}|_C = \Psi$. Hence $(\Phi(B), \circ)$ is an injective C^*-algebra. □

Definition 8.1.18 Let A be a unital C^*-algebra. An *extension* of A is a pair (C, k), where C is a unital C^*-algebra and k is a unital injective $*$-homomorphism of A into C. We call (C, k) an *injective envelope* of A if the following conditions are satisfied. First C is an injective C^*-algebra. Secondly, whenever Ψ is a unital, completely positive linear map from C into C such that $\Psi(k(a)) = k(a)$ for all a in A then $\Psi(c) = c$ for all $c \in C$.

Lemma 8.1.19 *Let A be a unital C^*-algebra and let B be any injective C^*-algebra such that A is a unital $*$-subalgebra of B. Let p be any minimal A-seminorm on B_{sa}. Let Φ be any completely positive idempotent linear map from B into itself such that $\Phi|_A = \iota_A$ and $\|\Phi(x)\| \le p(x)$ for all $x \in B_{sa}$. Then $((\Phi(B), \circ), \iota)$ is an injective envelope of A. That is, if Ψ is any completely positive map from $(\Phi(B), \circ)$ into itself that satisfies $\Psi|_A = \iota_A$, then Ψ must be the identity map.*

Proof Let Ψ be any completely positive mapping from $(\Phi(B), \circ)$ into $(\Phi(B), \circ)$ such that $\Psi|_A = \iota_A$. We only have to show that $\Psi = \iota_{(\Phi(B), \circ)}$.

For $x \in B_{sa}$, let

$$p_1(x) = \|\Psi \circ \Phi(x)\|$$

and

$$p_2(x) = \limsup_{n \to \infty} \left\| \frac{(\Psi + \cdots + \Psi^n) \circ \Phi(x)}{n} \right\|.$$

Then p_1 and p_2 are A-seminorms on B_{sa}. We claim that $p_1 \leq p$ and $p_2 \leq p$. Indeed, if $a \in A_{sa}$, then $\Psi(\Phi(a)) = a$ and so $p_1(a) = \|a\| = p_2(a)$. Next take any unitary u in A and $x \in B_{sa}$. Then, by Lemma 5.8.1 (3) we have

$$p_1(uxu^*) = \|\Psi(u\Phi(x)u^*)\|$$

$$= \|u\Psi \circ \Phi(x)u^*\|$$

$$= \|\Psi \circ \Phi(x)\| = p_1(x).$$

Similarly, we have $p_2(uxu^*) = p_2(x)$. So, p_1 and p_2 are A-seminorms such that $p_1 = p = p_2$, because $p_1, p_2 \leq p$ by the properties of Φ.

Then we have $\|\Phi(x)\| = \|\Psi \circ \Phi(x)\|$ and

$$\|\Phi(x)\| = \limsup_{n \to \infty} \left\| \frac{(\Psi + \Psi^2 + \cdots + \Psi^n) \circ \Phi(x)}{n} \right\|$$

for all $x \in B_{sa}$. Hence, for all $a \in \Phi(B)_{sa} = \Phi(B_{sa})$, we have $\|a\| = \|\Psi(a)\|$ and $\|a\| = \limsup_{n \to \infty} \left\| \frac{(\Psi + \Psi^2 + \cdots + \Psi^n)(a)}{n} \right\|$.

Since $a - \Psi(a) \in \Phi(B)_{sa}$ for all $a \in \Phi(B)_{sa}$, we have

$$\|a - \Psi(a)\| = \limsup_{n \to \infty} \left\| \frac{(\Psi + \Psi^2 + \cdots + \Psi^n)(a - \Psi(a))}{n} \right\|$$

$$= \limsup_{n \to \infty} \left\| \frac{\Psi(a) - \Psi^n(a)}{n} \right\| \leq \lim_{n \to \infty} \frac{2\|a\|}{n} = 0$$

which implies that $a = \Psi(a)$ if $a \in \Phi(B)_{sa}$. $\square$

The key result of Hamana, which we shall prove below, is that each unital C^*-algebra has an injective envelope. But first we show that injective envelopes are unique.

Lemma 8.1.20 Let (E, j) and (F, k) be two injective envelopes of A. Then there exists a $*$-isomorphism α from E onto F such that $\alpha \circ j = k$.

Proof Since F is injective and $k \circ j^{-1}$ maps $j(A)$ into F, there exists a completely positive linear map Φ from E into F such that $\Phi|_{j(A)} = k \circ j^{-1}$. Similarly, there is a completely positive linear map Ψ from F into E such that $\Psi\mid_{k(A)} = j \circ k^{-1}$. Then, $\Psi \circ \Phi$ is a completely positive contractive linear map such that $\Psi \circ \Phi\mid_{j(A)} = \Phi \circ k \circ j^{-1} = (j \circ k^{-1}) \circ (k \circ j^{-1}) = \iota_{j(A)}$. Since (E, j) is an injective envelope of A, it follows that $\Psi \circ \Phi$ is the identity on E. Similarly $\Phi \circ \Psi$ is the identity on F. So, Φ is a completely bipositive unital linear isomorphism from E onto F. Hence by Lemma 5.8.1 (2), we have $\Phi(a^2) = (\Phi(a))^2$ for all $a \in E_{sa}$. So, by Lemma 5.8.1 (3) Φ is a homomorphism. Hence Φ is a $*$-isomorphism from E onto F such that $\Phi \circ j = k$. $\square$

Theorem 8.1.21 *Let A be a unital C^*-algebra. Then there is a unique injective envelope $Inj(A)$ of A. The algebra $Inj(A)$ is monotone complete and the monotone closure $\hat{A}$ of A in $Inj(A)$ is the regular completion of A. That is, $\hat{A}$ is obtained by taking the monotone closure of A in $Inj(A)$ and for every $x \in \hat{A}_{sa}$, $x = \sup\{a \in A_{sa} : a \leq x\}$ in $Inj(A)_{sa}$. When $\hat{A}$ is wild, so is $Inj(A)$.*

Proof Let B be any injective C^*-algebra such that $\iota_A : A \mapsto B$ is an injective $*$-homomorphism. For example, we could let $B = L(H)$ where H is any Hilbert space on which A has a faithful representation.

By Lemma 8.1.11 the seminorm p is an A-seminorm on B_{sa}.

By Lemma 8.1.7 there is a minimal A-seminorm p_0 on B_{sa} such that $p_0 \leq p$.

By Lemma 8.1.10 there exists a completely positive map Ψ from B onto A such that Ψ is an idempotent A-projection and, for all $x \in B_{sa}$,

$$\|\Psi(x)\| \leq p_0(x).$$

It now follows from Lemma 8.1.17 that $(\Psi(B), \circ)$ is injective. By Lemma 8.1.19 $(\Psi(B), \circ)$ is the injective envelope of A.

We now apply Theorem 8.1.15 to show that $\tilde{A}$ is a regular completion of A. (As pointed out at the beginning of this subsection, the regular completion is unique.)

If λ is a normal state of $InjA$ then its restriction to $\tilde{A} = \hat{A}$ is also a normal state. If $\hat{A}$ is wild it has no normal states. So $InjA$ has no normal states, in other words, it is wild. $\square$

Corollary 8.1.22 *Let A be a unital C^*-algebra with the injective envelope $Inj(A)$. Let C be a unital C^*-algebra and suppose that $\Psi : Inj(A) \mapsto C$ is a unital completely positive linear map such that $\Psi\mid_A$ is a complete isometry. Then, Ψ is also a complete isometry.*

Proof Since, by our assumption, $(\Psi\mid_A)^{-1}$ is a completely positive map from the operator subsystem $\Psi(A)$, of C, into the injective algebra $Inj(A)$, there exists a unital completely positive linear map $\Psi' : C \mapsto Inj(A)$ such that $\Psi'\mid_{\Psi(A)} = (\Psi\mid_A)^{-1}$. Then $\Psi' \circ \Psi : Inj(A) \mapsto Inj(A)$ is a unital completely positive linear map such that $(\Psi' \circ \Psi)\mid_A = \iota_A$. Since $Inj(A)$ is the injective envelope of A, we must have that $\Psi' \circ \Psi = \iota_{Inj(A)}$. Hence Ψ is a complete isometry. $\square$

Corollary 8.1.23 *Let A be a unital C^*-algebra and let π be a $*$-automorphism of A. Then there exists a unique $*$-automorphism $Inj(\pi)$ of $Inj(A)$ which extends π such* that $Inj(\pi)(\hat{A}) = \hat{A}$. Hence there exists a unique $*$-automorphism $\hat{\pi}$ of $\hat{A}$ which is the restriction of $Inj(\pi)$ to $\hat{A}$. Let $Aut_A(\hat{A})$ be the automorphisms of $\hat{A}$ which restrict to automorphisms of A. Then the map $Aut(A) \ni \pi \mapsto \hat{\pi} \in Aut_A(\hat{A})$ is a group isomorphism

Proof Since $Inj(A)$ is injective and π is a completely positive map, there exists a completely positive extension π' of π to $Inj(A)$. Since π is completely isometric, by Corollary 8.1.22, π' is also completely isometric. By Lemma 5.8.1 (3), π' is an injective $*$-homomorphism of $Inj(A)$ into itself. By applying a similar argument to π^{-1}, we obtain that π' is surjective. Since $Inj(A)$ is an injective envelope of A, π' is unique. Moreover, the normality of π' tells us that $\pi'(\hat{A}) = \hat{A}$. □

Corollary 8.1.24 *Let A be a unital C^*-algebra with the injective envelope $Inj(A)$. Then the centre $Z_{Inj(A)}$ of $Inj(A)$ is $A' \cap Inj(A)$.*

Proof When u is any unitary element of $A' \cap Inj(A)$, Ad_u is a $*$-automorphism of $Inj(A)$ satisfying $Ad_u|_A = \iota_A$. Since $Inj(A)$ is an injective envelope of A, $Ad_u = \iota_{Inj(A)}$ and so, $ux = xu$ for all $x \in Inj(A)$, that is, u is in the centre of $Inj(A)$. Since every element of $A' \cap Inj(A)$ is a linear combination of at most four unitaries in $A' \cap Inj(A)$, $A' \cap Inj(A) \subset Z_{Inj(A)}$. Since the converse inclusion is clear, it follows that $A' \cap Inj(A) = Z_{Inj(A)}$. □

Exercise 8.1.25 Let B be an injective C^*-algebra and let e be a projection in B. Show that eBe is an injective C^*-algebra.

Corollary 8.1.26 *Let $Z_{\hat{A}}$ be the centre of the regular completion of a unital C^*-algebra A. Then $Z_{\hat{A}} = Z_{Inj(A)}$.*

Proof Since $Z_{\hat{A}} \subset A' \cap Inj(A) = Z_{Inj(A)}$, we only need to check the converse.

Assertion 1 Let A be a unital C^*-algebra. Then, for any $f \in Proj(Z_{Inj(A)})$, we have $fInj(A) = Inj(f\hat{A}) = Inj(fA)$.

Recall that $fInj(A)$ is an injective C^*-algebra which contains fA as a unital C^*-subalgebra. Take any completely positive linear map $\Phi : fInj(A) \mapsto fInj(A)$ such that $\Phi|_{fA} = \iota_{fA}$. We shall show that $\Phi = \iota_{fInj(A)}$. Let $\Psi(x) = \Phi(fx) + (1-f)x$ for $x \in Inj(A)$. Then, clearly Ψ is a completely positive linear map from $Inj(A)$ into $Inj(A)$. When $x \in A$, $\Psi(x) = \Phi(fx) + (1-f)x = fx + (1-f)x = x$, which implies that $\Psi = \iota_{Inj(A)}$ and $\Phi(fx) = f(\Phi(fx) + (1-f)x) = f(fx + (1-f)x) = fx$ for all $x \in Inj(A)$. So $\Phi = \iota_{fInj(A)}$. Hence $fInj(A)$ is an injective envelope of fA (and so of $f\hat{A}$).

Now take any projection $f \in Z_{Inj(A)}$. Let $F = \{e \in Proj(\hat{A}) : e \geq f\}$. Clearly $1 \in F$ and so $F \neq \emptyset$. Since $\hat{A}$ is monotone closed in $Inj(A)$, for any pair e_1 and e_2 in F, $e_1 \wedge_{\hat{A}} e_2 = e_1 \wedge_{Inj(A)} e_2$. So F is a decreasing net in $Proj(Inj(A))$. Let $f_1 = \bigwedge\{e : e \in F\}$ in $Proj(\hat{A})$ (and so in $Proj(Inj(A))$). Since for any unitary u in $\hat{A}$ and $e \in F$, we have $ueu^* \geq ufu^* = f$, and hence $uFu^* = F$ for all unitaries u in $\hat{A}$. So we have $uf_1u^* = f_1$ for all such u and it follows that $f_1 \in A' \cap Inj(A) = Z_{Inj(A)}$ and $f_1 \geq f$.

Let $\pi : f_1 Inj(A) \longmapsto fInj(A)$ be a (normal) $*$-homomorphism defined by $\pi(x) = fx$ for all $x \in f_1 Inj(A)$. Then we have the following.

Assertion 2 The restriction of π to $f_1 \hat{A}$, $\pi \mid_{f_1 \hat{A}}$, say, π' is a $*$-isomorphism from $f_1 \hat{A}$ into $fInj(A)$.

If the kernel $\pi'^{-1}(0) \neq \{0\}$, then there exists a non-zero projection e in $\hat{A}$ such that $e \leq f_1$ and $ef = \pi'(e) = \pi(e) = 0$. Hence we have $f \leq f_1 - e \lneq f_1$. Since $f_1 - e \in F$, this is a contradiction. So π' is injective and it is a complete isometry. By Assertion 1, $f_1 Inj(A)$ is the injective envelope of $f_1 A$ (and so, of $f_1 \hat{A}$), π is also completely isometric. So π is injective. Since $\pi(f_1 - f) = f(f_1 - f) = 0$, we have $f = f_1$. Hence $f = f_1 \in \hat{A}$. So it follows that $Z_{\hat{A}} = Z_{Inj(A)}$. $\qquad\qquad\square$

Until now we have been supposing A to be unital. We now drop that requirement. Let us recall that A^1 is the unital C^*-algebra obtained by adjoining a unit to A, when A is not unital, and $A^1 = A$ when A is unital. When A is not unital, we shall define $Inj(A)$ to be $Inj(A^1)$ and $\hat{A}$ to be $\widehat{A^1}$.

For any C^*-algebra, the positive elements of norm strictly less than 1, form an upward directed set which, following [121], we call the canonical approximate unit.

Compare the following result with Theorem 5.6.14.

Corollary 8.1.27 *Let A be a C^*-algebra and let I be any closed two-sided ideal of $\hat{A}$. If $A \cap I = \{0\}$, then $I = \{0\}$. More generally let B be a unital C^*-subalgebra of $Inj(A) = Inj(A^1)$ that contains A and let J be a closed ideal of B. Suppose that $A \cap J = \{0\}$. Then $J = \{0\}$.*

Proof First of all, we note that A is an essential ideal of A^1. For, by making use of the canonical approximate unit for A, we can easily check that if I is an ideal of A^1 such that $I \cap A = \{0\}$, then $I = \{0\}$.

Let $q : B \longmapsto B/J$ be the quotient $*$-homomorphism. If $J \cap A = \{0\}$, $J \cap A^1 = \{0\}$ as well by the above argument and so then $q \mid_{A^1}$ is injective. Hence

$$(q \mid_{A^1})^{-1} : q(A^1) \longmapsto A^1 \subset Inj(A)$$

is also an injective $*$-homomorphism. Since $Inj(A)$ is injective, there exists a completely positive linear map $\phi : B/J \mapsto Inj(A)$ such that $\phi \mid_{q(A^1)} = (q \mid_{A^1})^{-1}$. So, $\phi \circ q$ is a completely positive (unital) linear map from B into $Inj(A)$. Since $Inj(A)$ is injective, there is a completely positive linear map Φ from $Inj(A)$ into itself such that $\Phi \mid_B = \phi \circ q$. If $a \in A^1$, then

$$\Phi(a) = \phi \circ q(a) = (q \mid_{A^1})^{-1}(q(a)) = a$$

and so $\Phi \mid_{A^1} = \iota$. Since $Inj(A)$ is the injective envelope of A^1, $\Phi = \iota$ and when $b \in B$, $\phi(q(b)) = b$. In particular if $b \in J$, then $q(b) = 0$ and so $b = 0$ follows and $J = \{0\}$. The above proof is based on an idea due to Hamana. $\qquad\qquad\square$

Corollary 8.1.28 *For any C^*-algebra A, A is prime if, and only if $\hat{A}$ is a factor. Moreover, A is prime if, and only if, $Inj(A)$ is a factor.*

Proof Suppose that A is prime and that $\hat{A}$ is not a factor. Then there exists a non-trivial central projection z in $\hat{A}$, that is, $0 < z < 1$. Let $I = A \cap z\hat{A}$ and $J = A \cap (1 - z)\hat{A}$. Then both of them are non-zero closed two-sided ideals such that $IJ = \{0\}$. But this is a contradiction, because A is prime.

Conversely, suppose that $\hat{A}$ is a factor. Take any closed two-sided ideals K_1 and K_2 in A with $K_1 K_2 = \{0\}$. Let $\{a_\gamma\}$ be a bounded increasing approximate unit for K_1 and $\{b_\sigma\}$ be a bounded increasing approximate unit for K_2. Then there exist two central projections z_1 and z_2 such that $a_\gamma \uparrow z_1$ and $b_\sigma \uparrow z_2$ in $\hat{A}_{sa}$. Since $a_\gamma b_\sigma = 0$ for all γ and σ, we have $z_1 z_2 = 0$. But $\hat{A}$ is a factor, this implies that $z_1 = 0$ or $z_2 = 0$, that is, $K_1 = \{0\}$ or $K_2 = \{0\}$. Hence A is prime. By Corollary 8.1.24, $Z_{\hat{A}} = Z_{Inj(A)}$, which gives us the second statement. $\qquad\square$

Let A be a unital C^*-algebra and let I be a closed two-sided ideal of A. Let z be the least upper bound of the canonical approximate unit for I in the regular completion $\hat{A}$. Then, clearly z is a central projection in $\hat{A}$ and $\hat{A}z$ is the regular completion of $C^*(I, z)$ (which is $*$-isomorphic to I^1 when I is not unital and, when I is unital, then $z \in I$ is the unit of I). See [133] and [134].

Suppose that A is a *non-unital* C^*-algebra and let $\mathfrak{M}(A)$ be the multiplier algebra of A (see [121]). Since A is an essential ideal of $\mathfrak{M}(A)$ (see again [121]), if we apply the above result, we have:

Theorem 8.1.29 *The regular completion* $\widehat{\mathfrak{M}(A)}$ *of* $\mathfrak{M}(A)$ *is the regular completion of* A^1 *and the idealiser of* A *in* $\widehat{\mathfrak{M}(A)}$ *is the algebra* $\mathfrak{M}(A)$, *that is,*

$$\{a \in \widehat{\mathfrak{M}(A)} : aA \subset A \text{ and } Aa \subset A\} = \mathfrak{M}(A).$$

Since $z = 1$ in the above argument, the first statement follows directly. The second part is a well-known theorem of multiplier theory. Related bibliography can be found in [122] and [14]. But, by making use of the argument in [122], this is straightforward to prove. See also [133] for separable A.

When e is a projection in A, $\widehat{eAe} = e\hat{A}e$ by Lemma 5.9.1, it can also be shown that $Inj(eAe) = eInj(A)e$. See [63], where Hamana showed a more general theorem:

Theorem 8.1.30 *Let* B *be a hereditary* C^*-*subalgebra (see [121]) of a unital* C^*-*algebra* A. *Let* e *be the least upper bound of the canonical approximate units for* B *in* $\hat{A}$, *then* $eInj(A)e$ *is the injective envelope of* B^1 *and* $e\hat{A}e$ *is the regular completion of* B^1, *that is,* $Inj(B) = eInj(A)e$ *and* $\hat{B} = e\hat{A}e$.

By Corollary 8.1.23, for each $*$-automorphism γ of A, there exists a unique $*$-automorphism $Inj(\gamma)$ of $Inj(A)$ such that $Inj(\gamma)(\hat{A}) = \hat{A}$.

When A is *non-unital*, let us recall that γ is inner if there exists a unitary element u in $\mathfrak{M}(A)$ such that $\gamma(a) = uau^*$ for all $a \in A$. We call γ outer if γ is not inner in this sense. When A is unital, this definition coincides with the one already mentioned.

By making use of Theorems 8.1.29 and 8.1.30, we have

Theorem 8.1.31 *Let A be a simple C^*-algebra and let α be a $*$-automorphism of A. Then, α is outer if, and only if, $Inj(\alpha)$ is outer in $Inj(A)$. Let $\hat{\alpha} = Inj(\alpha)\big|_{\hat{A}}$ which is a $*$-automorphism of $\hat{A}$. If $\hat{\alpha}$ is outer in $\hat{A}$, then so is α in A. As a corollary, the regular completion $\hat{F}$ of the Fermion algebra F has an outer $*$-automorphism.*

See [138, 139, 142, 143]. For other topics on regular completions, see [135, 136].

8.1.3 Open Problem

The group von Neumann factor $A = M(\mathbb{F}_2)$ associated with the free group on two generators is monotone complete but is non-injective and so $Inj(A) \neq A = \hat{A}$.

When A is nuclear, does it hold that $Inj(A) = \hat{A}$?

If A is GCR (equivalently postliminary, see [34] and [121]), then $\hat{A}$ is a type I monotone complete C^*-algebra and so, it is injective. Hence it follows that $\hat{A} = Inj(A)$. See for example, [10, 11, 59, 61–63, 92, 133, 137, 154]. See also [6, 7, 17, 45–47, 49, 58].

We also note that for a unital C^*-algebra A, A is postliminary if, and only if, $\widehat{A/J}$ is a type I monotone complete C^*-algebra for every closed two-sided ideal J of A.

8.2 What Are AW^*-Algebras?

Kaplansky introduced AW^*-algebras as an algebraic generalisation of von Neumann algebras [13, 90–93]. In particular, he showed that the Murray-von Neumann classification of projections (in a von Neumann algebra) could be extended to these more general C^*-algebras.

For our purposes the most convenient characterisation is:

An AW^*-algebra is a (unital) C^*-algebra in which each maximal abelian $*$-subalgebra is monotone complete.

We showed in [146] that this is equivalent to Kaplansky's original definition. Since this is not obvious, indeed this assertion has been doubted by some, we shall give a proof here. Clearly this characterisation implies that every monotone complete C^*-algebra is an AW^*-algebra. So any results on AW^*-algebras can be applied to monotone complete C^*-algebras.

In this section, A will be a C^*-algebra which is assumed to have a unit element (unless we state otherwise). Let *ProjA* be the set of all projections in A. Let A_{sa} be the self-adjoint part of A. We recall that the positive cone $A^+ = \{zz^* : z \in A\}$ induces a partial ordering on A. Since each projection is in A^+, it follows that the partial ordering of A_{sa} induces a partial ordering on *ProjA*. When p and q are projections with $p \leq q$ then, by Lemma 2.2.1 $p = pq$. This implies that p and q commute. For

$$qp = (pq)^* = p^* = p = pq.$$

Kaplansky's original definition is:

Definition 8.2.1 The unital C^*-algebra A is an AW^*-*algebra* if (i) each maximal abelian self-adjoint subalgebra is (norm) generated by its projections and (ii) each family of orthogonal projections has a least upper bound in $ProjA$.

When A is an AW^*-algebra it can be proved that the partially ordered set of projections, $ProjA$, is a complete lattice.

Let B be a C^*-algebra which is not assumed to have a unit. For every non-empty subset S of B, the right annihilator of S is the set

$$\{x \in A : Sx = \{0\}\}.$$

Kaplansky [13, 90] showed:

Proposition 8.2.2 *A C^*-algebra B is an AW^*-algebra if, and only if, for every non-empty subset S of B there is a projection e such that eB is the right annihilator of S.*

Let B be an AW^*-algebra. Put $S = \{0\}$. Then Proposition 8.2.2 implies the existence of a projection $e \in B$ such that $B = eB$. So e is a unit of B. Hence B is unital.

When A is an AW^*-algebra it is straightforward to show that each maximal abelian $*$-subalgebra of A is monotone complete [13]. Also, as noted above, $ProjA$ is a complete lattice.

Recent work by Hamhalter [68], Heunen and others, see [74, 77, 99] investigate to what extent the abelian $*$-subalgebras of a C^*-algebra determine its structure. Also a number of interesting new results on AW^*-algebras have been discovered recently; for example [69] and [75, 76]. Also see [3–5] for more general algebras.

Lemma 8.2.3 *Let A be a unital C^*-algebra. Let every maximal abelian $*$-subalgebra of A be monotone complete. Let P be a family of commuting projections. Let L be the set of all projections in A which are lower bounds for P. Then (i) L is upward directed and (ii) P has a greatest lower bound in $ProjA$.*

Proof

(i) Let p and q be in L. Then each $c \in P$ commutes with both p and q and hence with $p + q$. So $P \cup \{p + q\}$ is a set of commuting elements. This set is contained in a maximal abelian $*$-subalgebra M_1. By spectral theory, $((\frac{p+q}{2})^{1/n})(n = 1, 2 \ldots)$ is a monotone increasing sequence whose least upper bound in M_1 is a projection f.

By operator monotonicity, see Sect. 2.1, for each positive integer n, and a, b in A_{sa}, $0 \leq a \leq b \leq 1$ implies $a^{1/n} \leq b^{1/n}$.

For any c in P, $c \geq p$ and $c \geq q$. So $c \geq \frac{p+q}{2}$.

So

$$c = c^{1/n} \geq \left(\frac{p+q}{2}\right)^{1/n}.$$

Hence $c \geq f$. Thus $f \in L$.

Also

$$f \geq \frac{p+q}{2} \geq \frac{1}{2}p.$$

So, by Lemma 2.2.2, $f \geq p$. Similarly, $f \geq q$. So L is upward directed.

(ii) Let C be an increasing chain in L. Then $C \cup P$ is a commuting family of projections. This can be embedded in a maximal abelian $*$-subalgebra M_2. Let e be the least upper bound of C in M_2. Clearly $1 \geq e \geq 0$.

To see that e is a projection we argue as follows. Since $e^{1/2}$ is an upper bound for C, $e^{1/2} \geq e$. So, by spectral theory, $e \geq e^2$. Since e commutes with each element of C, by spectral theory, e^2 is also an upper bound for C, so $e^2 \geq e$. It follows that $e^2 = e$.

For each $p \in P$, $p \geq e$. So $e \in L$. So every chain in L is upper bounded. So, by Zorn's Lemma, L has a maximal element. Since L is upward directed, a maximal element is a greatest element. In other words, P has a greatest lower bound in $Proj(A)$.   □

Proposition 8.2.4 *Let A be a unital C^*-algebra. Let every maximal abelian self-adjoint $*$-subalgebra of A be monotone complete. Then A is an AW^*-algebra.*

Proof Let $\{e_\lambda\}_{\lambda \in \Lambda}$ be a family of orthogonal projections. Let $P = \{1 - e_\lambda : \lambda \in \Lambda\}$. Since this is a commuting family of projections, it has a greatest lower bound f in $Proj(A)$. Hence $1 - f$ is the least upper bound of $\{e_\lambda\}_{\lambda \in \Lambda}$ in $Proj(A)$. Then, by Definition 8.2.1, A is an AW^*-algebra.   □

Theorem 8.2.5 *Let A be a C^*-algebra which is not assumed to be unital. Let each maximal abelian $*$-subalgebra be monotone complete. Then A is a (unital) AW^*-algebra.*

Proof All that is required is to show that A has a unit element. Then we can apply Proposition 8.2.4.

Given any $x \in A_{sa}$, there is a maximal abelian $*$-subalgebra M which contains x. Then the unit of M is a projection p such that $px = x = xp$. For any projection q, with $p \leq q$, $qx = qpx = px = x$. Taking adjoints, $xq = x$.

Let us argue as in Lemma 8.2.3 but with $L = ProjA$, and P the empty set. Then $Proj(A)$ has a maximal element e and $ProjA$ is upward directed. So e is a largest projection. In particular, $p \leq e$. So $ex = x = xe$.   □

No one has ever seen an AW^*-algebra which is not monotone complete. Are all AW^*-algebras monotone complete? Nobody knows but a number of positive results are known. We discuss this question in Sect. 8.3. In view of Theorem 8.2.5, this problem could be reformulated as: if every maximal abelian $*$-subalgebra of a C^*-algebra A is monotone complete is A also monotone complete?

The following technical lemma is used below. It is usually applied with $P = ProjB$ or with $P = B_{sa}$.

Lemma 8.2.6 *Let B be a unital C^*-algebra and let M be a maximal abelian $*$-subalgebra in B. Let P be a subset of B_{sa} such that $uPu^* = P$ whenever u is a unitary in B. Let Q be a subset of $P \cap M_{sa}$ which has a least upper bound q in P. Then q is in M.*

Proof Let u be any unitary in M. Then for any x in Q,

$$uqu^* \geq uxu^* = x.$$

Then uqu^* is in P and is an upper bound for Q. So $uqu^* \geq q$. Similarly $u^*qu \geq q$, that is $q \geq uqu^*$. Thus $uqu^* = q$. So q commutes with each unitary in M. But each element of M is a linear combination of at most four unitaries. So q commutes with each element of M. Hence, by maximality, $q \in M$.   □

Definition 8.2.7 Let A be an AW^*-algebra and let B be a C^*-subalgebra of A where B contains the unit of A. Then B is an AW^*-*subalgebra* of A if (i) B is an AW^*-algebra and (ii) whenever $\{e_\lambda : \lambda \in \Lambda\}$ is a set of orthogonal projections in B then its supremum in $ProjB$ is the same as its supremum in $ProjA$.

By Lemma 1 in [130], or see Exercise 27A of Section 4 page 27 and page 277 in [13], if B is an AW^*-subalgebra of A and Q is an upward directed set in $ProjB$ then the supremum of Q in $ProjB$ is the same as it is in $ProjA$.

In any C^*-algebra, each abelian C^*-subalgebra is contained in a maximal abelian $*$-subalgebra.

Our point of view, in this section, is to study AW^*-algebras in terms of their maximal abelian $*$-subalgebras. So the following proposition has its natural place here.

Proposition 8.2.8 *Let A be an AW^*-algebra. Let B be a C^*-subalgebra of A where B contains the unit of A. Suppose that whenever N is a maximal abelian $*$-subalgebra in B, M is a maximal abelian $*$-subalgebra in A and $N \subset M$ then N is monotone closed in M. Then B is an AW^*-subalgebra of A. The converse is also true.*

Proof Let N_1 be a maximal abelian $*$-subalgebra in B then it is a subalgebra of some maximal abelian $*$-subalgebra M_1 of A. Then M_1 is monotone complete because A is an AW^*-algebra. By hypothesis N_1 is a monotone closed subalgebra of M_1. So N_1 is monotone complete. Hence B is an AW^*-algebra.

Let C be a set of commuting projections in B such that C is upward directed. Let p be the supremum of C in $ProjA$.

Let N_2 be a maximal abelian $*$-subalgebra of B which contains C. Let u be any unitary in N_2. Then, for any $c \in C$,

$$upu^* \geq c.$$

So the projection upu^* is an upper bound for C in $ProjA$. Thus $upu^* \geq p$. On replacing u by u^*, we find that $u^*pu \geq p$. So $p \geq upu^*$. Thus $p = upu^*$. So

$pu = up$. Since each element of N_2 is the linear combination of four unitaries in N_2, it follows that p commutes with each element of N_2. So $N_2 \cup \{p\}$ is contained in a maximal abelian $*$-subalgebra M_2 of A.

Let q be the supremum of C in M_2. By spectral theory, q is a projection. Since p is the supremum of C in $ProjA$, $q \geq p$. But $p \in M_2$. So $q = p$. By hypothesis N_2 is a monotone closed subalgebra of M_2. So $p \in N_2 \subset B$. So p is the supremum of C in $ProjB$.

Now take $\{e_\lambda : \lambda \in \Lambda\}$ to be a set of orthogonal projections in B. Let

$$C = \{\textstyle\sum_{\lambda \in F} e_\lambda : F \text{ a finite, non-empty subset of } \Lambda\}.$$

It follows from the argument above that B is an AW^*-subalgebra of A.

Conversely suppose that B is an AW^*-subalgebra of A. Take any maximal abelian $*$-subalgebra N in B and any maximal abelian $*$-subalgebra M in A with $N \subset M$. We shall show that N is monotone closed in M.

Let (a_α) be any norm bounded increasing net in N_{sa} such that $a_\alpha \uparrow b$ in M_{sa}. We shall show $b \in N$. Suppose that $\|a_\alpha\| \leq k$ for all α. Since N is monotone complete, there exists $a \in N_{sa}$ such that $a_\alpha \uparrow a$ in N_{sa}. Clearly $b \leq a$. Suppose that $a - b \neq 0$. By spectral theory, there exist a non-zero projection p in M and a positive real number ε such that $\varepsilon p \leq (a - b)p$. Since $a_\alpha \uparrow a$ in N_{sa}, by Lemma 1.1 in [166] there exists an orthogonal family (e_γ) of projections in N with $\sup_\gamma e_\gamma = 1$ in $ProjN$ and a family $\{\alpha(\gamma)\}$ such that $\|(a - a_\alpha)e_\gamma\| \leq \frac{\varepsilon}{4}$ for all $\alpha \geq \alpha(\gamma)$ for each γ.

Since B is AW^*, $ProjB$ is a complete lattice. So (e_γ) has a least upper bound e in $ProjB$. By Lemma 8.2.6, $e \in N$. But $\sup_\gamma e_\gamma = 1$ in $ProjN$. So $e = 1$. Thus $\sup_\gamma e_\gamma = 1$ in $ProjB$. Since B is an AW^*-subalgebra of A, it follows that $\sup_\gamma e_\gamma = 1$ in $ProjA$.

Then

$$\varepsilon p \leq (a - b)p \leq (a - a_{\alpha(\gamma)})e_\gamma p + (a - a_{\alpha(\gamma)})p(1 - e_\gamma)$$

$$\leq \frac{\varepsilon}{4}p + 2k(1 - e_\gamma)$$

that is, $\frac{3\varepsilon}{4}p \leq 2k(1 - e_\gamma)$ for all γ. So

$$e_\gamma p e_\gamma = 0.$$

So $\|pe_\gamma\|^2 = 0$. Thus $1 - p \geq e_\gamma$ for all γ. So, in $ProjA$, $1 - p \geq 1$. Thus $p = 0$. This is a contradiction. So $b = a \in N$. □

Remark 8.2.9 Von Neumann algebras: originally Murray and von Neumann studied *Rings of Operators*. See [105–107, 110, 111, 117]. These were $*$-subalgebras of $L(H)$ which were closed in the weak operator topology or, equivalently, equal to their double commutants in $L(H)$. We use *von Neumann algebra* to mean an (abstract) C^*-algebra, A, which has a faithful $*$-representation π on a Hilbert space H where $\pi[A]$ is closed in the weak operator topology. Then a C^*-algebra, A, is a von Neumann algebra if, and only if, it is monotone complete and has a separating family of normal states. Equivalently, A has a (unique) Banach space predual, A_*. See [84, 87, 88, 147, 161].

J. Feldman [40] showed that when a finite AW^*-algebra (see Sect. 8.3 for "finite") has a separating set of completely additive states (states which are completely additive on families of orthogonal projections) it is $*$-isomorphic to a von Neumann algebra. Saitô [129] extended this to semi-finite AW^*-algebras. This culminated in Pedersen [119, 121] showing that every AW^*-subalgebra of $L(H)$ is closed in the strong operator topology.

8.3 Projections and AW^*-Algebras

In this section we give a brief survey avoiding proofs.

First let A be an arbitrary C^*-algebra. Then $ProjA$ is a partially ordered set, with the partial ordering induced by A_{sa}. In general, $ProjA$ need not be a lattice. Given projections p and q in A we define

$$p \sim q$$

to mean there exists a partial isometry $v \in A$ such that $v^*v = p$ and $vv^* = q$. Then $\sim$ is an equivalence relation (Murray-von Neumann equivalence) on $ProjA$. See Definition II.3.3.3 [15, p73].

We define another relation on $ProjA$ by

$$p \precsim q$$

meaning there exists a projection q_1 such that $p \sim q_1$ and $q_1 \leq q$. In general it is *NOT* true that $p \precsim q$ and $q \precsim p$ implies $p \sim q$. See [15, p73].

8.3.1 Projections in AW^*-Algebras

For the rest of this section, unless we specify otherwise, A is an AW^*-algebra with centre Z. Since every monotone complete C^*-algebra is an AW^*-algebra, all results for AW^*-algebras apply to monotone complete C^*-algebras. Just as for von Neumann algebras, A is a *factor* if its centre is one dimensional, that is, $Z = \mathbb{C}$.

We pointed out in Sect. 8.2 that $ProjA$ is a complete lattice. For any projection $p \in ProjA$, its *central cover,* $z(p)$, is the smallest projection in

$$\{z : z \in ProjZ \text{ and } z \geq p\}.$$

It can be seen that

$$z(p) = \bigvee_{ProjA} \{upu^* : u \text{ is a unitary in } A\}.$$

Kaplansky generalised a number of results of Murray and von Neumann to AW^*-algebras by using algebraic methods. Apart from the original papers of Kaplansky, an excellent source is [13]. Three results are fundamental for classifying projections.

First the Schroeder-Bernstein property:

Proposition 8.3.1 *Let p and q be projections in A. If $p \precsim q$ and $q \precsim p$ then $p \sim q$.*

Secondly, the generalised comparability of projections:

Proposition 8.3.2 *Let p and q be projections in A. Then there exists a central projection $z \in Z$ such that*

$$pz \precsim qz$$

and

$$q(1-z) \precsim p(1-z).$$

Thirdly, the complete additivity of projections:

Proposition 8.3.3 *Let $\{e_j\}_{j \in J}$ and $\{f_j\}_{j \in J}$ be two orthogonal families of projections in A. Suppose that there exists, for each $j \in J$, $v_j \in A$ such that $v_j^* v_j = e_j$ and $v_j v_j^* = f_j$. Then there exists $v \in A$ such that $v^* v = \bigvee_{j \in J} e_j$, $vv^* = \bigvee_{j \in J} f_j$ and $ve_j = v_j = f_j v$ for all $j \in J$. We sometimes write $v = \sum_{j \in J} v_j$.*

Proof The proof for general AW^*-algebras is fairly complicated but for monotone complete C^*-algebras we have the following straightforward argument.

We have $\{v_j\}_{j \in J}$ and $\{f_j\}$ are ℓ^2-summable over J and $f_j v_j = v_j$ for each $j \in J$. We apply Corollary 2.1.39. We put $v_F = \sum_{j \in F} v_j$ for each $F \in F(J)$. Then $\{v_F\}_{F \in F(J)}$ is convergent in A, say

$$v = LIM_{F \in F(J)} v_j = \sum_{j \in J} v_j$$

in A. The fact that $\left(\sum_{j \in F} v_j\right) e_{j_0} = v_{j_0}$ if $j_0 \in F$ and 0 if $j_0 \notin F$ tells us, by Proposition 2.1.22, that $ve_j = v_j$ for all $j \in J$. Similarly we have $f_j v = v_j$ for all $j \in J$. Since $v^* v_F = \sum_{j \in F} v_j^* v_j = \sum_{j \in F} e_j$, by Proposition 2.1.22 again, we have $v^* v = \bigvee_{j \in J} e_j$. Similarly, we have $vv^* = \bigvee_{j \in J} f_j$. It is clear, by Lemma 2.1.24, that such a v is uniquely determined.   □

For a proof of the above result for arbitrary AW^*-algebras see [13] or the original papers by Kaplansky.

These results are straightforward to prove for monotone complete C^*-algebras by adapting the classical Murray-von Neumann arguments. But since the AW^*-results are more general and easily accessible we do not need to spend time re-proving them for monotone complete C^*-algebras. We made an exception for Proposition 8.3.3 because the above argument is easy whereas the corresponding generalisation to AW^*-algebras is more intricate.

Types of AW^*-Algebras

We give a very brief account of the decomposition of AW^*-algebras into types. More detailed information can be found in [13] or [90–93].

Definition 8.3.4 An AW^*-algebra B is said to be *finite* if $x^*x = 1$ implies $xx^* = 1$. If B is not finite, it is said to be *infinite*. A projection $e \in B$ is said to be finite if eBe is finite and e is infinite if eBe is infinite. A projection e in B is said to be *abelian* if eBe is abelian. The algebra B is said to be *properly infinite* if the only finite central projection is 0. That is, the algebra B has no finite direct summand other than $\{0\}$. The algebra B is said to be *semi-finite* if it has a finite projection e such that $z(e) = 1$, where $z(e)$ is the central cover of e (see the second paragraph of the preceding subsection). The algebra is said to be *purely infinite* if it contains no finite projections other than 0. The algebra B is said to be *discrete* if it has an abelian projection e with $z(e) = 1$. The algebra is said to be *continuous* if it has no abelian projections other than 0. A central projection z in B is said to be finite (respectively infinite, purely infinite) if the direct summand Bz is finite (respectively infinite, purely infinite). The algebra B is said to be of *Type I* if it is discrete; *Type II_1* if it is continuous and finite; *Type II_∞* if it is continuous, semi-finite and properly infinite and *Type III* if it is purely infinite. A central projection z of B is said to be of type ν ($\nu = I, II_1, II_\infty, III$) if the direct summand Bz is of type ν.

Remark 8.3.5 We see from the above that a projection p is *finite* if $q \leq p$ and $q \sim p$ implies $q = p$.

Proposition 8.3.6 *Let A be an AW^*-algebra. There exist unique orthogonal central projections $z_I, z_{II_1}, z_{II_\infty}$ and z_{III} in A such that $z_\nu A$ is of Type ν ($\nu = I, II_1, II_\infty, III$) and $z_I + z_{II_1} + z_{II_\infty} + z_{III} = 1$.*

When A is monotone complete, Proposition 8.3.6 is a straightforward generalisation of the analogous result for von Neumann algebras. For general AW^*-algebras, see [13] and [90].

Monotone Completeness

Is every AW^*-algebra monotone complete? Kaplansky showed that all Type I factors are von Neumann, and so monotone complete. See [112, 113] concerning the complete classification of Type I algebras. In fact, all Type I AW^*-algebras are monotone complete (see [92] and [166]). But, in general, this is a difficult question. An impressive attack on it was made by Christensen and Pedersen [28] who showed that if A is a properly infinite AW^*-algebra then A is monotone σ-complete. So when A is properly infinite and has a faithful state then it is monotone complete. See also [141].

Let A be an AW^*-factor with a faithful state. Then, using Proposition 8.3.6 and [28], it follows immediately that A is monotone complete unless it is of Type II$_1$. By [175] a Type II$_1$ factor with a faithful state is always a von Neumann algebra. So any factor with a faithful state is monotone complete. In particular, all small factors are monotone complete. Moreover, when A is an AW^*-subalgebra of a type I AW^*-algebra with the centre Z, if A contains Z, then A must be monotone complete. See Theorem 8.3.10 and [130].

Quasi-Linear Maps

Definition 8.3.7 Let B be a unital C^*-algebra and Y a Banach space. A function $T : B \mapsto Y$ is *quasi-linear* if it has the following properties:

(i) $\{Tx : ||x|| \le 1\}$ is bounded in norm;
(ii) $T(a + ib) = Ta + iTb$ whenever a and b are self-adjoint;
(iii) Whenever C is a commutative $*$-subalgebra of B then the restriction of T to C is linear.

If we put $B = M_2(\mathbb{C})$ and $Y = \mathbb{C}$ there are many examples of quasi-linear T which are not linear.

(But it is known that if B is a von Neumann algebra without a Type I$_2$-direct summand then each quasi-linear T is linear. See [22]. This generalisation of Gleason's Theorem is the culmination of the work of many hands. See Hamhalter [67, Ch5] for its lucid, scholarly exposition of those results and its extensive bibliography.)

Finite AW^*-algebras

Let us now suppose the AW^*- algebra A to be finite that is, $x^*x = 1$ implies $xx^* = 1$. Then, see [13, Ch6], there exists a (unique) dimension function $d : ProjA \mapsto Z$ such that

(i) $e \sim f$ implies $d(e) = d(f)$.
(ii) $d(e) \ge 0$ and $d(e) = 0$ only if $e = 0$.
(iii) $d(z) = z$ if $z \in ProjZ$.
(iv) If e and f are orthogonal, then $d(e + f) = d(e) + d(f)$.
(v) The dimension function is completely additive on orthogonal projections.

In [16] there is a very lucid account of the "centre valued quasi-trace". More precisely, the dimension function d has a unique extension to a quasi-linear map $T : A \mapsto Z$ such that

(i) $Tz = z$ for each $z \in Z$.
(ii) $0 \le T(xx^*) = T(x^*x)$ for all x in A.
(iii) T is norm continuous, in particular, $\|T(a) - T(b)\| \le \|a - b\|$ for $a, b \in A_{sa}$.

It is natural to ask: is T linear? When A is a von Neumann algebra the answer is "yes". But, in general, this is unknown.

When A is a finite AW^*-factor then, see [175], if it has a faithful state then it has a faithful normal state. So it is a von Neumann algebra. (So then the quasi-trace, $T : A \mapsto \mathbb{C}$, is linear).

Now suppose that A is a Type II_∞ factor with a faithful state. Then, as remarked above, it is monotone complete. Since A is monotone complete we can apply [176] to show that A is a von Neumann algebra. See also [39, 128, 131, 132]. So, whenever a Type II factor has a faithful state it is von Neumann. It follows that *all small wild factors are of Type III*. Further more, in [39], it is shown that when A is an AW^*-algebra, if there exists a projection e with the central cover $z(e) = 1$ such that eAe is a von Neumann algebra, then A itself is a von Neumann algebra.

Embeddable Algebras

As before we suppose that A is an AW^*-algebra with centre Z.

Definition 8.3.8 We call A *embeddable* if there is a Type I AW^*-algebra, B, with centre Z and a $*$-isomorphism $\pi : A \mapsto B$ such that $\pi[A]$ equals its double commutant in B.

Ozawa [114] used model theory to establish a transfer principle from von Neumann algebras in Boolean-valued set theory to embeddable AW^*-algebras. See also [115]. Roughly speaking, he replaces the scalars by a commutative AW^*-algebra. This model theory approach makes the following generalisation of [175] seem plausible.

Theorem 8.3.9 *Let A be a finite AW^*-algebra with centre Z. Let $T : A \mapsto Z$ be a positive, faithful conditional expectation. Then A is embeddable in a type I AW^*-algebra B with the same centre Z.*

However this result was proved by Ozawa and Saitô [116] *using only standard set theory*.

Earlier Saitô [130] established the following:

Theorem 8.3.10 *If B is an AW^*-algebra of Type I and if A is an AW^*-subalgebra of B containing the centre of B, then A coincides with its bicommutant in B.*

A striking consequence of Theorems 8.3.9 and 8.3.10 is the following: When C is a *separable* unital C^*-algebra, its regular (σ-)completion $\hat{C}$ has no *type II* direct summand. See [116].

8.3.2 Open Problems

Let A be a Type II_1 AW^*-factor. Is A a von Neumann algebra? Or equivalently, is the quasi-trace T linear? This question has been unanswered for 60 years. By [175] a necessary and sufficient condition for a Type II_1 AW^*-factor to be von Neumann is for it to have a *faithful state*. So, an equivalent question is: When A is a finite type II AW^*-factor, does it have a *faithful state* ?

Since the quasi trace T restricts to a faithful normal state on each maximal abelian $*$-subalgebra of A, each maximal abelian $*$-subalgebra of A is a von Neumann algebra. So, this question can be reformulated: Let A be a C^*-algebra where each maximal abelian $*$-subalgebra is von Neumann (and so by Theorem 8.2.5 A is an AW^*-algebra). Suppose that A is simple, infinite dimensional and finite. Is A a von Neumann algebra ?

Handelman [70] showed that every stably finite C^*-algebra admits a $*$-homomorphism to a finite AW^*-factor. (A C^*-algebra B is stably finite, if $M_n(B)$ is finite, that is, if $xy = 1$ implies $yx = 1$ for $x, y \in M_n(B)$ for every $n \in \mathbb{N}$.) Hence the AW^*-problem becomes closely related to the following question in the K-theory of C^*-algebras. Does every stably finite C^*-algebra have a trace? In particular, Blackadar and Handelman [16] showed that the quasi-trace problem for C^*-algebras can be reduced to the AW^*-problem. See Theorem V.2.1.15 and Theorem V.2.2.15 in [15]. For a strong attack on this question see [56]. For a different approach to quasi-traces see [21]. We make no attempt to survey the many papers related to this problem. Instead we refer the reader to the excellent account in [15] and its extensive bibliography.

For a more general problem, consider a unital C^*-algebra B. Suppose there exists a $*$-homomorphism π from B onto $M_2(\mathbb{C})$. Let $\mu : M_2(\mathbb{C}) \mapsto \mathbb{C}$ be a quasi-linear map which is not linear. Then $\mu\pi : B \mapsto \mathbb{C}$ is a quasi-linear map which is not linear. The most optimistic conjecture is:

Let B be a unital C^*-algebra which does not have $M_2(\mathbb{C})$ as a quotient. Let T be a quasi-linear map from B into a Banach space Y. Then T is linear.

This problem can be reduced to the situation where $Y = \mathbb{C}$. It then turns out that T is linear if, and only if, it is uniformly weakly continuous on the closed unit ball of B, see [20]. When B is a von Neumann algebra, the Generalised Gleason Theorem, see above, gives a positive answer. But for general B this is a mystery.

8.4 Conclusions

There have been great advances in the investigation of monotone complete C^*-algebras. But many mysteries and challenges remain. The open problems listed throughout this book, suggest pathways which could lead to further progress.

The classification semigroup and the spectroid invariants introduced in Chap. 3 are far from giving a complete classification but they do produce some order out of chaos.

The following analogy may be helpful. Consider a vast city where each building contains a small monotone complete C^*-algebra. Suppose algebras in different buildings are never isomorphic and each building contains only one algebra. Also suppose each small monotone complete C^*-algebra is isomorphic to an algebra in one of the buildings. By Hamana's pioneering work there are 2^c-buildings. Our classification splits the whole city into parallel avenues, running west to east. At the centre is the 0th avenue, housing all the small von Neumann algebras. There are 2^c avenues, each one labelled by an element of the semigroup. Intersecting the avenues are streets running north to south. One of these streets is that where all the small commutative algebras are to be found. Many other streets remain to be explored before a complete map of the city can be made.

Bibliography

1. Akemann, C.A.: Separable representations of a W^*-algebra. Proc. Am. Math. Soc. **24**, 354–355 (1970)
2. Alfsen, E.M.: Compact Convex Sets and Boundary Integrals. Springer, Berlin/Heidelberg/New York (1971)
3. Ara, P.: Left and right projections are equivalent in Rickart C^*-algebras. J. Algebra **120**, 433–448 (1989)
4. Ara, P., Goldstein, D.: A solution of the matrix problem for Rickart C^*-algebras. Math. Nachr. **164**, 259–270 (1993)
5. Ara, P., Goldstein, D.: Rickart C^*-algebras are σ-normal. Arch. Math. (Basel) **65**, 505–510 (1995)
6. Ara, P., Mathieu, M.: Local Multipliers of C^*-Algebras. Springer, London/Tokyo (2003)
7. Ara, P., Mathieu, M.: Maximal C^*-algebras as quotients and injective envelopes of C^*-algebras. Houst. J. Math. **34**, 827–872 (2008)
8. Araki, H., Woods, E.J.: A classification of factors. Publ. Res. Inst. Math. Sci. **3**, 51–130 (1968)
9. Archbold, R.J.: Prime C^*-algebras and antilattices. Proc. Lond. Math. Soc. **24**, 669–680 (1972)
10. Argerami, M., Farenick, D.R.: Injective envelopes of separable C^*-algebras. ArXiv:math/0506309v1. [math.G.A] 15 June 2005
11. Argerami, M., Farenick, D.R.: Local multiplier algebras, injective envelopes, and type I W^*-algebras. J. Oper. Theory **59**, 237–245 (2008)
12. Balcar, B., Simon, P.: Part III (Appendix on general topology). In: Monk, J.D., Bonnet, R. (eds.) Handbook of Boolean Algebras. North Holland, Amsterdam/New York/Oxford/Tokyo (1989)
13. Berberian, S.K.: Baer*-Rings. Springer, Berlin/Hidelberg/New York (1972)
14. Berberian, S.K.: Isomorphisms of large ideals of AW^*-algebras. Math. Z. **182**, 81–86 (1983)
15. Blackadar, B.: Operator Algebras. Encyclopedia of Mathematical Sciences, vol. 122. Springer, Berlin/New York (2006)
16. Blackadar, B., Handelman, D.: Dimension functions and traces on C^*-*algebras*. J. Funct. Anal. **45**, 297–340 (1982)
17. Blecher, D.P., Paulsen, V.I.: Multipliers of operator spaces and the injective envelope. Pac. J. Math. **200**, 1–17 (2001)
18. Blecher, D.P., Le Merdy, C.: Operator Algebras and Their Modules. London Mathematical Society Monographs Series, vol. 30. Oxford University Press, London (2004)
19. Brown, L.: Large C^*-algebras of universally measurable operators. Q. J. Math. **65**, 857–867 (2014)

© Springer-Verlag London 2015

K. Saitô, J.D.M. Wright, *Monotone Complete C*-algebras and Generic Dynamics*,
Springer Monographs in Mathematics, DOI 10.1007/978-1-4471-6775-4

20. Bunce, L.J., Wright, J.D.M.: The quasi-linearity problem for C^*-algebras. Pac. J. Math. **12**, 271–280 (1994)
21. Bunce, L.J., Wright, J.D.M.: A topological characterization of linearity for quasi-traces. Proc. Am. Math. Soc. **124**, 2377–2381 (1996)
22. Bunce, L.J., Wright, J.D.M.: The Mackey-Gleason problem for vector measures on projections in von Neumann algebras. J. Lond. Math. Soc. **49**, 133–149 (1994)
23. Bures, D.: Abelian Subalgebras of von Neumann Algebras. Memoirs of the American Mathematical Society, vol. 110. American Mathematical Society, Providence (1971)
24. Choi, M.-D.: A Schwarz inequality for positive linear maps on C^*-algebras. Ill. J. Math. **18**, 565–574 (1974)
25. Choi, M.-D., Effros, E.G.: Injectivity and operator spaces. J. Funct. Anal. **24**, 156–209 (1977)
26. Choquet, G.: Lectures on Analysis, II Representation Theory. W.A. Benjamin, London (1969)
27. Christensen, E.: Non-commutative integration for monotone sequentially closed C^*-algebras. Math. Scand. **31**, 171–190 (1972)
28. Christensen, E., Pedersen, G.K.: Properly infinite AW^*-algebras are monotone sequentially complete. Bull. Lond. Math. Soc. **16**, 407–410 (1984)
29. Chu, C.-H.: Prime faces in C^*-algebras. J. Lond. Math. Soc. **7**, 175–180 (1973)
30. Connes, A.: Classification of injective factors. Ann. Math. **104**, 73–115 (1976)
31. Connes, A., Feldman, J., Weiss, B.: An amenable equivalence relation is generated by a single transformation. Ergod. Theory Dyn. Syst. **1**, 431–450 (1981)
32. Davies, E.B.: On the Borel structure of C^*-algebras (with an appendix by R.V. Kadison). Commun. Math. Phys. **8**, 147–163 (1968)
33. Dixmier, J.: Sur certains espaces considérés par M.H. Stone. Summa Bras. Math. **11**(2), 151–182 (1951)
34. Dixmier, J.: Les C^*-algèbres et leurs représentations, 2nd edn. Gauthier-Villars, Paris (1969)
35. Dougherty, R., Foreman, M.: Banach-Tarski decompositions using sets with the property of Baire. J. Am. Math. Soc. **7**, 75–124 (1994)
36. Dyer, J.A.: Concerning AW^*-algebras. Not. Am. Math. Soc. **17**, 788 (1970)
37. Edwards, D.A.: A class of Choquet boundaries that are Baire spaces. Q. J. Math. **17**, 282–284 (1966)
38. Effros, E.G., Ruan, Z.: Operator Spaces. Oxford University Press, Oxford (2000)
39. Elliott, G.A., Saitô, K., Wright, J.D.M.: Embedding AW^*-algebras as double commutants in type I algebras. J. Lond. Math. Soc. **28**, 376–384 (1983)
40. Feldman, J.: Embedding of AW^*-algebras. Duke Math. J. **23**, 303–308 (1956)
41. Feldman, J., Moore, C.C.: Ergodic equivalence relations, cohomology, and von Neumann algebras I. Trans. Am. Math. Soc. **234**, 289–324 (1977)
42. Feldman, J., Moore, C.C.: Ergodic equivalence relations, cohomology, and von Neumann algebras II. Trans. Am. Math. Soc. **234**, 325–359 (1977)
43. Figà-Talamanca, A., Picardello, M.A.: Harmonic Analysis on Free Groups. Lecture Notes in Pure and Applied Mathematics, vol. 87. Marcel Dekker, New York/Basel (1983)
44. Floyd, E.E.: Boolean algebras with pathological order properties. Pac. J. Math. **5**, 687–689 (1955)
45. Frank, M.: Elements of Tomita-Takesaki theory for embeddable AW^*-algebras. Ann. Glob. Anal. Geom. **7**, 115–131 (1989)
46. Frank, M.: C^*-modules and related subjects-a guided reference overview I (1996). arXiv preprint funct-an/9605003
47. Frank, M.: Injective envelopes and local multiplier algebras of C^*-algebras(1999). arxiv preprint math/9910109
48. Frank, M., Paulsen, V.I.: On Hahn-Banach type theorems for Hilbert C^*-modules. Int. J. Math. **13**, 675–693 (2002)
49. Frank, M., Paulsen, V.I.: Injective envelopes of C^*-algebras as operator modules. Pac. J. Math. **212**, 57–69 (2003)
50. Gelfand, I.M., Naimark, M.A.: On the imbedding of normed rings into the rings of operators in Hilbert space. Mat. Sb. **12**, 197–213 (1943)

51. Gillman, L., Jerison, M.: Rings of Continuous Functions. van Nostrand, Princeton (1960)
52. Golodets, V.Y., Kulagin, V., Sinel'shchikov, S.D.: Orbit properties of pseudo-homeomorphism groups and their cocycles. Lond. Math. Soc. Lect. Note Ser. **227**, 211–229 (2000)
53. Golodets, V.Y., Kulagin, V.: Weak equivalence of cocycles and Mackey action in generic dynamics. Qual. Theory Dyn. Syst. **4**, 39–57 (2003)
54. Goodner, D.B.: Projections in normed linear spaces. Trans. Am. Math. Soc. **69**, 89–108 (1950)
55. Grätzer, G.: General Lattice Theory. Birkhäuser Verlag, Basel/Stuttgart (1978)
56. Haagerup, U., Thorbjornsen, S.: Random matrices and K-theory for exact C^*-algebras. Doc. Math. **4**, 341–450 (1999) (electronic)
57. Halmos, P.R.: Lectures on Boolean Algebras. Van Nostrand, Toronto/New York/London (1963)
58. Halpern, H.: The maximal GCR ideal in an AW^*-algebra. Proc. Am. Math. Soc. **17**, 906–914 (1966)
59. Hamana, M.: Injeetive envelopes of C^*-algebras. J. Math. Soc. Jpn. **31**, 181–197 (1979)
60. Hamana, M.: Injective envelopes of operator systems. Publ. Res. Inst. Math. Sci. **15**, 773–785 (1979)
61. Hamana, M.: Regular embeddings of C^*-algebras in monotone complete C^*-algebras. J. Math. Soc. Jpn. **33**, 159–183 (1981)
62. Hamana, M.: The centre of the regular monotone completion of a C^*-algebra. J. Lond. Math. Soc. **26**, 522–530 (1982)
63. Hamana, M.: Tensor products for monotone complete C^*-algebras, I. Jpn. J. Math. **8**, 259–283 (1982)
64. Hamana, M.: Tensor products for monotone complete C^*-algebras, II. Jpn. J. Math. **8**, 285–295 (1982)
65. Hamana, M.: Modules over monotone complete C^*-algebras. Int. J. Math. **3**, 185–204 (1992)
66. Hamana, M.: Infinite, σ-finite, non-W^*, AW^*-factors. Int. J. Math. **12**, 81–95 (2001)
67. Hamhalter, J.: Quantum Measure Theory. Fundamental Theories of Physics, vol. 134. Kluwer, Dordrecht/Boston/London (2003)
68. Hamhalter, J.: Isomorphisms of ordered structures of abelian C^*-subalgebras of C^*-algebras. J. Math. Anal. Appl. **383**, 391–399 (2011)
69. Hamhalter, J.: Dye's theorem and Gleason's theorem for AW^*-algebras. J. Math. Anal. Appl. **422**, 1103–1115 (2015)
70. Handelman, D.: Homomorphisms of C^*-algebras to finite AW^*-algebras. Mich. Math. J. **28**, 229–240 (1981)
71. Hardy, G.H., Wright, E.M.: An Introduction to the Theory of Numbers, 6th edn. Oxford University Press, Oxford (2008)
72. Hasumi, M.: The extension property of complex Banach spaces. Tôhoku Math. J. **10**, 135–142 (1958)
73. Hervé, M.: Sur les représentations intégral à l'aide des points extremaux dans un ensemble compact métrisable. C. R. Acad. Sci. Paris **253**, 366–368 (1961)
74. Heunen, C.: The many classical faces of quantum structures. arXiv preprint arXiv:1412.2177, 2014-arxiv.org
75. Heunen, C., Reyes, M.L.: Diagonalizing matrices over AW^*-algebras. J. Funct. Anal. **264**, 1873–1898 (2013)
76. Heunen, C., Reyes, M.L.: Active lattices determine AW^*-algebras. J. Math. Anal. Appl. **416**, 289–313 (2014)
77. Heunen, C., Reyes, M.L.: On discretization of C^*-algebras. ArXiv preprint arXiv:1412.1721, 2014-arxiv.org
78. Hewitt, E.: A remark on density characters. Bull. Am. Math. Soc. **52**, 641–643 (1946)
79. Hjorth, G.: Classification and Orbit Equivalence Relations of Generic Dynamics. Mathematical Surveys and Monographs, vol. 75. American Mathematical Society, Providence (2000)
80. Hjorth, G., Kechris, A.S.: Rigidity Theorems for Actions of Product Groups and Countable Equivalence Relations. Memoirs of the American Mathematical Society, vol. 177. American Mathematical Society, Providence (2005)

81. Kadison, R.V.: Order properties of bounded self-adjoint operators. Proc. Am. Math. Soc. **2**, 505–510 (1951)
82. Kadison, R.V.: A Representation Theory for Commutative Topological Algebra. Memoirs of the American Mathematical Society, vol. 7. American Mathematical Society, New York City (1951)
83. Kadison, R.V.: A generalized Schwarz inequality and algebraic invariants for operator algebras. Ann. Math. **56**, 494–503 (1952)
84. Kadison, R.V.: Operator algebras with a faithful weakly closed representation. Ann. Math. **64**, 175–181 (1956)
85. Kadison, R.V.: Unitary invariants for representations of operator algebras. Ann. Math. **66**, 304–379 (1957)
86. Kadison, R.V., Pedersen, G.K.: Equivalence in operator algebras. Math. Scand. **27**, 205–222 (1970)
87. Kadison, R.V., Ringrose, J.R.: Fundamentals of the Theory of Operator Algebras, I Elementary Theory. Academic, New York (1983); II Advanced Theory. Academic, New York (1986)
88. Kadison, R.V., Ringrose, J.R.: Fundamentals of the Theory of Operator Algebras Special Topics, III Elementary Theory–An Exercise Approach. Quinn-Woodbine (1991); IV Advanced Theory–An Exercise Appoach. Quinn-Woodbine, New Jersey (1992)
89. Kakutani, S.: Concrete representation of abstract (M)-spaces (A characterization of the space of continuous functions). Ann. Math. **42**, 994–1024 (1941)
90. Kaplansky, I.: Projections in Banach algebras. Ann. Math. **53**, 235–249 (1951)
91. Kaplansky, I.: Algebras of type I. Ann. Math. **56**, 460–472 (1952)
92. Kaplansky, I.: Modules over operator algebras. Am. J. Math. **75**, 839–858 (1953)
93. Kaplansky, I.: Rings of Operators. Benjamin, New York (1968)
94. Kelly, J.L.: Banach spaces with the extension property. Trans. Am. Math. Soc. **72**, 323–326 (1952)
95. Koppelberg, S.: General theory of Boolean algebras. In: Donald Monk, J., Bonnet, R., Koppelberg, S. (eds.) Handbook of Boolean Algebras, vol. 1. North-Holland, Amsterdam/New York/Oxford/Tokyo (1989)
96. Krieger, W.: On ergodic flows and the isomorphism of factors. Math. Ann. **223**, 19–70 (1976)
97. Kuratowski, C.: Topology I. Academic, New York (1966)
98. Lacey, H.E.: The Isometric Theory of Classical Banach Spaces. Springer, Berlin/Heidelberg/New York (1974)
99. Lindenhovius, A.J.: Classifying finite dimensional C^*-algebras by posets of their commutative C^*-algebras. Int. J. Theor. Phys. **54**(12), 4615–4635 (2015)
100. McDuff, D.: Uncountably many II$_1$-factors. Ann. Math. **90**, 372–377 (1969)
101. Mercer, R.: Convergence of fourier series in discrete cross products of von Neumann algebras. Proc. Am. Math. Soc. **94**, 254–258 (1985)
102. Mokobodski, G.: Quelques propriétés des fonctions numériques convexes sur ensemble convexe compact, Sém. Brelot-Choqet-Deny de la théorie du potential, 6e année, Exposé No.9(1962)
103. Monk, J.D., Bonnet, R. (ed.): Handbook of Boolean Algebras, vol. 3. North-Holland, Amsterdam/New York/Oxford/Tokyo (1989)
104. Monk, J.D., Solovay, R.: On the number of complete Boolean algebras. Algebra Univers. **2**, 365–368 (1972)
105. Murray, F.J., von Neumann, J.: On rings of operators. Ann. Math. **37**, 116–229 (1936)
106. Murray, F.J., von Neumann, J.: On rings of operators, II. Trans. Am. Math. Soc. **41**, 208–248 (1937)
107. Murray, F.J., von Neumann, J.: On rings of operators, IV. Ann. Math. **44**, 716–808 (1943)
108. Nachbin, L.: A theorem of the Hahn-Banach type for linear transformations. Trans. Am. Math. Soc. **68**, 28–46 (1950)
109. Nakai, M.: Some expectations in AW^*-algebras. Proc. Jpn. Acad. Ser. A Math. Sci. **34**, 411–416 (1958)

110. von Neumann, J.: Zur Algebra der Funktionalen Operationen und Theorie der normalen Operatoren. Math. Ann. **102**, 370–427 (1929)

111. von Neumann, J.: On rings of operators, III. Ann. Math. **41**, 94–161 (1940)

112. Ozawa, M.: A classification of type I AW^*-algebras and Boolean valued analysis. J. Math. Soc. Jpn. **36**, 589–608 (1984)

113. Ozawa, M.: Non-uniqueness of the cardinality attached to homogeneous AW^*-algebras. Proc. Am. Math. Soc. **93**, 681–684 (1985)

114. Ozawa, M.: A transfer principle from von Neumann algebras to AW^*-algebras. J. Lond. Math. Soc. **32**, 141–148 (1985)

115. Ozawa, M.: Boolean-valued interpretation of Banach space theory and module structures of von Neumann algebras. Nagoya Math. J. **117**, 1–36 (1990)

116. Ozawa, M., Saitô, K.: Embeddable AW^*-algebras and regular completions. J. Lond. Math. Soc. **34**, 511–523 (1986)

117. Pedersen, G.K.: Measure theory for C^*-algebras, III. Math. Scand. **25**, 71–93 (1969)

118. Pedersen, G.K.: On weak and monotone σ-closures of C^*-algebras. Commun. Math. Phys. **11**, 221–226 (1969)

119. Pedersen, G.K.: Operator algebras with weakly closed abelian subalgebras. Bull. Lond. Math. Soc. **4**, 171–175 (1972)

120. Pedersen, G.K.: Isomorphisms of UHF algebras. J. Funct. Anal. **30**, 1–16 (1978)

121. Pedersen, G.K.: C^*-Algebras and Their Automorphism Groups. Academic, London/New York/San Francisco (1979)

122. Pedersen, G.K.: Multipliers of AW^*-algebras. Math. Z. **187**, 23–24 (1984)

123. Peressini, A.L.: Ordered Topological Vector Spaces. Harper and Row, New York (1967)

124. Phelps, R.R.: Extreme positive operators and homomorphisms. Trans. Am. Math. Soc. **108**, 265–274 (1963)

125. Phelps, R.R.: Lectures on Choquet's Theorems. Van Nostrand, Princeton (1966)

126. Powers, R.T.: Representations of uniformly hyperfinite algebras and their associated von Neumann rings. Ann. Math. **86**, 138–171 (1967)

127. Royden, H.L.: Real Analysis, 3rd edn. Macmillan, New York/Collier Macmillan, London (1988)

128. Saitô, K.: Non-commutative extension of Lusin's theorem. Tôhoku Math. J. **19**, 332–340 (1967)

129. Saitô, K.: A non-commutative theory of integration for a semi-finite AW^*-algebra and a problem of Feldman. Tôhoku Math. J. **22**, 420–461 (1970)

130. Saitô, K.: On the embedding as a double commutator in a type I AW^*-algebra II. Tôhoku Math. J. **26**, 333–339 (1974)

131. Saitô, K.: AW^*-algebras with monotone convergence property and type III, non W^*, AW^*-factors. Lecture Notes in Mathematics, vol. 650, pp. 131–134. Springer, Berlin/Heidelberg/New York (1978)

132. Saitô, K.: AW^*-algebras with monotone convergence property and examples by Takenouchi and Dyer. Tôhoku Math. J. **31**, 31–40 (1979)

133. Saitô, K.: A structure theory in the regular σ-completion of C^*-algebras. J. Lond. Math. Soc. **22**, 549–558 (1980); A correction to "A structure theory in the regular σ-completion of C^*-algebras". J. Lond. Math. Soc. **25**, 498 (1982)

134. Saitô, K.: A structure theory in the regular monotone completion of C^*-algebras. Proc. Symp. Pure Math. Am. Math. Soc. **38**, Part 2, 601–604 (1982)

135. Saitô, K.: On C^*-algebras with countable order dense sets and tensor products of monotone complete C^*-algebras. Q. J. Math. **43**, 349–360 (1992)

136. Saitô, K.: Wild, type III, monotone complete, simple C^*-algebras indexed by cardinal numbers. J. Lond. Math. Soc. **49**, 543–554 (1994)

137. Saitô, K.: The smallness problem for C^*-algebras. J. Math. Anal. Appl. **360**, 369–376 (2009)

138. Saitô, K., Wright, J.D.M.: Outer automorphisms of regular completions. J. Lond. Math. Soc. **27**, 150–156 (1983)

139. Saitô, K., Wright, J.D.M.: Outer automorphisms of injective C^*-algebras. Math. Scand. **54**, 40–50 (1984)

140. Saitô, K., Wright, J.D.M.: On tensor products of monotone complete algebras. Q. J. Math. **35**, 209–221 (1984)

141. Saitô, K., Wright, J.D.M.: All AW^*-factors are normal. J. Lond. Math. Soc. **44**, 143–154 (1991)

142. Saitô, K., Wright, J.D.M.: A continuum of discrete group actions on a fermionic factor. Q. J. Math. **44**, 339–343 (1993)

143. Saitô, K., Wright, J.D.M.: Ergodic actions on the fermionic factor. Q. J. Math. **44**, 493–496 (1993)

144. Saitô, K., Wright, J.D.M.: On classifying monotone complete algebras of operators. Ric. Mat. **56**, 321–355 (2007)

145. Saitô, K., Wright, J.D.M.: Monotone complete C^*-algebras and generic dynamics. Proc. Lond. Math. Soc. **107**, 549–589 (2013)

146. Saitô, K., Wright, J.D.M.: On defining AW^*-algebras and Rickart C^*-algebras (2015). arxiv.org/abs/1501.02434

147. Sakai, S.: A characterization of W^*-algebras. Pac. J. Math. **6**, 763–773 (1956)

148. Sakai, S.: An uncountable number of II_1 and II_∞-factors. J. Funct. Anal. **5**, 236–246 (1970)

149. Schaefer, H.H.: Banach Lattices and Positive Operators. Springer, Berlin/Heidelberg/New York (1974)

150. Segal, I.E.: Irreducible representations of operator algebras. Bull. Am. Math. Soc. **53**, 73–88 (1947)

151. Sherman, S.: The second adjoint of a C^*-algebra. In: Proceedings of the International Congress of Mathematicians, Cambridge, vol. 1, p. 470 (1950)

152. Sherman, S.: Order in operator algebras. Am. J. Math. **73**, 227–232 (1951)

153. Sikorski, R.: Boolean Algebras, 3rd edn. Springer, Berlin/Heidelberg/New York (1969)

154. Somerset, D.W.B.: The local multiplier algebra of a C^*-algebra, II. J. Funct. Anal. **171**, 308–330 (2000)

155. Stinespring, W.F.: Positive functions on C^*-algebras. Proc. Am. Math. Soc. **6**, 211–216 (1955)

156. Stone, M.H.: Boundedness properties in function-lattices. Can. J. Math. **1**, 176–186 (1949)

157. Sullivan, D., Weiss, B., Wright, J.D.M.: Generic dynamics and monotone complete C^*-algebras. Trans. Am. Math. Soc. **295**, 795–809 (1986)

158. Takeda, Z.: Conjugate spaces of operator algebras. Proc. Jpn. Acad. Ser. A Math. Sci. **30**, 90–95 (1954)

159. Takenouchi, O.: A Non-W^*, AW^*-Factor. Lecture Notes in Mathematics, vol. 650, pp. 135–139. Springer, Berlin/Heidelberg/New York (1978)

160. Takesaki, M.: On the Hahn-Banach type theorem and the Jordan decomposition of module linear mapping over some operator algebras. Kōdai Math. Sem. Rep. **12**, 1–10 (1960)

161. Takesaki, M.: Theory of Operator Algebras, I. Springer, Berlin/Heidelberg/New York (1979)

162. Takesaki, M.: Theory of Operator Algebras, II, III. Springer, Berlin/Heidelberg/New York (2001)

163. Tomiyama, J.: On the projection of norm one in W^*-algebras. Proc. Jpn. Acad. Ser. A Math. **33**, 608–612 (1957)

164. Vincent-Smith, G.F.: The Hahn-Banach theorem for modules. Proc. Lond. Math. Soc. **17**, 72–90 (1967)

165. Weiss, B.: A survey of generic dynamics. London Math. Soc., Lecture Note Series. **277**, 273–291 (2000)

166. Widom, H.: Embedding in algebras of type I. Duke Math. J. **23**, 309–324 (1956)

167. Willard, S.: General Topology. Addison-Wesley, Reading (1970)

168. Wright, J.D.M.: An extension theorem and a dual proof of a theorem of Gleason. J. Lond. Math. Soc. **43**, 699–702 (1968)

169. Wright, J.D.M.: A spectral theorem for normal operators on a Kaplansky-Hilbert module. Proc. Lond. Math. Soc. **19**, 258–268 (1969)

170. Wright, J.D.M.: Measures with values in a partially ordered vector space. Proc. Lond. Math. Soc. **25**, 675–688 (1972)

171. Wright, J.D.M.: Every monotone σ-complete C^*-algebra is the quotient of its Baire* envelope by a two sided σ-ideal. J. Lond. Math. Soc. **6**, 210–214 (1973)

172. Wright, J.D.M.: On minimal σ-completions of C^*-algebras. Bull Lond. Math. Soc. **6**, 168–174 (1974)

173. Wright, J.D.M.: Regular σ-completions of C^*-algebras. J. Lond. Math. Soc. **12**, 299–309 (1976)

174. Wright, J.D.M.: On von Neumann algebras whose pure states are separable. J. Lond. Math. Soc. **12**, 385–388 (1976)

175. Wright, J.D.M.: On AW^*-algebras of finite type. J. Lond. Math. Soc. **12**, 431–439 (1976)

176. Wright, J.D.M.: On semi-finite AW^*-algebras. Math. Proc. Camb. Philos. Soc. **79**, 443–445 (1976)

177. Wright, J.D.M.: Wild AW^*-factors and Kaplansky-Rickart algebras. J. Lond. Math. Soc. **13**, 83–89 (1976)

178. Wright, J.D.M.: On C^*-algebras which are almost separably representable. J. Lond. Math. Soc. **18**, 147–150 (1978)

179. Wright, J.D.M.: On some problems of Kaplansky in the theory of rings of operators. Math. Z. **172**, 131–141 (1980)

180. Wright, J.D.M.: Hyperfiniteness in wild factors. J. Lond. Math. Soc. **38**, 492–502 (1988)

181. Wright, J.D.M.: Paradoxical decompositions of the cube and injectivity. Bull. Lond. Math. Soc. **22**, 18–24 (1989)

182. Wright, J.D.M.: On classifying monotone complete algebras of operators. Contemp. Math. **503**, 307–317 (2009)

Index

$A(K; \mathcal{V})$, 93
A^1, 10
A^∞, 34
A^b, 107
F_σ-set, 79
G_δ-set, 71
$L(f)$, 184
$LP(z)$, 26
$ProjA$, 24
$RP(z)$, 26
$\mathcal{M}(E)$, 185
$\mathfrak{M}_g$, 73
$\partial_{(T,N)}$, 58
$\precsim$, 51

Action
    free action, 142, 158, 175, 194
    group action, 5
    pseudo free, 158
Admissible, 82
Almost separably representable, 50, 214
Anti-lattice, 8, 9
Approximately finite dimensional, 215
AW^*-algebra, 4, 204, 219, 234

Baire measurability, 79, 151
Baire measurable, 79
Baire Property, 67, 195
Baire space, 4
Big Cantor space, 80
Birkhoff-Ulam Theorerm, 67, 113
Boolean algebra, 5, 63
Borel
    Borel measurable, 2

Borel measure, 5
Brown, 107, 114
    Brown-Borel envelope, 93, 107, 109

Cantor space, 79
Central cover, 239, 241
Centre, 214
Choi, 127, 129, 219
Choquet, 94
    Choquet's Lemma, 95, 99
Christensen, 242
Classification semigroup, 51
Completely positive, 2, 49, 91, 126, 219
Completely regular, 5, 38
Concave function, 93
Conditional expectation, 5, 42, 190, 202
Cone
    hereditary cone, 8
Connes, 3, 215
Convex, 42
    compact convex, 91
    convex function, 93
Cross-product, 196

Davies, 99
    Davies-Baire envelope, 100, 102
Dedekind cut, 3
Direct product, 52, 184
Direct sum, 34
Dixmier algebra, 77, 89, 176
Downward directed, 8
Dyadic Group, 143, 166
Dyer, 4, 217

© Springer-Verlag London 2015
K. Saitô, J.D.M. Wright, *Monotone Complete C*-algebras and Generic Dynamics*,
Springer Monographs in Mathematics, DOI 10.1007/978-1-4471-6775-4